Advances in Intelligent Systems and Computing

Volume 1356

AF847063

The series "Advances in Intelligent Systems and Computing" contains publications on theory, applications, and design methods of Intelligent Systems and Intelligent Computing. Virtually all disciplines such as engineering, natural sciences, computer and information science, ICT, economics, business, e-commerce, environment, healthcare, life science are covered. The list of topics spans all the areas of modern intelligent systems and computing such as: computational intelligence, soft computing including neural networks, fuzzy systems, evolutionary computing and the fusion of these paradigms, social intelligence, ambient intelligence, computational neuroscience, artificial life, virtual worlds and society, cognitive science and systems, Perception and Vision, DNA and immune based systems, self-organizing and adaptive systems, e-Learning and teaching, human-centered and human-centric computing, recommender systems, intelligent control, robotics and mechatronics including human-machine teaming, knowledge-based paradigms, learning paradigms, machine ethics, intelligent data analysis, knowledge management, intelligent agents, intelligent decision making and support, intelligent network security, trust management, interactive entertainment, Web intelligence and multimedia.

The publications within "Advances in Intelligent Systems and Computing" are primarily proceedings of important conferences, symposia and congresses. They cover significant recent developments in the field, both of a foundational and applicable character. An important characteristic feature of the series is the short publication time and world-wide distribution. This permits a rapid and broad dissemination of research results.

Indexed by DBLP, EI Compendex, INSPEC, WTI Frankfurt eG, zbMATH, Japanese Science and Technology Agency (JST).

All books published in the series are submitted for consideration in Web of Science.

More information about this series at http://www.springer.com/series/11156

Susanta Kumar Paikray · Hemen Dutta ·
John N. Mordeson
Editors

New Trends in Applied Analysis and Computational Mathematics

Proceedings of the International Conference on Advances in Mathematics and Computing (ICAMC 2020)

Springer

Editors
Susanta Kumar Paikray
Department of Mathematics
Veer Surendra Sai University of Technology
Burla, Odisha, India

Hemen Dutta
Department of Mathematics
Gauhati University
Guwahati, India

John N. Mordeson
Department of Mathematics
Center for Mathematics of Uncertainty
Creighton University
Omaha, NE, USA

ISSN 2194-5357 ISSN 2194-5365 (electronic)
Advances in Intelligent Systems and Computing
ISBN 978-981-16-1401-9 ISBN 978-981-16-1402-6 (eBook)
https://doi.org/10.1007/978-981-16-1402-6

This Springer imprint is published by the registered company Springer Nature Singapore Pte Ltd.
The registered company address is: 152 Beach Road, #21-01/04 Gateway East, Singapore 189721, Singapore

Preface

The book contains original research papers as the Proceedings of the International Conference on Advances in Mathematics and Computing, held at Veer Surendra Sai University of Technology, Odisha, India, during 7–8 February, 2020. It contains research works on new trends in applied analysis, computational mathematics and related areas. The readers shall also find new models, image analysis technique, fluid flow problems, in particular and several new topics of computational mathematics. The book should be a valuable resource for researchers, graduate students, teachers, scientists and engineers interested in recent advances in applied analysis and computational mathematics necessary for understanding, analyzing and solving problems arise in various branches of physical sciences and engineering. The book consists of 21 chapters and they are organized as follows.

The chapter "A Fuzzy Two-Warehouse Inventory Model of Deteriorating Items with Exponentially Demand and Backlogged Condition" aims to develop an optimization inventory policy for deteriorating items under two ware-house systems in fuzzy environment assuming the demand for the products depends upon selling price and shortages are partially backlogged. The model assumes that the rate of backlogging and demand are exponential functions, and it is formulated to optimize the total average cost using Graded Mean Integration Method. It also includes two numerical examples for feasibility of the model as well as sensitivity analysis has been carried out to identify the model parameters.

The chapter "A Comparative Study of Nonlinear Convection in a Confined Fluid Overlying a Porous Layer" uses the alternating direction implicit method for a comparative study among the Darcy, Brinkman-extended Darcy and Brinkman-Forchheimer-extended Darcy models of free convection in a cavity containing a fluid layer overlying a porous layer saturated with the same fluid. It also contains numerical simulations revealing that all three models have yielded almost the same results for Darcy numbers up to about 10^{-4}. It observes that for higher Darcy numbers, the Brinkman-Forchheimer-extended Darcy model is useful to simulate momentum transfer as it accounts for the effects of inertia.

The chapter "Persistence Barcoded Vehicular Traffic Videos in a Topology of Data Approach to Shape Tracking" introduces a computational CW topology of data approach to track the persistence of image object shapes that appear in triangulated video frames. Fermi energy and Betti numbers have been used to construct persistence barcodes derived from nested cycles inherent in triangulated video frame shapes. An application is also included in terms of Ghrist persistence barcoding of vehicular traffic videos.

The chapter "Heat Transfer of MHD Channel Flow of Viscoelastic (PTT) Fluid" presents heat transfer aspect of MHD channel flow of Phan-Thien-Tanner (PTT) conducting flow accounting for the viscous dissipation. The important findings include the role of magnetic parameter is to enhance the temperature across the flow domain whereas Deborah number and other parameters act adversely. It also observes that contribution of viscous dissipative heat as insignificant due to linear variation across the temperature field in the present PTT model indicates the preservation of thermal energy loss.

The chapter "Properties of Characteristic Polynomials of Oriented Graphs" introduces some spectral properties of oriented graphs. It determines characteristic polynomials of several oriented graph classes, the effect of edge addition to characteristic polynomial, and also includes several recursive results for the characteristic polynomial of an oriented graph by means of cut vertices, bridges, paths and pendant edges.

The chapter "An Inventory Model for Linear Deteriorating Item with Shortages Under Partial-Backlogged Condition" studies an inventory planning problem for decaying commodities with price-dependent linear demand. It considers allowable shortage which is partial-backlogged and uses price-dependent demand and holding up time subordinate build-up rates in a general structure to develop the model. The solution of the optimization model is adorned by a numerical example. It also includes a convexity checkup of the total average cost function by plotting a 3D graph. Sensitivity analysis and managerial insights are further incorporated to examine the impact of various system parameters of the model.

The chapter "On the Local Convergence of a Sixth-Order Iterative Scheme in Banach Spaces" applies the Lipschitz continuity of the first-order Fréchet derivative to describe the local convergence of a sixth-order convergent nonlinear system solver in Banach spaces. It extends the algorithm applicability through the use of a theory based on the first-order derivative only. The analysis also includes the radius of convergence ball and computable error bounds together with the uniqueness of the solution. It further incorporates numerical experiments to demonstrate that the technique adopted is beneficial when prior studies fail to solve problems.

The chapter "New Results on Chromatic Polynomials" provides some short-cut results enabling to calculate the chromatic polynomial of a relatively large graph by dividing it into smaller graphs. Coloring is an important study area in graph theory and related fields, especially when the combinatorial calculations are needed. The Birkhoff-Lewis Theorem gives a step-by-step reduction method to calculate the chromatic polynomial of any given graph as the difference of the chromatic polynomials of two smaller graphs, one is edge deleted, and the other is edge contracted.

The chapter "Recurrence Relations of Edge-Zagreb and Sum-Edge Characteristic Polynomials of Some Graphs" recalls two graph matrices and obtains recurrence relations for the graph energies corresponding to the matrices making it easy to calculate the energy of large graphs by means of smaller graphs. Spectral graph theory is an important area of graph theory and it has applications related to molecular energy in mathematical chemistry. Classical molecular energy was defined by means of the adjacency matrix of the graph modeling the given molecular structure.

The chapter "Effects of Radiation on MHD Flow with Induced Magnetic Field" aims to analyze numerically the impact of radiation and viscous dissipation on steady MHD flow over a stretching sheet with induced magnetic field. It also includes comparisons with earlier results and claims to have found good agreements. Surface transport phenomena such as skin friction and Nusselt number are discussed besides the three boundary layers. The important results obtained include the magnetic parameter β decelerates velocity, and accelerates temperature profiles; the radiation parameter (R) enhances the thermal boundary layer thickness for both cases $a/c < 1$ and $a/c > 1$.

The chapter "Various Approximate Multiplicative Inverse Lie $\star$-Derivations" determines various approximate multiplicative inverse Lie $\star$-derivations in the framework of normed $\star$-algebras pertinent to Ulam-Hyers stability theory. The results are applied to achieve other classical stabilities by taking different upper bounds.

The chapter "Analysis and Computation of Reactive Second-Grade Fluid Flow with Variable Viscosity Within Porous Couette Device" examines the influence on velocity and temperature profiles of an unsteady, reactive and incompressible second grade fluid with low electrical conductivity and variable viscosity within a porous channel with asymmetric cooling at the walls. The system is solved by damped-Newton method. The numerical method observes second-order convergence and high stability with reduction of error after iterations depending on the choice of damping applied. The effects of different physical parameters on velocity and temperature profile are also discussed and demonstrated graphically.

The chapter "Hopf Bifurcation and Stability Analysis of Delayed Lotka–Volterra Predator–Prey Model Having Disease for Both Existing Species" proposes a delay predator-prey model with logistic growth in the prey population. The model assumes to include an SIS infection in both prey and predator species. It first deals with disease in the prey population and then analyzes the disease predator population. It also proves the existence of Hopf bifurcation for this system by analyzing characteristic equations, and identifies the important threshold quantities.

The chapter "An EOQ Model Without Shortages with Uncertain Cost Associated with Some Fuzzy Parameters and Interval Parameters" presents Economic Order Quantity (EOQ) models without shortages for single item and multi-items in which the holding cost of the item is a continuous function of the order quantity. The proposed model is discussed in two cases by describing the model in an uncertain environment. In the first case, an EOQ model with ordering cost, holding cost and unit product cost has been considered in a fuzzy environment with certain limitations. In the second case, it considers parameters like ordering cost, holding cost, unit product cost and the total money investment for the quantities as interval numbers. The proposed model is then discussed by obtaining the best and the worst optimum values of the objective function. It also includes numerical examples to support the proposed model.

The chapter "Free Poisson Elements Induced by Orthogonal Projections" aims to study free (weighted-) Poisson elements in a topological noncommutative free *-probability space generated by countable infinitely many mutually free, (weighted-) semicircular elements which are induced by mutually orthogonal countable infinitely many projections in a fixed C*-probability space.

The chapter "On Approximation of Functions in the Generalized Zygmund Class Using $(E, r)(N, q_n)$ Mean Associated with Conjugate Fourier Series" aims to establish a result on degree of approximation of conjugate Fourier series of functions in the generalized Zygmund class by using Euler-Nörlund product mean.

The chapter "Electrification Effect of Nanoparticles on Nanofluid Flow over a Continuous Stretching Sheet" concerns with the study of nano fluid flow past a continuous stretching sheet including electrification of nano particles along with Brownian diffusion and thermophoresis to show the impact on the enhancement of thermal conductivity and cooling process. It investigates electrification effect of nano particles on dimensionless velocity, normalized temperature, dimensionless nano particle concentration as well as the non-dimensional heat and mass transfer coefficients. It observes that the electrification of nano particles is a possible mechanism for heat transfer enhancement of base fluids.

The chapter "Determinantal Formulas and Recurrent Relations for Bi-Periodic Fibonacci and Lucas Polynomials" finds closed formulas and recurrent relations for bi-periodic Fibonacci polynomials and for bi-periodic Lucas polynomials in terms of the Hessenberg determinants. It further derives closed formulas and recurrent relations for the Fibonacci, Lucas, bi-periodic Fibonacci and bi-periodic Lucas numbers in terms of the Hessenberg determinants.

The chapter "PBIB Designs and Their Association Schemes Arising from Minimum Bi-Independent Dominating Sets of Circulant Graphs" obtains the total number of γ_{bi}-sets along with partially balanced incomplete block (PBIB) designs with their association schemes arising from the γ_{bi}-sets in different jump sizes of some circulant graphs.

The chapter "Skew-Harmonic and Skew-Sum Connectivity Energy of Some Digraphs" calculates the skew-harmonic energy and skew-sum connectivity energy of some digraphs. Graph energies are the backbone of chemical graph theory.

The chapter "Multi-valued Analysis of *CR* Iterative Process in Banach Spaces" aims to define multi-valued CR iteration scheme involving quasi-nonexpansive mapping by relaxing the compactness of the domain. It also establishes the multi-valued CR iteration scheme in a new sense.

The editors are thankful to contributors for their cooperation and patience while the chapters were being reviewed and processed. Reviewers deserve gratitude for their help voluntarily offered to us. The editors thankfully acknowledge the encouragement received from many colleagues in completing this book.

Burla, India
Guwahati, India
Omaha, USA
June, 2021

Susanta Kumar Paikray
Hemen Dutta
John N. Mordeson

Contents

About the Editors

Dr. Susanta Kumar Paikray has held the position of Associate Professor in the Department of Mathematics Veer Surendra Sai University of Technology, Burla, since 2016, having joined the faculty there in 2014, first as Reader (2014–2016) and then as Associate Professor (2016–till date). Also, he has served as a lecturer in the Department of Mathematics, Ravenshaw University Cuttack (2011–2014). He earned his Ph.D. degree in 2010 at Berhampur University, while he was a full-time member of the teaching faculty at the DRIEMS in Odisha. His current research interests include several areas of pure and applied mathematical sciences like summability theory, Fourier series, approximation theory, functional analysis, statistical convergence, operations research and inventory optimization. He has published 02 books, 10 book chapters, 12 papers in international conference proceedings and more than 60 scientific research articles in peer-reviewed national and international journals of repute, as well as Forewords and Prefaces to many books and journals, and so on.

Dr. Hemen Dutta is a regular faculty member at the Department of Mathematics, Gauhati University, India. He has more than 14 years of teaching and research experiences. His primary research areas are in the field of mathematical analysis covering both abstract and applied aspects. He has over 100 research papers for journals, 16 chapters in books and contributed 12 papers in conference proceedings so far. He has to credit 10 published books, 6 books in press and 3 conference proceedings. He has delivered several invited talks at national and international level events and visited some foreign institutions on invitations. He has organized 5 events for academicians and researchers. He is the recipient of some grants for conference organization, research project and travelling. He has reviewed research papers for journals and databases and associated with editing special issues in reputed journals. He has also authored several general articles for newspaper, popular books, magazines and science portals.

Dr. John N. Mordeson is Professor Emeritus of Mathematics at Creighton University, USA. He received his B.S., M.S. and Ph.D. from Iowa State University. He is a member of Phi Kappa Phi. He is President of the Society for Mathematics of Uncertainty. He has published 15 books and 200 journal articles. He is on the editorial board of numerous journals. He has served as an external examiner of Ph.D. candidates from India, South Africa, Bulgaria and Pakistan. He has refereed for numerous journals and granting agencies. He is particularly interested in applying mathematics of uncertainty to combat the problem of human trafficking.

A Fuzzy Two-Warehouse Inventory Model of Deteriorating Items with Exponentially Demand and Backlogged Condition

A. K. Sahoo, S. K. Indrajitsingha, P. N. Samanta, and U. K. Misra

Abstract The objective of this study is to develop an inventory policy for deteriorating items under a two-warehouse system in fuzzy environment under selling price-dependent demand with shortages which are partially backlogged. It is assumed that the retailer has two warehouses of which one is treated as owned warehouse (OW) of finite capacity and the other treated as rented warehouse (RW) with large capacity to keep the surplus products for uncertain demand with a reasonable distance from marketplace. It is also assumed that the holding cost for RW is inversely proportional to the distance of RW from market. As the retailer normally wants to sell the inventory in RW first, the inventory in RW decreases due to demand and deteriorates until it reaches zero level, while the inventory in OW is depleted due to deterioration only. In this model, it is assumed that shortages are allowed and backlogged exponentially. The demand is also assumed to be an exponential function. The model is formulated to optimize the total average cost using Graded Mean Integration Method (GMIR). Two numerical examples are given for testing the feasibility of the model and sensitivity analysis has been carried out to identify the model parameters.

A. K. Sahoo (✉) · P. N. Samanta
Department of Mathematics, Berhampur University, Bhanja Bihar, Berhampur 760007, Odisha, India
e-mail: ajit2ksahoo@gmail.com

P. N. Samanta
e-mail: pns.math@buodisha.edu.in

S. K. Indrajitsingha
Department of Mathematics, Saraswati Degree Vidya Mandir, Neelakantha Nagar, Berhampur 760002, Odisha, India
e-mail: skismath86@gmail.com

U. K. Misra
NIST, Palur Hills, Berhampur 761 008, Odisha, India
e-mail: umakanta_misra@yahoo.com

S. K. Paikray et al. (eds.), *New Trends in Applied Analysis and Computational Mathematics*, Advances in Intelligent Systems and Computing 1356,
https://doi.org/10.1007/978-981-16-1402-6_1

1 Introduction

In contrast to the past, it is quite impossible to have a big showroom or market complex at the heart of a city due to scarcity of space and very high rents. Hence, the retailers usually are taking two warehouses, one in the marketplace named as owned warehouse (OW) with a finite capacity, and the other just nearby the market with a large capacity in rent named as rented warehouse. In actual practice, the management goes for a huge purchase of products once at a time, when either the seasonal demand is high or attractive discount in price for bulk. The actual service to the costumer is provided at owned warehouse only and after the items are exhausted in owned warehouse, items are transferred from rented warehouse to owned warehouse. Consequently, the inventory will diminish to zero at RW first and then OW.

During the last two decades, some inventory models for two warehouses have been developed and widely applied in the business world. The first model was developed by Hartely [10] where the transportation cost from RW to OW was not considered. In 1983, Sharma [25] modified Hartley's model by introducing the transportation cost. Murdeswar and Sathe [24] developed a combined work on finite refilling rate, which was corrected by Sathe in the year 1988 by mentioning the case of bulk unleash pattern for each finite and infinite refilling rate. In 1992, an EOQ model was introduced by Goswami and Choudhuri [9] for the items with two levels of shortages, where the demand is linear constant. Subsequently, many prominent models are developed such as by Bhunia et al. [1, 2], Huang [11], Jaggi et al. [16, 17], Lee and Hsu [19], Liang and Zhou [20], Mandal and Giri [23], Sheikh and Patel [26], Xu et al. [27], Yang [29], Yang and Chang [30], etc.

Demand has a great role in inventory management and in real life. It is observed that the demand for uncertain commodities not only depends upon selling price but it also depends upon some other factors like time horizon, inventory level, etc. Many researchers have been published their work on price-dependent demand rates in the last few decades. Usually, in the inventory models, resource constraints are assumed to be deterministic. But practically, the parameters engaged in the system fluctuate. Priory, the concept of probability had been used to handle such situations in the inventory models, but it was not more efficient to give the accuracy. To overcome such situations, we can use the parameters as fuzzy variables. Hence, the theoretical approach of fuzzy sets may be used to formulate the models. In 1970, Zadeh and Bellman [31] developed an inventory model on decision making with fuzzy approach. Henceforth, so many research papers were published on inventory models with fuzzy approach.

In 1999, using triangular-fuzzy-number, Chang [5] proposed a model for fuzzy production quantity. Subsequently, Chen and Ouyang [6] have developed a model for deteriorating-items allowing delay-payment, under a fuzzy environment. However, De and Rawat [7] established a model without shortages using triangular fuzzy number, which was extended for shortages by Indrajitsingha et al. [12] have established an EOQ model on time-dependent rate. They have also established a fuzzy model allowing shortages which are fully backlogged. A time-varying demand model for deteriorating items with shortages was developed by Jaggi et al. [15]. Many more papers are published by different researchers like Dutta and Kumar [8], Indrajitsingha et al. [13, 14], Kumar and Rajput [18], etc.

The works that have been developed in the last decade on two-warehouse model using fuzzy set theory are Matti [21] developed a fuzzy model with two warehouses under constraints and Mallick et al. [22] found a fuzzy mixture two-warehouse inventory model with linear demand. Yadav et al. [28] and Boina et al. [3, 4] developed a fuzzy-based two-warehouse inventory model in which the deterioration is instantaneous. Recently, Indrajitsingha et al. [14] established a fuzzy model for single deteriorating items, under two-warehouse system in which demand is selling price-dependent.

A large number of research papers have been published for two-warehouse system with demand as linear function or constant in crisp approach. But sometimes the demand behaves like an exponential function during some instances and it is very difficult to manage the inventory part in the business. Since the management has to keep stock more products for a high demand, a model is developed in such a way that it will give us more profit during a short period of time when the demand will rise exponentially. Presently, we try to develop such a model allowing shortages which are partially backlogged. Triangular fuzzy numbers are used for all the parameters like demand, deterioration, holding cost for both RW and OW, shortages cost, and Lost sale cost and for defuzzification, GMIR method has been employed. The total model is being illustrated by two numerical examples. Using Mathematica 11.1 software sensitivity analysis has been carried out.

2 Notations

OW	Owned warehouse.
RW	Rented warehouse.
$I_R(t)$	Stock amount in RW at time t.
$I_O(t)$	Stock amount in OW at time t.
α	Rate of deterioration.
β	Initial demand rate.
Θ	Positive demand parameter.
t_R	Time at which the inventory level in rented warehouse depletes to zero.
t_O	Time at which the inventory level in owned warehouse depletes to zero.
W	Storage capacity of OW.
s	Selling price (\$/unit/year).
$D(p)$	Demand rate depending upon selling price.
q	Order quantity.
PC	Purchasing cost per unit.
HC	Holding cost per unit.

DC	Deterioration cost per unit.
SC	Shortage cost per unit.
LC	Lost sale cost per unit.
S	Initial stock level.
q_1	Backorder quantity during stock out.
T	Length of the replenishment cycle.
p	Purchasing cost ($/unit/day).
k	Backlogging-rate.
h_R	Holding cost ($/unit/year) in Rented Warehouse.
h_O	Holding cost ($/unit/year) in Owned Warehouse.
d	Unit deterioration cost ($/unit/day).
A	Shortage cost per unit ($/unit/day).
B	Unit lost sale cost.($/unit/day).
$TAC(t_R, t_O)$	Total average cost ($/unit/day).
$\tilde{\alpha}$	Fuzzy deterioration rate.
$\tilde{h}_R$	Fuzzy holding cost ($/unit/day) in RW.
$\tilde{h}_O$	Fuzzy holding cost ($/unit/day) in OW.
$\tilde{A}$	Fuzzy shortage cost ($/unit/day).
$\tilde{B}$	Fuzzy opportunity cost due to lost sale ($/unit/day).
$\widetilde{TAC}(t_R, t_O)$	Fuzzy total cost ($/unit/day).
$\widetilde{TAC}_G(t_R, t_O)$	Defuzzified value of $\widetilde{TAC}(t_R, t_O)$ by applying GMIR method.

3 Assumptions

Throughout the paper, we assume that

(i) The model is based on a finite planning horizon.
(ii) The demand rate is of exponential function and depending upon selling price i.e. $D(s) = \beta s^{-\theta}, \theta, \beta > 0$.
(iii) Lead time is negligible.
(iv) The partially backlogging shortages are allowed.
(v) The capacity of OW is limited.
(vi) The RW has unlimited capacity.
(vii) The holding cost per unit in RW is more than that of OW.
(viii) Deterioration is considered to be by nature.
(ix) Higher powers of α are neglected.
(x) Items are kept in OW first.
(xi) The priority has given to the RW for first consumption.

4 Mathematical Formulation

Let q be the initial order quantity of an inventory discussed in the present model. Let 'W' units of inventory be kept in OW and the rest be dispatched into the RW. Since we are taking the maximum possible inventory in a bulk for this model, the order quantity must be greater than the initial stock level ($q > W$). Hence, it is necessary to take a RW for keeping the excess stock. Let $(0, t_R)$ be the time interval during which period the inventory level in RW reduces to zero level due to demand and deterioration. However, a portion of the inventory in OW is depleted due to deterioration only. Suppose at time $t = t_O$, the inventory level in OW becomes zero because of demand and deterioration. Thus, during the interval (t_R, t_O) the inventory in OW decreases due to demand and deterioration. After $t = t_O$ shortages will start. In this model, it is assumed that the rate of backlogging is a negative exponential function with respect to time. The pictorial observation of the above model is presented below in Fig. 1.

4.1 Crisp Model

The behavior of the problem is instructed by the differential equations given below.

$$\frac{dI_R(t)}{dt} = -\alpha I_R(t) - \beta s^{-\theta}, 0 \le t \le t_R \tag{1}$$

with $I_R(t_R) = 0$.

$$\frac{dI_O(t)}{dt} = -\alpha I_O(t), 0 \le t \le t_R \tag{2}$$

with $I_O(0) = W$.

$$\frac{dI_O(t)}{dt} = -\alpha I_O(t) - \beta s^{-\theta}, t_R \le t \le t_O \tag{3}$$

with $I_O(t_O) = 0$.
The solutions of (1)–(3) are as follows:

$$I_R(t) = \frac{\beta s^{-\theta}}{\alpha}\left\{e^{\alpha(t_R - t)}\right\} 0 \le t \le t_R I_R(t) \tag{4}$$

$$I_O(t) = We^{-\alpha t}, 0 \le t \le t_R, \tag{5}$$

$$I_O(t) = \frac{-\beta s^{-\theta}}{\alpha} + \frac{\beta e^{\alpha t_O}.s^{-\theta}}{\alpha} e^{-\alpha t} t_R \le t \le t_O. \tag{6}$$

From (5), we have

$$I_R(0) = S - W$$

$$S = \frac{\beta s^{-\theta}}{\alpha}\left(e^{\alpha t_R} - 1\right) \tag{7}$$

At $t = t_R$, Eqs. (5) and (6) yield.

$$W = \frac{\beta s^{-\theta}}{\alpha} e^{\alpha t_R}\left\{e^{\alpha(t_O - t_R)} - 1\right\} \tag{8}$$

$$S = \frac{\beta s^{-\theta}}{\alpha}\left\{e^{\alpha t_O} - 1\right\} \tag{9}$$

With the above data following parameters are calculated:
$PC = (S + q_1)p$,
where
$q_1 = \int_{t_O}^{T} k\beta s^{-\theta} dt$.
Then

$$PC = \left\{\beta s^{-\theta}\left\{\frac{e^{\alpha t_O} - 1}{\alpha} + k(T - t_O)\right\}\right\}p. \tag{10}$$

$$HC = HC_R + HC_O,$$

where

$$HC_R = h_R \int_0^{t_R} I_R(t)dt = \frac{h_R \beta s^{-\theta}}{\alpha}\left\{\left(\frac{e^{\alpha t_R} - 1}{\alpha}\right) - t_R\right\}. \tag{11}$$

and

$$\begin{aligned} HC_O &= h_O\left\{\int_0^{t_R} I_O(t)dt + \int_{t_R}^{t_O} I_O(t)dt\right\} \\ &= \frac{h_O}{\alpha}\left\{W\left(1 - e^{\alpha t_R}\right) - \frac{\beta s^{-\theta}}{\alpha}\left(1 - e^{\alpha(t_O - t_R)}\right) - \frac{\beta s^{-\theta}}{\alpha}(t_O - t_R)\right\} \end{aligned} \tag{12}$$

$$DC = DC_R + DC_O,$$

where

$$DC_R = d\left\{\beta s^{-\theta}\left\{\frac{e^{\alpha t_R} - 1}{\alpha} - t_R\right\}\right\}. \tag{13}$$

and

$$DC_O = d\{W - \beta s^{-\theta}(t_O - t_R)\}. \tag{14}$$

$$SC = A\int_{t_O}^{T} \beta s^{-\theta} dt = A\beta s^{-\theta}(T - t_O) \tag{15}$$

$$LC = B\int_{t_O}^{T} (1-k)\beta s^{-\theta} dt = B(1-k)\beta s^{-\theta}(T - t_O). \tag{16}$$

Total average cost $TAC(t_R, t_O)$ for this model during a cycle is given by

$$\begin{aligned} TAC(t_R, t_O) &= \frac{1}{T}[PC + HC + DC + SC + LC] \\ &= \frac{1}{T}\Bigg[\left\{\beta s^{-\theta}\left\{\frac{e^{\alpha t_O} - 1}{\alpha} + k(T - t_O)\right\}\right\}p \\ &\quad + \frac{h_R \beta s^{-\theta}}{\alpha}\left\{\left(\frac{e^{\alpha t_R} - 1}{\alpha}\right) - t_R\right\} \\ &\quad + \frac{h_O}{\alpha}\left\{W\left(1 - e^{\alpha t_R}\right) - \frac{\beta s^{-\theta}}{\alpha}\left(1 - e^{\alpha(t_O - t_R)}\right) - \beta s^{-\theta}(t_O - t_R)\right\} \\ &\quad + d\left\{\beta s^{-\theta}\left\{\frac{e^{\alpha t_R} - 1}{\alpha} - t_R\right\}\right\} \\ &\quad + d\{W - \beta s^{-\theta}(t_O - t_R)\} + A\beta s^{-\theta}(T - t_O) \\ &\quad + B(1-k)\beta s^{-\theta}(T - t_O)\Bigg] \end{aligned} \tag{17}$$

To minimize the total average cost function $TAC(t_R, t_O)$ per unit time, the values of t_R and t_O can be obtained by solving the equations

$$\frac{\partial TAC(t_R, t_O)}{\partial t_R} = 0 \text{ and } \frac{\partial TAC(t_R, t_O)}{\partial t_O} = 0. \tag{18}$$

satisfying

$$\frac{\partial^2 TAC(t_R, t_O)}{\partial t_R^2} > 0 \frac{\partial^2 TAC(t_R, t_O)}{\partial t_R^2} > 0, \frac{\partial^2 TAC(t_R, t_O)}{\partial t_O^2} > 0 \text{ and}$$

$$\left(\frac{\partial^2 TAC(t_R, t_O)}{\partial t_R^2}\right)\left(\frac{\partial^2 TAC(t_R, t_O)}{\partial t_O^2}\right) - \left(\frac{\partial^2 TAC(t_R, t_O)}{\partial t_R^2 \partial t_O^2}\right)^2 > 0$$

4.2 Fuzzy Model

It is very difficult to define all the system parameters exactly because of uncertainty in nature. Let us suppose that some of these parameters namely $\tilde{\alpha}, \tilde{h}_R, \tilde{h}_O, \tilde{A}, \tilde{B}$, may change within some limits.

Suppose $\tilde{\alpha} = (\alpha_1, \alpha_2, \alpha_3)$, $\tilde{h}_R = (R_1, R_2, R_3)$, $\tilde{h}_o = (O_1, O_2, O_3)$, $\tilde{A} = (a_1, a_2, a_3)$, and $\tilde{B} = (b_1, b_2, b_3)$ are considered as triangular fuzzy numbers.

Then the total average cost for the proposed model in fuzzy sense is given by

$$\begin{aligned}\widetilde{TAC}(t_R, t_O) = \frac{1}{T}\Bigg[&\left\{\beta s^{-\theta}\left\{\frac{e^{\tilde{\alpha} t_O} - 1}{\tilde{\alpha}} + k(T - t_O)\right\}\right\}p + \frac{\widetilde{h_R}\beta s^{-\theta}}{\tilde{\alpha}}\left\{\left(\frac{e^{\tilde{\alpha} t_R} - 1}{\tilde{\alpha}}\right) - t_R\right\} \\ &+ \frac{\widetilde{h_O}}{\tilde{\alpha}}\left\{W\left(1 - e^{\tilde{\alpha} t_R}\right) - \frac{\beta s^{-\theta}}{\tilde{\alpha}}\left(1 - e^{\tilde{\alpha}(t_O - t_R)}\right) - \beta s^{-\theta}(t_O - t_R)\right\} \\ &+ d\left\{\beta s^{-\theta}\left\{\frac{e^{\tilde{\alpha} t_R} - 1}{\tilde{\alpha}} - t_R\right\}\right\} + d\{W - \beta s^{-\theta}(t_O - t_R)\} \\ &+ \tilde{A}\beta s^{-\theta}(T - t_O) + \tilde{B}(1 - k)\beta s^{-\theta}(T - t_O)\Bigg] \end{aligned} \tag{19}$$

Using GMIR method for defuzzification of total average cost $\widetilde{TAC}(t_R, t_O)$, we get

$$\widetilde{TAC}_G(t_R, t_O) = \frac{1}{6}\left[\widetilde{TAC}_{G1}(t_R, t_O) + 4\widetilde{TAC}_{G2}(t_R, t_O) + \widetilde{TAC}_{G3}(t_R, t_O)\right] \tag{20}$$

where

$$\begin{aligned}\widetilde{TAC}_{G1}(t_R, t_O) = \frac{1}{T}\Bigg[&\left\{\beta s^{-\theta}\left\{\frac{e^{\alpha_1 t_O} - 1}{\alpha_1} + k(T - t_O)\right\}\right\}p + \frac{R_1\beta s^{-\theta}}{\alpha_1}\left\{\left(\frac{e^{\alpha_1 t_R} - 1}{\alpha_1}\right) - t_R\right\} \\ &+ \frac{O_1}{\alpha_1}\left\{W(1 - e^{\alpha_1 t_R}) - \frac{\beta s^{-\theta}}{\alpha_1}\left(1 - e^{\alpha_1(t_O - t_R)}\right) - \beta s^{-\theta}(t_O - t_R)\right\} \\ &+ d\left\{\beta s^{-\theta}\left\{\frac{e^{\alpha_1 t_R} - 1}{\alpha_1} - t_R\right\}\right\} + d\{W - \beta s^{-\theta}(t_O - t_R)\} + a_1\beta s^{-\theta}(T - t_O) \\ &+ b_1(1 - k)\beta s^{-\theta}(T - t_O)\Bigg] \end{aligned}$$

$$\begin{aligned}\widetilde{TAC}_{G2}(t_R, t_O) = \frac{1}{T}\Bigg[&\left\{\beta s^{-\theta}\left\{\frac{e^{\alpha_2 t_O} - 1}{\alpha_2} + k(T - t_O)\right\}\right\}p + \frac{R_2\beta s^{-\theta}}{\alpha_2}\left\{\left(\frac{e^{\alpha_2 t_R} - 1}{\alpha_2}\right) - t_R\right\} \\ &+ \frac{O_2}{\alpha_2}\left\{W(1 - e^{\alpha_1 t_R}) - \frac{\beta s^{-\theta}}{\alpha_2}\left(1 - e^{\alpha_1(t_O - t_R)}\right) - \beta s^{-\theta}(t_O - t_R)\right\} \\ &+ d\left\{\beta s^{-\theta}\left\{\frac{e^{\alpha_2 t_R} - 1}{\alpha_2} - t_R\right\}\right\} + d\{W - \beta s^{-\theta}(t_O - t_R)\} + a_2\beta s^{-\theta}(T - t_O) \\ &+ b_1(1 - k)\beta s^{-\theta}(T - t_O)\Bigg] \end{aligned}$$

$$\widetilde{TAC}_{G3}(t_R, t_O) = \frac{1}{T}\Bigg[\left\{\beta s^{-\theta}\left\{\frac{e^{\alpha_3 t_O}-1}{\alpha_3} + k(T-t_O)\right\}\right\}p + \frac{R_3\beta s^{-\theta}}{\alpha_3}\left\{\left(\frac{e^{\alpha_3 t_R}-1}{\alpha_3}\right) - t_R\right\}$$
$$+ \frac{O_3}{\alpha_3}\left\{W\left(1-e^{\alpha_1 t_R}\right) - \frac{\beta s^{-\theta}}{\alpha_3}\left(1-e^{\alpha_1(t_O-t_R)}\right) - \beta s^{-\theta}(t_O-t_R)\right\}$$
$$+ d\left\{\beta s^{-\theta}\left\{\frac{e^{\alpha_3 t_R}-1}{\alpha_3} - t_R\right\}\right\} + d\left\{W - \beta s^{-\theta}(t_O-t_R)\right\} + a_3\beta s^{-\theta}(T-t_O)$$
$$+ b_1(1-k)\beta s^{-\theta}(T-t_O)\Bigg]$$

To minimize the value of total average cost $\widetilde{TAC}_G(t_R, t_O)$ per unit time, the optimum value of t_R and t_O can be obtained by solving the equations
$\frac{\partial \widetilde{TAC}_G(t_R,t_O)}{t_R} = 0$ and $\frac{\partial \widetilde{TAC}_G(t_R,t_O)}{t_O} = 0$
satisfying

$$\frac{\partial^2 \widetilde{TAC}_G(t_R, t_O)}{\partial t_R^2} > 0,\ \frac{\partial^2 \widetilde{TAC}_G(t_R, t_O)}{\partial t_O^2} > 0 \text{ and}$$

$$\left(\frac{\partial^2 \widetilde{TAC}_G(t_R, t_O)}{\partial t_R^2}\right)\left(\frac{\partial^2 \widetilde{TAC}_G(t_R, t_O)}{\partial t_O^2}\right) - \left(\frac{\partial^2 \widetilde{TAC}_G(t_R, t_O)}{\partial t_R^2 \partial t_O^2}\right)^2 > 0$$

5 Numerical Example

For better illustration, we consider an inventory system with the following parametric values.

5.1 *Crisp Model*

Let us take the values of the system parameters as $\beta = 80$ units, $\theta = 0.001$, $s = \$30$/unit/day, $k = 0.7$ unit, $A = \$12$/unit/day, $B = \$16$/unit/day, $p = \$15$/unit/day, $\alpha = 0.006$, $W = 100$ units, $d = 0.5$ unit, $h_R = \$.0.07$/unit/day, $h_O = \$0.06$/unit/day, T = 365 days. Corresponding to these input values, $t_R = 33.9843$ days, $t_O = 81.6802$ days will minimize *TAC* and the minimum value is $TAC(t_R, t_O) = \$2056.61$ (Fig. 2).

5.2 *Fuzzy Model*

Let us take the values of the system parameters as $\beta = 80$ units, $\theta = 0.001$, $p = \$15/\text{unit/day}$, $s = \$30/\text{unit/day}$, $k = 0.7$ unit, $\tilde{\alpha} = (0.004, 0.006, 0.008)$, $\tilde{h}_R = (0.05, 0.07, 0.09)$, $\tilde{h}_O = (0.04, 0.06, 0.08)$ $\tilde{A} = (10, 12, 14)$, $\tilde{B} = (14, 16, 18)$, $W = 100$ units, $d = 0.5$ unit T = 365 days. Corresponding to these input values, $\tilde{t}_R = 34.3552$ days, $\tilde{t}_O = 83.1179$ days will minimize *TAC* and the minimum value is $\widetilde{TAC}(t_R, t_O) = \$2055.96$.

6 Sensitivity Analysis

See Tables 1, 2, 3, 4, 5 and Figs. 3, 4, 5, 6, 7.

The sensitivity analysis is performed by changing five parameters $\alpha, h_R, h_O, A, and B$ taking one parameter at a time and keeping the other parameters unchanged. It is observed that the parameters α, $A, and B$ are highly sensitive and the parameters h_R *and* h_O are slightly sensitive on total average cost and it is directly proportional to all the above parameters.

7 Conclusion

The article presented is a two-warehouse inventory model for deteriorating items with price-dependent demand and shortages with partially backlogging under fuzzy environment, where the demand is an exponential function. This is analyzed and verified by both crisp and fuzzy surroundings. For uncertainty in the parameters like deterioration rate, holding cost (both in RW and OW), shortage cost and lost sale cost, which are implemented by triangular fuzzy number and defuzzified by GMIR method. Here the total average cost decreases in fuzzy model as compared to crisp model. It is also observed that the parameters α, $A, and B$ are highly sensitive and the parameters h_R *and* h_O are slightly sensitive on total average cost and it is directly proportional to all the above parameters. This proposed model will be more effective and will give optimum results for a business plan in which the demand will grow exponentially in a certain time. The problem can be extended further either by neglecting the deterioration cost or using the preservation technology.

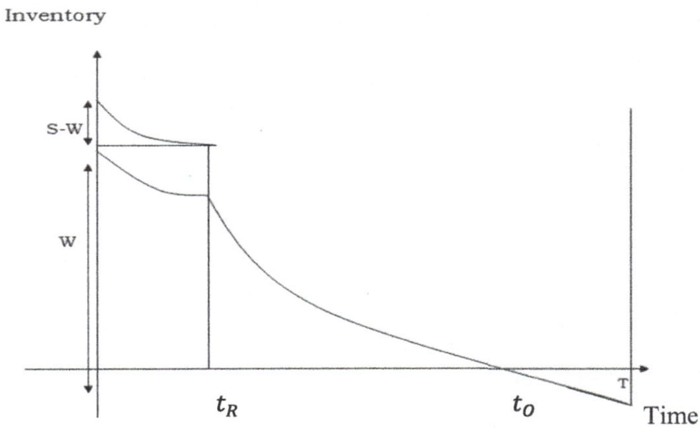

Fig. 1 Inventory level versus time graph for two warehouses

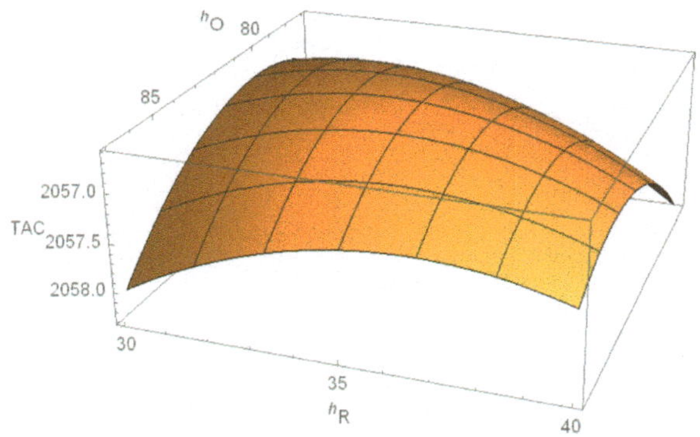

Fig. 2 Graphical representation of convexity of the function

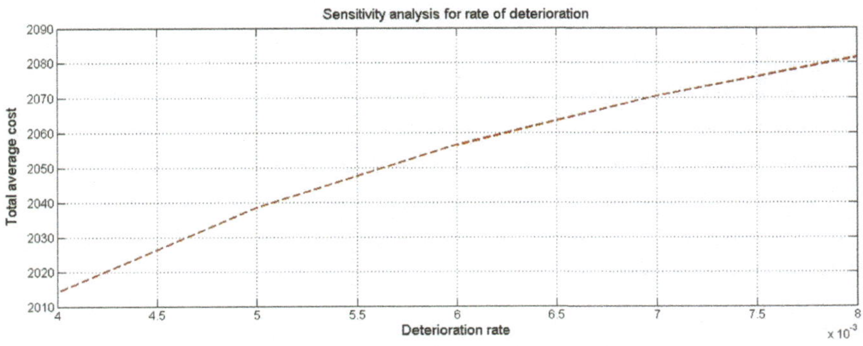

Fig. 3 TAC versus deterioration rate

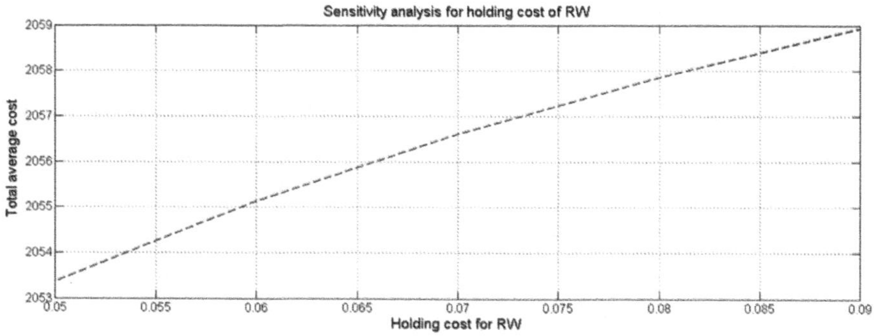

Fig. 4 TAC versus Holding cost for RW

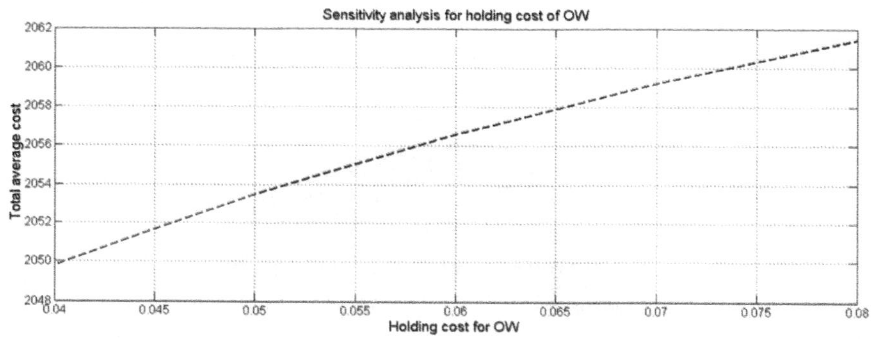

Fig. 5 TAC versus Holding cost for OW

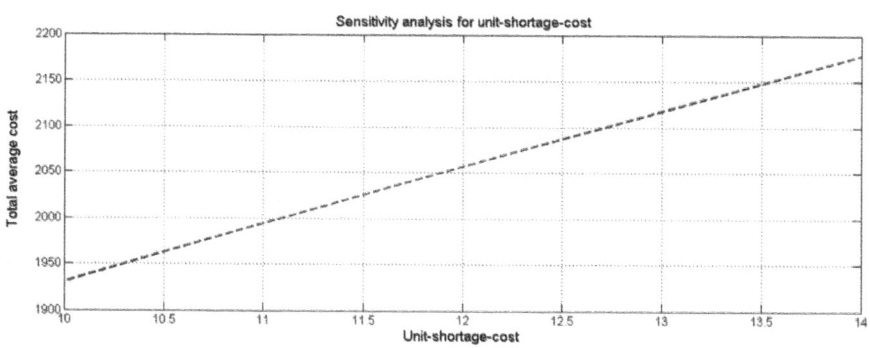

Fig. 6 TAC versus Unit shortage cost

Fig. 7 TAC versus Lost sale cost

Table 1 Sensitivity analysis for the rate of deterioration $(\tilde{\alpha})$

$\tilde{\alpha}$	t_R	t_O	$\widetilde{TAC}_G$
(0.004, 0.005, 0.006)	40.2038	95.1125	2037.4983
(0.005, 0.006, 0.007)	34.2147	82.1779	2055.9166
(0.006, 0.007, 0.008)	29.6605	72.3421	2070.02

Table 2 Sensitivity analysis for holding cost in RW $(\tilde{h}_R)$

$\tilde{h}_R$	t_R	t_O	$\widetilde{TAC}_G$
(0.05, 0.06, 0.07)	37.0007	82.7340	2055.09
(0.06, 0.07, 0.08)	34.0587	81.7047	2056.57
(0.07, 0.08, 0.09)	31.5705	80.8249	2057.83

Table 3 Sensitivity analysis for holding cost in OW $(\tilde{h}_O)$

$\tilde{h}_O$	t_R	t_O	$\widetilde{TAC}_G$
(0.04, 0.05, 0.06)	31.1339	83.4473	2053.41
(0.05, 0.06, 0.07)	33.9026	81.7237	2056.53
(0.06, 0.07, 0.08)	36.1700	80.2663	2059.15

Table 4 Sensitivity analysis for shortage cost $(\tilde{A})$

$\tilde{A}$	t_R	t_O	$\widetilde{TAC}_G$
(10, 11, 12)	31.4375	76.2882	1993.94
(11, 12, 13)	33.9728	81.6553	2056.41
(12, 13, 14)	36.4397	86.873	2117.74

Table 5 Sensitivity analysis for Lost sale cost $(\tilde{B})$

$\tilde{B}$	t_R	t_O	$\widetilde{TAC}_G$
(14, 15, 16)	33.2301	80.0843	2037.98
(15, 16, 17)	33.9832	81.6779	2056.59
(16, 17, 18)	34.7301	83.2580	2075.11

References

1. A.K. Bhunia, C.K. Jaggi, A. Sharma, R.A. Sharma, Two-warehouse inventory model for deteriorating items under permissible delay in payment with partial backlogging. Appl. Math. Comput. **232**(1), 1125–1137 (2014)
2. A.K. Bhunia, A.A. Shaikh, R.K. Gupta, A study on two-warehouse partially backlogged deteriorating inventory models under inflation via particle swarm optimization. Int. J. Syst. Sci. **46**(8), 1036–1050 (2015)
3. A.K. Boina, S.K. Paikray, H. Dutta, Cost optimization model for items having fuzzydemand and deterioration with two-warehouse facility under the trade credit financing. AIMS Math. **5**, 1603–1620 (2020)
4. A.K. Boina, S.K. Paikray, U. Misra, Two-storage fuzzy inventory model with time dependent demand and holding cost under acceptable delay in payment. Math. Model. Anal. **25**, 441–460 (2020)
5. S.C. Chang, Fuzzy production inventory for fuzzy product quantity with triangular fuzzy number. Fuzzy Sets Syst. **107**, 37–57 (1999)
6. L.H. Chen, L.Y. Ouyang, Fuzzy inventory model for deteriorating items with permissible delay in payment. Appl. Math. Comput. **182**(1), 711–726 (2006)
7. P.K. De, A. Rawat, A fuzzy inventory model without shortages using triangular fuzzy number. Fuzzy Inf. Eng. **1**, 59–68 (2011)
8. D. Dutta, P. Kumar, Fuzzy inventory model for deteriorating items with shortages under fully backlogged condition. Int. J. Soft Comput. Eng. **3**(2), 393–398 (2012)
9. A. Goswami, K.S. Chaudhuri, An economic order quantity model for items with two levels of storage for a linear trend in demand. J. Oper. Res. Soc. **43**(2), 155–167 (1992)
10. V.R. Hartely, *Operations Research-A Managerial Emphasis'* (Good Year Publishing Company, Santa Monica, California, 1976), pp. 315–317
11. Y.F. Huang, An inventory model under two levels of trade credit and limited storage space derived without derivatives. Appl. Math. Model. **30**(5), 418–436 (2006)
12. S.K. Indrajitsingha, S.S. Routray, S.K. Paikray, U. Misra, Fuzzy economic production quantity model with time dependent demand rate. LogForum **12**(3), 193–198 (2016)
13. S.K. Indrajitsingha, P.N. Samanta, U.K. Misra, A production fuzzy inventory model for deteriorating items defuzzified by graded mean integration representation method. J. Ultra Sci. Phys. Sci. Sect. A Math. **29**(9), 388–400 (2017)
14. S.K.Indrajitsingha, P.N. Samanta, U.K. Misra, A fuzzy two ware-house Inventory model for singledeteriorating items with selling price-dependent demand and shortage under partial backlogged condition. Int. J. Appl. Appl. Math. (AAM) **14**, 511–536 (2019)
15. C.K. Jaggi, S. Pareek, A. Sharma, Nidhi, , Fuzzy inventory model for deteriorating items with time-varying demand and shortages. Am. J. Oper. Res. **2**(6), 81–92 (2012)
16. C.K. Jaggi, S. Pareek, P. Verma, R. Sharma, Ordering policy for deteriorating items in a two-warehouse environment with partial backlogging. Int. J. Logist. Syst. Manag. **16**(1), 16–40 (2013)
17. C.K. Jaggi, L.E. Cardenas-Barron, S. Tiwari, A. Shafi, A Two-warehouse inventory model for deteriorating items with imperfect quality under the conditions of permissible delay in payments. Sci. Iran. **157**(2), 344–356 (2017)
18. S. Kumar, U.S. Rajput, Fuzzy inventory model for deteriorating items with time dependent demand and partial backlogging. Appl. Math. **6**, 496–509 (2015)
19. C.C. Lee, S.L. Hsu, A two-warehouse production model for deteriorating inventory items with time dependent demands. Eur. Oper. Res. **194**(3), 700–710 (2009)
20. Y. Liang, F. Zhou, A two-warehouse inventory model for deteriorating items under conditionally permissible delay in payments. Appl. Math. Model. **35**(1), 2221–2231 (2011)
21. M.M. Maiti, M. Maiti, Fuzzy inventory model with two-warehouse under possibility constraints. Fuzzy Sets Syst. **157**(1), 52–73 (2006)
22. A.K. Malik, Y. Singh, A fuzzy mixture two warehouse inventory model with linear demand. Int. J. Appl. Innov. Eng. Manag. **2**(1), 180–186 (2013)

23. P. Mandal, B.C. Giri, A two-warehouse integrated inventory model with imperfect production process under stock dependent demand quantity discount offer. Int. J. Syst. Sci. Oper. Logist. **4**(4), 1–12 (2017)
24. T.M. Murdeswar, Y.S. Sathe, Some aspects of lot size model with two levels of storage. Opsearch **22**(4), 255–262 (1985)
25. K.V. Sharma, A deterministic inventory model with two levels of storage and optimum release rule. Opsearch **20**(1), 175–180 (1983)
26. S.R. Sheikh, R. Patel, Two-warehouse inventory model with different deterioration rates under time dependent demand and shortages. Glob. J. Pure Appl. Math. **13**(8), 3951–3960 (2017)
27. X. Xu, Q. Bai, M. Chen, A comparison of different dispatching policies in two-warehouse inventory systems for deteriorating items over a finite time horizon. Appl. Math. Model. **41**(1), 1–16 (2016)
28. A.S. Yadav, S. Sharma, A. Swami, A fuzzy based two-warehouse inventory model for non-instantaneous deteriorating items with conditionally permissible delay in payment. Int. J. Control. Theory Appl. **10**(10), 107–123 (2017)
29. H.L. Yang, Two-warehouse inventory models for deteriorating items with shortages under inflation. Eur. J. Oper. Res. **157**(2), 344–356 (2004)
30. H.L. Yang, C.T. Chang, A two-warehouse partial backlogging inventory model for deteriorating items with permissible delay in payment under inflation. Appl. Math. Model. **37**(1), 2717–2726 (2013)
31. L.A. Zadeh, R.E. Bellman, Decision making in a fuzzy environment. Manag. Sci. **17**, 140–164 (1970)

A Comparative Study of Nonlinear Convection in a Confined Fluid Overlying a Porous Layer

Atul K. Srivastava

Abstract The alternating direction implicit method (ADI method) has been used to investigate a comparative study between the Darcy (DM), Brinkman-extended Darcy (BM) and Brinkman–Forchheimer-extended Darcy models (BFM) of free convection in a cavity containing a fluid layer overlying a porous layer saturated with the same fluid. The two-dimensional enclosure is heated from below and cooled from above, while the other two vertical sides are adiabatic. The Beavers–Joseph empirical boundary condition at the fluid/porous layer interface is employed, while the Darcy model is used to simulate momentum transfer in a porous medium. In case of the BM and BFM models, the two regions are coupled by matching the velocity and stress components at the interface. Numerical simulations have revealed that all three models have yielded almost same results for Darcy numbers up to about 10^{-4}. For higher Darcy numbers, the BFM model can be used to simulate momentum transfer as it accounts for the effects of inertia.

Keywords Darcy model · Brinkman-extended Darcy model · Brinkman–Forchheimer-extended Darcy model · Natural convection

A. K. Srivastava (✉)
Department of Mathematics, Central University of Jharkhand, Ranchi, India
e-mail: atulshaswat@gmail.com

S. K. Paikray et al. (eds.), *New Trends in Applied Analysis and Computational Mathematics*, Advances in Intelligent Systems and Computing 1356,
https://doi.org/10.1007/978-981-16-1402-6_2

List of Symbols

Latin Symbols

Ar	aspect ratio in x' direction $\frac{H}{L}$.
C	inertia coefficient in the Brinkman–Forchheimer-extended Darcy modal.
D_f	$\frac{D'_f}{L}$.
D'_f	depth of the fluid layer from upper surface, m.
Da	Darcy number.
g	gravitational acceleration, m/s^2.
H	height of the enclosure, m.
k	permeability of porous layer, m^2.
L	length of the enclosure in y'-direction, m.
Nu_av	overall average Nusselt number.
$Nu(x)$	local Nusselt number.
Pr	Prandtl number.
Ra	Rayleigh number.
Rc	ratio of thermal conductivities of porous and fluid layers.
T'	temperature, ^0K.
$u, v,$	dimensionless velocity components in x- and y-directions.
x, y	non-dimensional coordinates.
x', y'	coordinates, m.

Greek symbols

α	thermal diffusivity, m/s^2.
$\overline{\alpha}$	coefficient in the Beavers–Joseph matching condition.
β	coefficient of thermal expansion, $^0K^{-1}$.
ϵ	porosity of porous region.
κ	thermal conductivity, w/m^0K.
ψ	dimensionless stream function.
ψ'	stream function, m^2/s.
μ	viscosity, kg/ms.
$\Delta T'$	temperature difference, ^0K.
θ	dimensionless temperature.
ν	viscous diffusivity, m^2/s.
ζ	non-dimensional vorticity.
ζ'	vorticity, $1/s$.

Subscripts

c	cold wall.
f	fluid layer.
h	hot wall.
p	porous layer.

1 Introduction

Transport phenomena due to natural convection in a composite system consisting of fluid and a porous medium saturated with the same fluid are of considerable interest due to their numerous applications in different fields. It occurs in a variety of fields, such as industrial applications (multi-component solidification, thermal insulation, drying process, cooling and fumigation of stored food grains, cooling of aluminum billets, filtration, etc.) or in the environment (benthic boundary layers, groundwater pollution, etc.). A typical example of this configuration is the liquid alloys above the frozen alloys at the bottom, separated by a mushy zone, when the mold containing concentrated alloys is cooled from below for directional solidification.

Most of the investigations performed assumed a two-domain approach and stability analysis have been carried out. In these cases, conservation equations in the fluid and porous regions are coupled by inter-facial boundary conditions which can depend on partial differential equations. The pioneering investigation on an exact solution for the thermal stability of superposed porous and fluid layers with rigid top and bottom boundaries for the case in which constant heat flux is applied at the bottom is performed by Nield [20]. Nield [21], Chen and Chen [10], Poulikakos et al. [24], Carr [8] and Carr and Straughan [9] have performed numerical studies for natural convection. The analysis presented in this paper is based on assumption that for moment transport, most of the studies use Darcy's law in porous region and Navier–Stokes equation in fluid region. Under these circumstances, the coupling between the two homogeneous regions is obtained using a slip boundary condition [2], where the slip coefficient depends on the local nature and the position of the interface [3]. Brinkman [6], used the Brinkman-extended Darcy models (termed as Brinkman model) to describe the flow in the porous layer and equate all components of the velocity and stress at the interface. Chen and Chen [10] produced a classical paper in which they studied thermal convection in a two-layer system composed of a porous layer saturated with fluid over which was a layer of the same fluid. The system was heated from below and the depth of both the layers considered to be fixed. These results were qualitatively and quantitatively verified by the experimental work of Chen and Chen [11].

Chen and Chen [12] extended his previous work [11] by assuming that the motion of the fluid in porous layer is governed by Darcy's equation with the Brinkman terms for viscous effects and the Forchheimer term for inertial effects. Numerical results were obtained using a combination of Galerkin and finite difference method. Nishimura et al. [23] and Sathe et al. [26] used this model in their studies of natural convection in rectangular enclosures containing a porous/fluid composite system by considering the interia term as used in the Navier-stokes equation. As proposed by Forchheimer [14], Beckermann et al. [4, 5] investigated their model by considering inertia term, therefore, due to this modification, this is known as the Brinkman–Forchheimer-extended Darcy model (termed as Brinkman–Forchheimer model) in the literature. Kim and Choi [16] investigated convective and diffusive phenomena between porous and overlying fluid layer, for Brinkman–Forchheimer-extended

Darcy model for Porous media. His study provided a fundamental framework for predicting heat transfer, chemical diffusion, and fluid flow between a fluid and a saturated porous medium.

The problem of Chen and Chen [10] has been solved numerically by Bukhari [7] using first- and second-order Chebyshev tau methods. Mckay [18] studied a similar porous-fluid layer problem to [10], but he allowed chemical reactions in the layers. Valencia-Lopez and Ochoa-Tapia [32] presented a comparison of two models, Darcy and Brinkman, to study the buoyancy-driven flow in a confined fluid overlying a porous layer. Straughan [28, 29] used a D^2 Chebyshev tau method to obtain the surface tension-driven convection results for the effect of porosity and modeling, respectively, in a fluid overlying a porous layer.

Two-domain approach consists of using Brinkman correction in Darcy's law allowing to satisfy the continuity of both velocity and stress at the fluid/porous interface [6, 19]. Only one stability analysis has been performed using this modeling. This comparison with the results obtained using the Beavers and Joseph condition [2] shows a quantitative agreement depending on the values of the slip coefficient [15]. Recently, [1] investigated a linear stability analysis in horizontal fluid layer overlying a layer of porous media saturated with the same fluid, with uniform heating from below in the presence of vertical magnetic field. The flow is assumed to be governed by Darcy' s law. The Beavers–Joseph condition is applied at the interface between two layers. He used first-order Chebyshev tau method to solve linear stability equations including magnetic field.

The above discussion illustrates that the momentum transfer in the porous region can be modeled by either the Darcy, Brinkman, or Forchheimer models. Application of the Darcy model with the Beavers–Joseph condition results in a slip velocity condition at the interface, as well as at the impermeable walls. On the other hand, the use of the Brinkman or Brinkman–Forchheimer models with the continuity of the velocity and stress components at the interface satisfies the no-slip criteria at the interface as well as the impermeable surfaces. Singh and Thorpe [27] have presented a comparative study between these models for confined fluid overlying a porous layer when it is heated from one vertical side and cooled on opposite side. The other two sides are taken to be adiabatic.

In the present work, our focus is to compare the above three momentum transport models when they are applied to free convection processes in a rectangular enclosure containing fluid and porous layers when heated from below and cooled from above. The governing equations are formulated in terms of derived variables such as vorticity and a stream function. The set of coupled, nonlinear PDEs concerned with the behavior of the system is solved by the Samarskii and Andreyev [25] alternating direction implicit method by adding false transient terms as suggested by Mallinson and Davis [17]. Here, the thermal conductivities of the fluid and porous medium are assumed to be same.

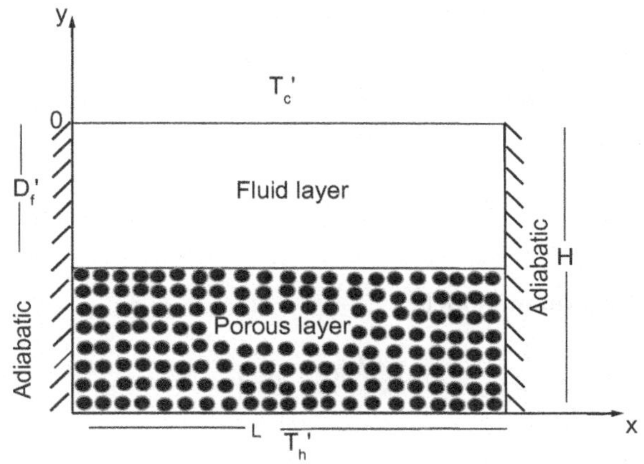

Fig. 1 Schematic diagram of physical model

2 Analysis

The schematic configuration is composed of a horizontal porous bed underlying a fluid layer as shown in Fig. 1. Overlying fluid which is assumed to be Newtonian and suits Boussinesq's approximation can flow from one layer to another as the interface is considered to be permeable. The horizontal walls of the composite system are impermeable and are maintained at different temperatures T_h' and T_c' such that $T_h' > T_c'$, whereas the other two walls are adiabatic (perfectly thermally insulated). Due to heating from below, there is a density difference and thus arises the phenomena of natural convection within cavity. We have derived the equation of motion in terms of the vorticity and stream function defined by

$$u = \frac{\partial \psi}{\partial y}, v = -\frac{\partial \psi}{\partial x}, \zeta = \frac{\partial v}{\partial x} - \frac{\partial u}{\partial y}, \tag{1}$$

for each domain.

The governing equations for each region, in non-dimensional form (in vector form; see the) are given (as described by Nield and Bejan [22]) in terms of vorticity, stream function, and temperature under mentioned assumptions as

Fluid region:

$$u_f \frac{\partial \zeta_f}{\partial x} + v_f \frac{\partial \zeta_f}{\partial y} = Pr \left(\frac{\partial^2 \zeta_f}{\partial x^2} + \frac{\partial^2 \zeta_f}{\partial y^2} \right) + Ra Pr \frac{\partial \theta_f}{\partial y} \tag{2}$$

$$\frac{\partial^2 \psi_f}{\partial x^2} + \frac{\partial^2 \psi_f}{\partial y^2} = -\zeta_f \tag{3}$$

$$u_f \frac{\partial \theta_f}{\partial x} + v_f \frac{\partial \theta_f}{\partial y} = \left(\frac{\partial^2 \theta_f}{\partial x^2} + \frac{\partial^2 \theta_f}{\partial y^2} \right) \tag{4}$$

Momentum equation for porous region in terms of the vorticity for different models,

Darcy Model:
Darcy model contains no viscous term and therefore the corresponding equation is derived directly in terms of the stream function as

$$\frac{\partial^2 \psi_p}{\partial x^2} + \frac{\partial^2 \psi_p}{\partial y^2} = -DaRa\frac{\partial \theta_p}{\partial y} \tag{5}$$

Brinkman model:

$$u_p\frac{\partial \zeta_p}{\partial x} + v_p\frac{\partial \zeta_p}{\partial y} = Pr\left(\frac{\partial^2 \zeta_p}{\partial x^2} + \frac{\partial^2 \zeta_p}{\partial y^2}\right) + RaPr\frac{\partial \theta_p}{\partial y} - \frac{Pr}{Da}\zeta_p \tag{6}$$

Brinkman–Forchheimer model:

$$\left(\frac{\partial^2 \zeta_p}{\partial x^2} + \frac{\partial^2 \zeta_p}{\partial y^2}\right) + RaPr\frac{\partial \theta_p}{\partial y} - \left(\frac{Pr}{Da} + \frac{C\,|V_p|}{Da^{\frac{1}{2}}}\right)\zeta_p - \frac{C}{Da^{\frac{1}{2}}}\left(v_p\frac{\partial |V_p|}{\partial x} - u_p\frac{\partial |V_p|}{\partial y}\right) \tag{7}$$

The derived equation in terms of the stream function is the same for both models and it is given by

$$\frac{\partial^2 \psi_p}{\partial x^2} + \frac{\partial^2 \psi_p}{\partial y^2} = -\zeta_p \tag{8}$$

The energy equation is same for all the models in the porous region and is expressed by the following equation:

$$u_p\frac{\partial \theta_p}{\partial x} + v_p\frac{\partial \theta_p}{\partial y} = Rc\left(\frac{\partial^2 \theta_p}{\partial x^2} + \frac{\partial^2 \theta_p}{\partial y^2}\right) \tag{9}$$

As suggested by Neale and Nader [19], we have considered $\mu_p = \mu_f$ in deriving the above equations due to the fact that this provides good agreement with experimental data. The equations have been rendered dimensionless by using the following transformations:

$$(x, y) = \frac{(x', y')}{L},\ (u, v) = \frac{(u', v')}{\left(\frac{\alpha_f}{L}\right)}, \tag{10}$$

$$\zeta = \frac{\zeta'}{\left(\frac{\alpha_f}{L^2}\right)},\ \psi = \frac{\psi'}{\alpha_f},\ \theta = \frac{(T' - T_0')}{\Delta T'} \tag{11}$$

The non-dimensional physical parameters are Darcy number (Da), Prandtl number (Pr), Rayleigh number (Ra), and ratio of thermal conductivities of porous and fluid layers (Rc) corresponding to each domain are defined as

$$Da = \frac{K}{L^2},\ Pr = \frac{\nu_f}{\alpha_f},\ Ra = \frac{g\beta\Delta T' L^3}{\nu_f \alpha_f},\ Rc = \frac{\kappa_p}{\kappa_f} \tag{12}$$

Using the empirical formula [13], the inertia coefficient C appearing in Eq. (7) can be evaluated, i.e.,

$$C = \frac{1.75}{\sqrt{(175\varepsilon^3)}} \tag{13}$$

Following boundary conditions of the two-dimensional region bounded by the planes $x = 0$, Ar and $y = 0, 1$ are imposed on the temperature field:

$$\frac{\partial \theta}{\partial x} = 0 \quad on \quad x = 0 \quad and \quad 1, \tag{14}$$

$$\theta = 1 \quad on \quad y = 0 \quad and \quad \theta = 0 \quad y = Ar. \tag{15}$$

At the boundaries, vanishing of velocity components implies that

$$\psi = 0. \tag{16}$$

Boundary conditions for the vorticity in terms of the stream function are taken as

$$\zeta = -\frac{\partial^2 \psi}{\partial x^2} \quad on \quad x = 0 \quad and \quad 1, \tag{17}$$

$$\zeta = -\frac{\partial^2 \psi}{\partial y^2} \quad on \quad y = 0 \quad and \quad Ar. \tag{18}$$

Matching condition at the fluid/porous interface

By considering the continuity of temperature and heat flux, the respective matching conditions, at the interface $y = D_f$ are

$$\theta_f = \theta_p, \quad \frac{\partial \theta_f}{\partial y} = Rc \frac{\partial \theta_p}{\partial y}, \tag{19}$$

However, it cannot be possible to specify directly the boundary conditions on the vorticity and stream function and it is necessary to derive from matching conditions of velocity and stress components at the interface which are obtained as

$$\zeta_f = \zeta_p, \quad \frac{\partial \zeta_f}{\partial y} = \frac{\partial \zeta_p}{\partial y} \tag{20}$$

$$\psi_f = \psi_p, \quad \frac{\partial \psi_f}{\partial y} = \frac{\partial \psi_p}{\partial y} \tag{21}$$

that are applicable only in the cases of the Brinkman and Brinkman–Forchheimer models. In the case of Darcy model, the stream function at the interface is obtained by using the matching conditions

$$v_f = v_p, \quad \frac{\partial u_f}{\partial y} = \frac{\overline{\alpha}\left(u_f - u_p\right)}{\sqrt{Da}} \tag{22}$$

proposed by Beavers and Joseph [2]. The vorticity at the interface is calculated by using the expression

$$\zeta_f = \frac{\partial v_f}{\partial x} - \frac{\partial u_f}{\partial y}. \quad (23)$$

3 Numerical Procedure

Due to the nonlinear behavior of autonomous differential equations, we have to solve it using the numerical method. By adding the false transient method as introduced by Mallinson and Davis [17], the resulting parabolic transformed equations are derived by using central difference for spatial derivative and backward difference for time derivative. Using the alternating direction implicit method (ADI) introduced by Samarskii and Andreyev [25], the resulting finite difference equations in tridiagonal form are solved by Thomas algorithm [31].

We have done the numerical integration for each dependent variable in the composite system separately and moved according to the following steps. In the first step, using the finite difference form of the fluid layer equations, our calculation proceeds until the grid point situated immediately above the horizontal interface $x = D_f$ is reached, whereas every dependent variable is obtained from the upper horizontal surface. In the next step, the finite difference equation obtained with respect to the porous layer equations is used to advanced the solution process from the grid points situated immediately below the interface down to the lower horizontal surface. In the last step, the value of the dependent variables at the inter-facial grid points is obtained using the matching conditions at the interface.

The solution procedure is iterated until a quasi-steady state is approached by satisfying the following convergence criterion with respect to each variable:

$$\frac{\sum_i \sum_j \left| \tau_{i,j}^{n+1} - \tau_{i,j}^{n} \right|}{m \, |\tau|_{max}} < \tau_{err}, \quad (24)$$

where m is the number of the interior grid points, $|\tau|_{max}$ the maximum magnitude of τ, τ_{err} some pre-determined error criterion determined by numerical experiment and τ may stand for temperature, vorticity and stream function. The superscripts denote the values of the dependent variables after the nth and $(n+1)$th integrations, respectively, whereas Indies i and j indicate grid location in the (x, y) plane. In the case of $Ra = 10^4$ and 10^5, τ_{err} is set at 10^{-5}, whereas for $Ra = 10^6$, it is set as 10^{-4} because of the large number of iterations required to satisfy the first criterion.

For the validity of our results, we have compared the computed results with published work of Poulikakos et al. [24], Chen and Chen [12], and Kim and Choi [16] in the case of cavities, considering with a single-phase fluid, a porous layer and a fluid-porous layer. The correctness of the numerical code was found to somewhat insensitive to the range of grid size. Therefore, the investigation has been performed on a uniform grid, and results are obtained for 51×51 grid point in all three cases.

Calculations were made for the parameters used by Poulikakos et al. [24] for the purpose of comparison. As [24, 30] observed, we found that when $Da = 10^{-3}$ for all Rayleigh numbers the solutions based on the B-J condition became unstable when the mesh size was reduced from that obtaining on a 31×31 grids. As α that is B-J coefficient changes from 1.0 to 2.0, the solution is stable.

4 Results and Discussions

There are six non-dimensional parameters namely Da, Ra, Rc, Pr, D_f, and Ar, that control the behavior of transport phenomena in the composite system. It is very difficult to study the effect of all parameters in a single study because it makes the computational work a formidable task. Hence, we have taken $Pr = 0.71$, $Ar = 1.0$, and $Rc = 1.0$ and concentrated to investigate the effects of Ra, Da and D_f.

We found from this study that BM and BFM are predicting almost the same results. We have plotted the Figs. 2 and 3 for the isotherms and streamlines for the DM and BM. In Fig. 2, the composite system is divided equally into fluid and porous regions $(D_f = 0.5)$ and in Fig. 3, the fluid layer is thicker than the porous layer $(D_f = 0.75)$. In these figs., as the Rayleigh number increases by one order of magnitude at a time from 10^4 to 10^6. There is symmetric cellular flow whose intensity increases with Rayleigh number for both the models and the shape of two-dimensional rolls changes from circular to elliptic. These figures clearly show the presence of kinks in the streamlines that cross the plane of the fluid–bed interface. The kinks are disappearing for a higher Rayleigh number. This indicates that the flow is progressively forced out of the porous bed. From Fig. 3, it is clear that when porous medium is less, the visibility of kinks goes down. Also from these figures, we can clearly see that isotherms and streamlines indicate, convection increases for increasing Ra. These result for DM is similar to the results of Poulikakos et al. [24]. Comparative study shows that DM is predicting more convection than BM and BFM. This is also confirmed by tables.

We have also done the comparative study for the maximum absolute values of the stream function and average Nusselt numbers representing the overall heat transfer rate at the heated wall. The average Nusselt number is obtained by using the following expression:

$$Nu_{av} = \frac{1}{Ar} \int_0^{Ar} Nu\,(x)\,dx, \tag{25}$$

where $Nu(y)$ is the local Nusselt number on the hot wall defined by

$$Nu\,(y) = -Rc\frac{\partial \theta_p}{\partial x}. \tag{26}$$

For the three models, the average Nusselt number on the heated wall and the maximum absolute value of the stream function corresponding to $D_f = 0.5$ and $D_f = 0.75$ are calculated for a range of Rayleigh and Darcy numbers and are shown in Tables 1 and 2, respectively. From these tables, we can conclude that changes in the values of

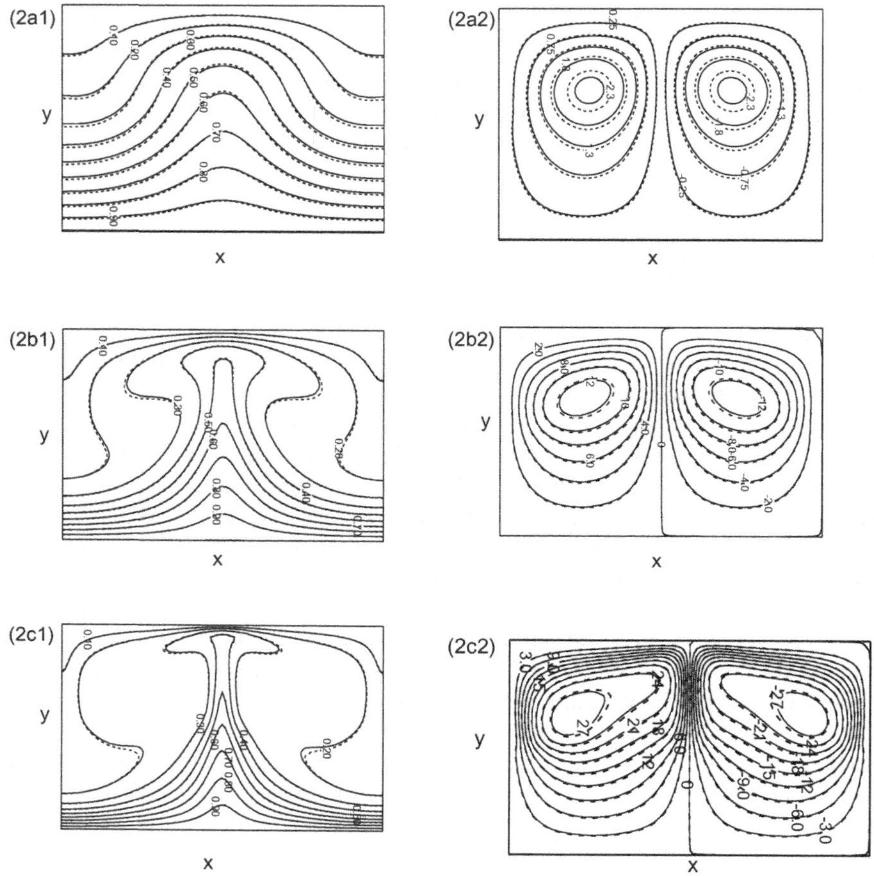

Fig. 2 Isotherms and streamlines for $Da = 10^{-5}$ and $D_f = 0.5$; (2a1, 2a2) $Ra = 10^4$, (2b1, 2b2) $Ra = 10^5$ and (2c1, 2c2) $Ra = 10^6$; $\cdots$ Darcy model, – Brinkman model

Nu_{av} and $|\psi|_{max}$ are less than 1% and lower in the case when Ra increases. This suggests that for a comparative study between DM, BM, and BFM, there is no significant effect of resistance due to viscous as well as interia term on the computed velocity and temperature fields. Numerically, BM and BFM are different for higher values of Darcy number(10^{-3}) and for different inertia terms in the models. The parameter C in the interia term is considered a fixed value for comparative study, but for specific physical systems, the value of parameter may assume different values.

For high Rayleigh number, DM gives more differences in Nu_{av} and $|\psi|_{max}$ than BM and BFM because of the lack of inertia and viscous terms. Although for the best of authors's knowledge, very little work has been done on calculating the Nu_{av} and $|\psi|_{max}$ in superposed fluid when heated from below in porous cavity, we found results very similar to the published work of Poulikakos et al. [24].

Table 1 Comparison of the average Nusselt number and maximum absolute value of stream function obtained by using the Dracy, Brinkman, and Brinkman–Forchheimer models when $D_f = 0.5$

Ra	Da	Nu_{av}			ψ		
		DM	BM	BFM	DM	BM	BFM
10^4	10^{-3}	1.3625,$\bar{\alpha} = 2$	1.44	1.44	2.51	2.89	2.87
10^4	10^{-4}	1.27	1.27	1.28	2.2	2.21	2.21
10^4	10^{-5}	1.27	1.24	1.25	2.2	2.07	2.08
10^4	10^{-6}	1.27	1.24	1.24	2.21	2.05	2.06
10^5	10^{-3}	3.4822,$\bar{\alpha} = 2$	3.2	3.18	14.14	13.95	13.78
10^5	10^{-4}	2.8797,$\bar{\alpha} = 1$	2.82	2.88	12.82	12.66	12.78
10^5	10^{-5}	2.8093,$\bar{\alpha} = 1$	2.78	2.8	12.68	12.48	12.53
10^5	10^{-6}	2.8038,$\bar{\alpha} = 1$	2.78	2.79	12.67	12.47	12.48
10^6	10^{-3}	6.8338,$\bar{\alpha} = 2$	6.23	6.22	35.32	30.01	28.54
10^6	10^{-4}	4.74	4.7	4.44	26.9	28.01	26.37
10^6	10^{-5}	4.48	4.41	4.47	29.33	28.83	28.81
10^6	10^{-6}	4.43	4.38	4.4	29.26	28.81	28.81

Table 2 Comparison of the average Nusselt number and maximum absolute value of stream function obtained by using the Dracy, Brinkman, and Brinkman–Forchheimer models when $D_f = 0.75$

Ra	Da	Nu_{av}			ψ		
		DM	BM	BFM	DM	BM	BFM
10^4	10^{-3}	2.3957,$\bar{\alpha} = 2$	2.35	2.38	7.46	7.48	7.54
10^4	10^{-4}	2.37	2.34	2.35	7.38	7.25	7.31
10^4	10^{-5}	2.36	2.34	2.35	7.35	7.23	7.25
10^4	10^{-6}	2.36	2.34	2.35	7.34	7.23	7.23
10^5	10^{-3}	4.4596,$\bar{\alpha} = 2$	4.21	4.3	23.34	23.78	23.56
10^5	10^{-4}	4.31	4.22	4.32	23.1	22.93	23.09
10^5	10^{-5}	4.3	4.27	4.3	23.15	22.84	22.91
10^5	10^{-6}	4.3	4.27	4.29	23.19	22.83	22.86
10^6	10^{-3}	7.6894,$\bar{\alpha} = 2$	7.52	7.82	40.48	53.95	50.66
10^6	10^{-4}	7.79	7.45	7.76	50.1	48.48	49.59
10^6	10^{-5}	7.69	7.52	7.66	52.22	47.21	46.94
10^6	10^{-6}	7.7	7.53	7.57	52.76	47.15	46.5

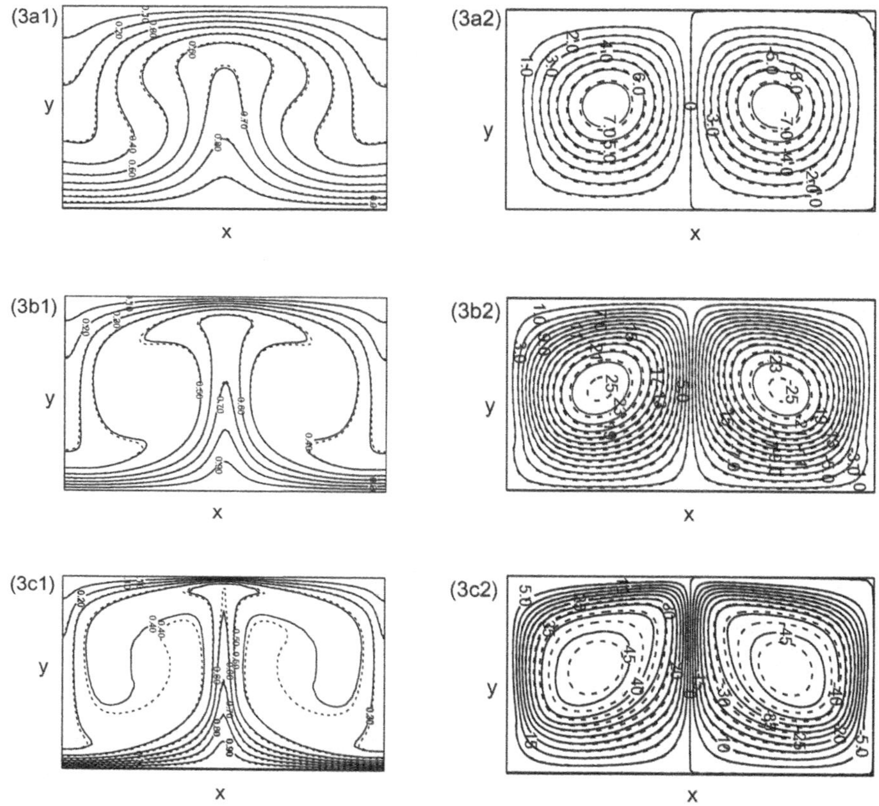

Fig. 3 Isotherms and streamlines for $Da = 10^{-5}$ and $D_f = 0.75$; (3a1, 3a2) $Ra = 10^4$, (3b1, 3b2) $Ra = 10^5$ and (3c1, 3c2) $Ra = 10^6$; ··· Darcy model, – Brinkman model

5 Conclusions

In the present work, we have compared three different models in the sense of streamlines, isotherms and by calculating Nu_{av} and $|\psi|_{max}$. There is a convective flow in an enclosure containing a fluid overlying a porous media heated from below and other two sides are adiabatic. Comparative study shows that BM and BFM are producing almost same results. DM also predicts almost same results when Da and Ra are small. However, when the value of Da and Ra increases, DM predicts more convection.

Appendix

Mathematical Modeling of Governing Equations in Vector Form

Following paper of Kim and Choi [16].
For fluid region:
Conservation equation of mass, momentum, and energy equation for fluid region

$$\nabla \cdot \vec{q} = 0, \tag{27}$$

$$\vec{q} \cdot \nabla \vec{q} = -\frac{1}{\rho_f} \nabla p + \nu_f \nabla^2 \vec{q} + \vec{g} \tag{28}$$

$$\vec{q} \cdot \nabla T = \alpha_f \nabla^2 T \tag{29}$$

where $\vec{q}$ represents velocity similarly T, temperature, p, pressure, and all other symbols are defined above.
For porous region:
Brinkman–Forchheimer-extended Darcy model

$$\nabla \cdot \vec{q} = 0, \tag{30}$$

$$\vec{q} \cdot \nabla \vec{q} = -\frac{1}{\rho_f} \nabla p \nu_{eff} \nabla^2 \vec{q} - \frac{\nu_f}{K} \vec{q} - \frac{F}{\sqrt{K}} |\vec{q}| \vec{q} + g \tag{31}$$

$$\vec{q} \cdot \nabla T = \alpha_{eff} \nabla^2 T \tag{32}$$

where the symbols used defined above, $\alpha_{eff} = k_{eff} \nabla^2 T$ is effective thermal diffusivity.
Following [22]
Darcy model: It is enough to show only momentum equation (other two are same)

$$\vec{q} = -\frac{K}{\mu} \nabla p \tag{33}$$

Brinkman's Model:

$$\nabla p = -\frac{\mu}{K} \vec{q} + \tilde{\mu} \nabla^2 \vec{q}, \tag{34}$$

where μ is the dynamic viscosity of the fluid, $\tilde{\mu}$ is effective viscosity, K is intrinsic permeability of porous medium.

References

1. H.M. Banjer, A.A. Abdullah, Convection in superposed fluid and porous layers in the presence of a vertical magnetic field. WSEAS Trans. Fluid Mech. **5**(3), 175–185 (2010)

2. G.S. Beavers, D.D. Joseph, Boundary conditions at a naturally permeable wall. J. Fluid Mech. **30**, 197–207 (1967)
3. G.S. Beavers, E.M. Sparrow, R.A. Magnuson, Experiments on coupled parallel flows in a channel and a bounding porous medium. J. Basic Eng. **92**, 843–848 (1970)
4. C. Beckermann, S. Ramadhyan, R. Viskanta, Natural convection flow and heat transfer between a fluid layer and a porous layer inside a rectangular enclosure. ASME J. Heat Trans. **109**, 363–370 (1987)
5. C. Beckermann, R. Viskanta, S. Ramadhyani, Natural convection in vertical enclosures containing simultaneously fluid and porous layers. J. Fluid Mech. **186**, 257–284 (1988)
6. H.C. Brinkman, A calculation of the viscous force exerted by flowing fluid on a dense swarm of particles. Appl. Sci. Res. **A1**, 27–34 (1947)
7. A.F. Bukhari, A spectral eigenvalue method for multi-layered continua. Ph.D. thesis, University of Glasgow (1997)
8. M. Carr, Penetrative convection in a superposed porous-medium-fluid layer via internal heating. J. Fluid Mech. **509**, 305–329 (2004)
9. M. Carr, B. Straughan, Penetrative convection in a fluid overlying a porous layer. Adv. Water Res. **26**, 263–276 (2003)
10. F. Chen, C.F. Chen, Onset of finger convection in a horizontal porous layer underlying a fluid layer. J. Heat Transf. **110**, 403–409 (1988)
11. F. Chen, C.F. Chen, Experimental investigation of convective instability in a superposed fluid and porous layer when heated from below. J. Fluid Mech. **207**, 311–321 (1989)
12. F. Chen, C.F. Chen, Convection in superposed fluid and porous layers. J. Fluid Mech. **234**, 97–119 (1992)
13. S. Ergun, Fluid flow through packed columns. Chem. Eng. Progr. **48**(2), 89–94 (1952)
14. P. Forchheimer, Wasserbewegung durch Boden. Zeits. V. deutsch. Ing. **45**, 1782–1788 (1901)
15. S.C. Hirata, B. Goyeau, D. Gobin, M. Carr, R.M. Cotta, Stability analysis of natural convection in adjacent fluid and porous layer: influence of interfacial modeling. Int. J. Heat Mass Trans. **50**, 1356–1367 (2007)
16. S.J. Kim, C.Y. Choi, Convective heat transfer in porous and overlying fluid layers heated flow. Int. J. Heat Mass Trans. **39**(2), 319–329 (1996)
17. G.D. Mallinson, G. De Vahl Davis, The method of the false transient for the solution of coupled elliptic equations. J. Comput. Phys. **12**, 435–461 (1973)
18. G. McKay, Onset of buoyancy-driven convection in superposed reacting fluid and porous layers. J. Eng. Math. **33**, 31–46 (1998)
19. G. Neale, W. Nader, Practical significance of Brinkman's extension of Darcy's law: coupled parallel flow within a channel and a bounding porous medium. Can. J. Chem. Eng. **52**, 475–478 (1974)
20. D.A. Nield, Onset of convection in a fluid layer overlying a layer of a porous medium. J. Fluid Mech. **81**, 513–522 (1977)
21. D.A. Nield, The boundary correction for the Rayleigh-Darcy problem: limitations of the Brinkman equation. J. Fluid Mech. **128**, 37–46 (1983)
22. D.A. Nield, A. Bejan, *Convection in Porous Media* (Springer, New York, 2013)
23. T. Nishimura, T. Takumi, M. Shiraishi, Y. Kawamura, H. Ozoe, Numerical analysis of natural convection in a rectangular horizontal cavity divided into fluid and porous region. Int. J. Heat Mass Transf. **26**, 889–898 (1986)
24. D. Poulikakos, A. Bajan, B. Selimos, K. Blake, High Rayleigh number convection in a fluid overlaying a porous bed. Int. J. Heat Fluid Flow **7**(2), 109–116 (1986)
25. A.A. Samarskii, V.B. Andreyev, On a high-accuracy difference scheme for elliptic equation with several space variables. USSR Comput. Math. Math. Phys. **3**, 1373–1382 (1963)
26. S.B. Sathe, W.Q. Lin, T.W. Tang, Natural convection in enclosures containing an insulation with permeable fluid-porous interface. Int. J. Heat Fluid Flow **9**, 389–395 (1988)
27. A.K. Singh, G.R. Thorpe, Natural convection in a confined fluid overlying a porous layer-a comparison study of different models. Indian J. Pure Appl. Math. **5**(3), 81–95 (1994)

28. B. Straughan, Surface-tension-driven convection in a fluid overlying a porous layer. J. Comput. Phys. **170**, 320–337 (2001)
29. B. Straughan, Effect of property variation and modeling on convection in a fluid overlying a porous layer. Int. J. Numer. Anal. Meth. Geomech. **26**, 75–97 (2002)
30. M.W. Taslim, U. Narusawa, Thermal stability of horizontally superposed porous and fluid layers. ASME J. Heat Transf. **111**, 357–362 (1989)
31. L.H. Thomas, *Watson Scientific Computing Laboratory* (Columbia University, New York, 1949)
32. J. Valencia-Lopez, J. Ochoa-Tapia, A study of buoyancy-driven flow in a confined fluid overlying a porous layer. Int. J. Heat Mass Transf. **44**, 4725–4736 (2001)

Persistence Barcoded Vehicular Traffic Videos in a Topology of Data Approach to Shape Tracking

Arjuna P. H. Don, James F. Peters, and Sheela Ramanna

Abstract This paper introduces a computational CW topology of data approach to tracking the persistence of image object shapes that appear in triangulated video frames. Shapes are cell complexes are viewed in the context of an Alexandroff–Hopf–Whitehead CW (Closure finite Weak) topological space. Fermi energy and Betti numbers are used to construct persistence barcodes derived from nested cycles (optical vortexes) inherent in triangulated video frame shapes. An application of this approach is given in terms of Ghrist persistence barcoding of vehicular traffic videos.

Keywords Betti number · CW topology · Fermi energy · Optical vortex · Triangulated frame · Video

2010 Mathematics Subject Classification 54C56 (shape theory) · 55R40 (homology of classifying shapes) · 55U10 (cell complexes)

1 Introduction

This paper tackles the problem of tracking the persistence of moving shapes in a video by reducing each triangulated video frame to a collection of elementary cells. Each video frame is a finite, bounded region of the Euclidean plane. A *cell* in the Euclidean plane is either a 0-cell (vertex) or 1-cell (edge) or 2-cell (filled triangle). A

A. P. H. Don · J. F. Peters (✉)
Computational Intelligence Laboratory, University of Manitoba, WPG, Winnipeg, MB R3T 5V6, Canada
e-mail: james.peters3@umanitoba.ca

J. F. Peters
Department of Mathematics, Faculty of Arts and Sciences, Adiyaman University, 02040 Adiyaman, Turkey

S. Ramanna
Applied Computer Science, University of Winnipeg, Winnipeg, MB R3B 2E9, Canada
e-mail: s.ramanna@uwinnipeg.ca

S. K. Paikray et al. (eds.), *New Trends in Applied Analysis and Computational Mathematics*, Advances in Intelligent Systems and Computing 1356,
https://doi.org/10.1007/978-981-16-1402-6_3

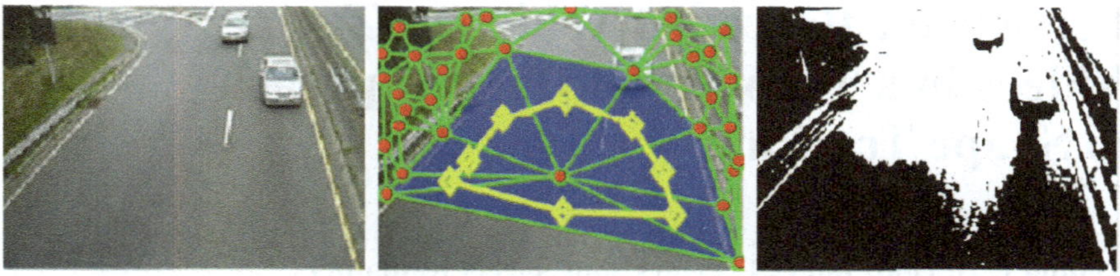

1.1: Traffic video frame 1.2: Traffic video frame vortex 1.3: Frame binary image

Fig. 1 Sample video frame barycentric vortex

cell complex is a collection of cells attached to each other by edges or by having one or more common vertices. A nonvoid collection of cell complexes K has a *Closure finite Weak* (CW) topology, provided K is Hausdorff (every pair of distinct cells is contained in disjoint neighbourhoods [1, Sect. 5.1, p. 94]) and the collection of cell complexes in K satisfy the Alexandroff–Hopf–Whitehead [2, Sect. III, starting on p. 124], [3, pp. 315–317], [4, Sect. 5, p. 223] conditions, namely, containment (the closure of each cell complex is in K) and intersection (the nonempty intersection of cell complexes is in K). The focus on this work on detecting nesting cycles in optical vortexes on the barycenters of triangles covering video frame shapes.

A **vortex** is a collection of nesting, usually non-concentric, path-connected, barycentric, intersecting cycles. A **cycle** E (denoted by $\text{cyc}E$) contains vertexes so that each pair of vertexes $p, q \in \text{cyc}E$ in the cycle is path-connected, i.e., there is a sequences of edges leading from vertex p to vertex q in the cycle. The vertexes in barycentric cycles are the barycenters (intersection of the median lines) of triangles. An **optical vortex** is a vortex constructed from vertexes that are picture elements (pixels), snapshots of reflected light from surfaces record in a video frame (for a sample optical vortex, see Example 1).

Example 1 A barycentric optical vortex containing a single cycle (in yellow) is shown in the triangulation of the traffic video frame in Fig. 1a is shown in Fig. 1b. ■

Vortex cycles are examples of nerve structures (called an optical vortex nerve). A ***vortex nerve*** is a collection of nesting, possibly overlapping filled vortexes attached to each other and have nonempty intersection [5–9]. A ***filled vortex*** has a boundary that is a simple closed curve and a nonempty interior. Due to the fact that the optical cycles intersect, i.e., the cycles in such vortexes have one or more common vertexes. Because the paths between vertexes are on intersecting cycles in an optical vortex are bidirectional, we obtain the following result.

Theorem 1 [10, Sect. 4.13, p. 212] *An optical vortex has a free Abelian group representation.*

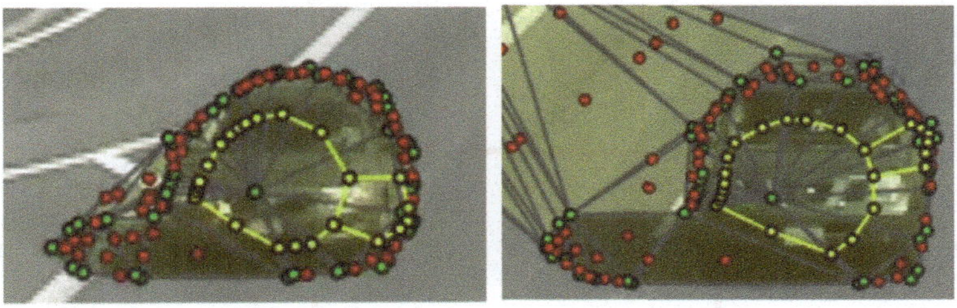

2.1: Traffic optical vortex nerve 1 2.2: Traffic optical vortex nerve 2

Fig. 2 Sample traffic video frame optical vortex nerves

Recall that a group is a nonempty set V (for vortex) equipped with a binary operation (represented here with a + (traverse or move)), so that each vertex p of V has an inverse $-p$ with $p + (-p) = 0$ (i.e., no traversal or movement occurs) and $p + q = q + p$ (Abelian property). That is, traversing the edges from p to q in the vortex can always be followed by a traversal of the edges from q to p, which takes back where we started. A zero move is the identity element of the group. For example, $p + 0$ reads 'no traversal occurs at p'. In a free Abelian group representation of a vortex nerve, each vertex in the nerve can be written as a summation of the generating elements. The number of generators in such a group is the **rank** of the group.

Example 2 Sample traffic video frame optical vortex nerves are shown in Fig. 2. The vertexes on the edges attached between the inner yellow cycles are the generators of the free Abelian group representations of these nerves. ■

A ***Ghrist barcode***, usually called a ***persistence barcode***, is a topology-of-data pictograph that represents that appearance and disappearance of consecutive sequences of video frames having a particular feature value [11], [12, Sect. 5.13, pp. 104–106]. The origin of topology-of-data barcodes can be traced back to Edelsbrunner et al. [13, 14]. For a complete view of the landscape for a topology-of-data barcode viewed as a multiset of intervals,[1] see Perea [15].

Example 3 An overview of the steps leading to barcode for traffic is given in Fig. 3. These steps with the construction of an optical vortex nerve resulting from the triangulation of a moving traffic video frame shape. Two measures of shape structure (Fermi energy and counts of the basic parts of each nerve) are reflected in varying length horizontal bars in a 2D barcode for each video. From shape energy and shape part counts, a 3D barcode is constructed for each shape vortex nerve. ■

2 Preliminaries

This section briefly introduces Betti numbers and Fermi energy.

[1]Many thanks to Vidit Nanda for pointing this out.

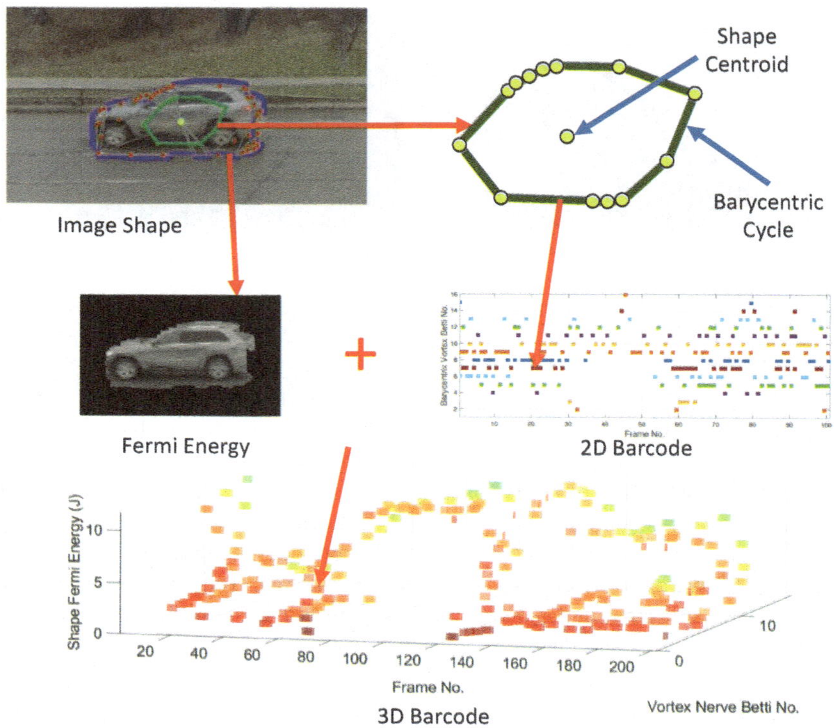

Fig. 3 Barycentric cycle

2.1 Betti Numbers

There are two forms of Betti numbers, an **algebraic Betti number** (number of generators of a free Abelian group, denoted simply by β), introduced in Munkres [16, Sect. 1.4, p. 24] and three **geometric Betti numbers** that give us the cardinality of geometric structures in a CW complex formed by the triangulation of a video frame shape, introduced by Zomorodian [17, Sect. 4.3.2, p. 55]. The focus here is on geometric Betti numbers, which are more informative in characterizing triangulated video frame shapes. The important thing to notice is that we isolate and triangulate moving vehicle shapes in traffic video frames. That is, we restrict Betti numbers to triangulated shapes [18] as opposed to triangulation of an entire video frame.

In the context of finite CW complexes found on triangulated finite bounded planar regions, triangulated video frame shapes are characterized by the geometric Betti numbers, namely, $\mathcal{B}_0$ (cell or CW complex components count), $\mathcal{B}_1$ (vortex nerve cycle count) and $\mathcal{B}_2$ (hole or void count). In addition, the ***Betti number***. By contrast, from an algebraic perspective, the Betti number of a vortex nerve is a count of the number of generating elements that define a free Abelian group representation of the nerve.

In triangulated traffic video frames, vertexes are the centroids of dark frame regions. Each frame dark region absorbs sunlight and hence the dark regions are considered holes with corresponding β_2 counts. In that case, β_0 counts the number of centroidal vertexes, edges, and filled triangles in a video frame. And β_1 is a count of the number of barycentric cycles on a triangulated video frame.

Example 4 In Fig. 1b, we have the following Betti numbers:

Geometricviews :

$$\beta_0 = 8 + 40(\text{filled triangle count}) + 34\text{vertexes} + 34\text{edges} = 116.$$
$$\beta_1 = 1(\text{cycle count}).$$
$$\beta_2 = 75(\text{hole count}).$$

Algebraicview :.

$$\beta = 2,\ i.e., 2 \text{ cycles, shape boundary cycle, and shape barycentric cycle.}$$

In the binary image for Fig. 1b given in Fig. 1c, there are a total of 150 bounded regions (75 of the bounded regions are holes). ■

2.2 *Fermi Energy*

The Hummel form of Fermi Energy [19, Sect. 6, p. 69, Eq. (6.11)] derived from the pixel population of an object shape is used in this paper.

$$E_F = \frac{k}{2m_0}\left(\frac{3\pi^2 N}{V}\right)^{2/3}, \tag{1}$$

where N is the number of pixels in a moving object shape, V, the total area of the moving objects, and m_0, the average intensity of the object shape pixels (in greyscale). A scaling factor k was used to scale the values appropriately. This form of Fermi energy for structures in digital images appears in Pradikar et al. [20, p. 807].

To isolate moving objects in each video frame, background subtraction was used, i.e., an initial video frame that displays no moving objects is subtracted from the remaining frames in the video. In effect, subtracting an initial frame from the remaining frames in a video makes it possible to remove all background (dark) regions in a video. Morphological operations were performed on the binary images to further isolate the object shapes in the remaining frames. Algorithm 1 gives the basic pseudocode to construct optical vortex nerves on triangulated video frame shapes.

Visualization of shape pixel intensities is represented by a 3D projection of isolated frame object shapes.

Algorithm 1: Vortex Energy Barcode

```
input  : bkGray (initial greyscale frame), f video frames
output: A stationary background frame of size w × l.
for count ← 1 to f do
    frameDif ← frameGray- bkGray;
    BW ← Convert the frameDif to binary image.;
    shNo ← Number of moving objects in BW;
    for shNo ← 1 to f do
        shapeBW ← Extract shape shNo from BW;
        shCentroid ← Get the centroid of white blob area;
        shGray ← frameGray- shapeBW;
        shPoints ← Find strongest SWIFT points on shGray;
        seedPoints ← shCentroid + shPoints;
        triDelaunay ← Perform Delaunay triangulation on seedPoints.;
        ncNode ← Calculate the nodes associated with shCentroid;
        matBarycenters ← Calculate barycenters of triangles in triDelaunay;
        ncBarycenters ← matBarycenters(ncNode);
        bettyNo = Number of nodes in Barycentric Cycle;
        matEnergy ← Save bettyNo and Energy associated with shGray;
plot(matEnergy);
```

Fig. 4 Shape Fermi energy in traffic video frame

Example 5 Shape Fermi energy in a video frame is shown in Fig. 4 for a traffic video of Portage Avenue in Winnipeg. From Algorithm 1, Fermi energy results from extracting the number of pixels N in a 3D projection of a 2D vehicle shape. ■

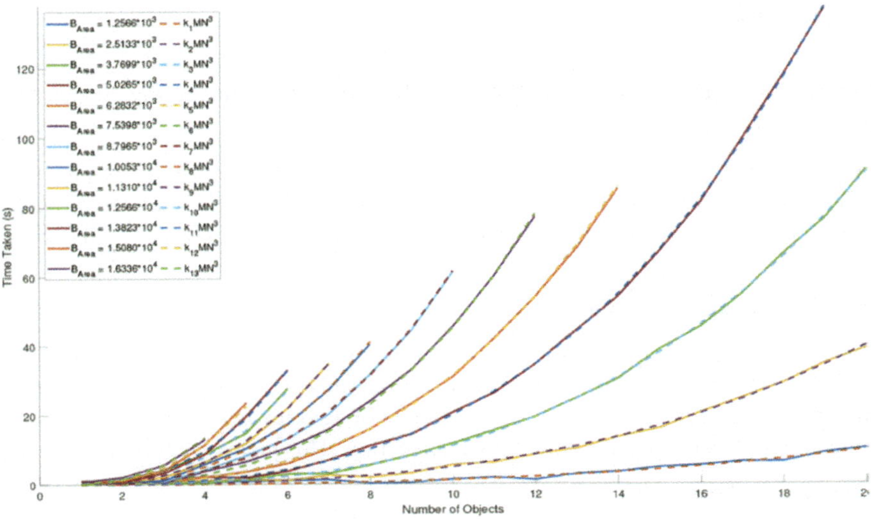

Fig. 5 Time complexity

3 Time Complexity Analysis

Figure 5 shows the results of time complexity analysis for Algorithm 1. In estimating the complexity of Algorithm 1, both theoretical and actual time values were calculated. The time taken for the inbuilt functions (e.g., video frame extraction, video frame exporting, triangulation) were not considered since these functions are optimized. When calculating the theoretical time complexity addition, subtraction, multiplication, and division were considered as four different calculations.

The video frame shape triangulation method introduced in this paper has theoretical time complexity $O\left(mn^3\right)$, which is computed in terms of m (pixel area of a moving object) and n (number of moving objects) in a video frame. An experimentally derived scaling factor of k is used to align each theatrical graph with the actual time graph. The value of k is in the order of 1×10^{-6} and varies slightly depending on each time curve shown in Fig. 5.

It is evident from Fig. 5 that the theoretical and the actual time complexity curves follow each other very closely. This is especially the case as the number of object shapes and the object shape area increase. There are some deviations from this observation, when the object area and the number are small, is to be expected. This anomaly results from not considering the impact of some functions, which at work in the background of the system, affecting the time taken for processing. But when the computation time increases, the influence of the background functions becomes minimal compared to the overall computation time.

4 Results

Two of the results of this research reflect the influence of increasing shapes in a sequence of video frames.

Lemma 1 *Let* shA, shA' *be object shapes in a pair of adjacent video frames with* shA' *appearing after* shA*. Also, let* N, V, E, N', V', E' *be the number of shape pixels, shape area, and Fermi energy in* shA, shA'*, respectively. Then* $N' > V', N' > N$ *and* $N > V$ *implies* $E' > E$*.*

Proof From Eq. 1, we obtain the desired result. □

Observe that nesting, usually non-concentric cycles covering all or the principal part of the interior of a video frame shape, form a ribbon. A *ribbon* is a collection of such nesting-filled cycles, which is an example of a vortex nerve. A *vortex nerve* is collection of cycle that have nonempty intersection. This yields the following result.

Theorem 2 [21] *A ribbon is a vortex nerve.*

Theorem 3 *Let* shA, shA' *be object shapes covered by nesting, non-concentric filled cycles in a pair of adjacent video frames with* shA' *appearing after* shA*. Let* β_1 *be the cycle count Betti number for shA,* $\beta_1^{'}$ *for* shA'*. Assume that shape* shA' *has* $N_{pixels}^{'} > V_{area}^{'}$ *and* shA *has* $N_{pixels} > V_{area}$*. Then* $\beta_1^{'} > \beta_1$ *implies* $E' > E$*.*

Proof From Theorem 2, sh$A$, shA' are vortex nerves. $\beta_1^{'} > \beta_1$ indicates that the number of nesting, non-concentric cycles in shA' is greater than the number of cycles in sh$A$. The outer cycle (call it cycA') on the interior of shA' has the remaining cycles of shA' nested inside cycA'. Also, from $\beta_1^{'} > \beta_1$, the number of pixels $N' > V'$ in the interior of cycA' is greater than the number of pixels $N > V$ in the interior of cyc$A$. Hence, from Lemma 1, $E' > E$. □

Barcoding traffic videos provides a concise means of tracking the persistences of shapes across sequences of video frames. Each occurrence of a frame shape is represented by a horizontal bar in 2D Ghrist pictographs called barcodes, introduced by Ghrist [11, 12] and others [13–15, 22] and later extended to 3D persistence barcodes in [23] (Fig. 6).

The basis for picture proof of Theorem 3 is given in the 3D persistence barcode in Example 6.

Example 6 In the 3D persistence barcode in Fig. 4, the cycles Betti number β_1 is also vortex Betti number. Uniformly in Fig. 4, as the cycles Betti number β_1 increases, there is a corresponding increase in shape Fermi energy across a sequence of frames. This tends to corroborate the observation in Theorem 3. ■

A 2D persistence barcode structured in terms of barycentric cycle counts in the vortex Betti number β_1, exhibits which of the β_1 vortex cycle counts tend to be more persistent over a sequence of vehicular traffic video frames. A *barycentric cycle* is a simple, closed curve in which its vertexes are triangle barycenters that are path-connected.

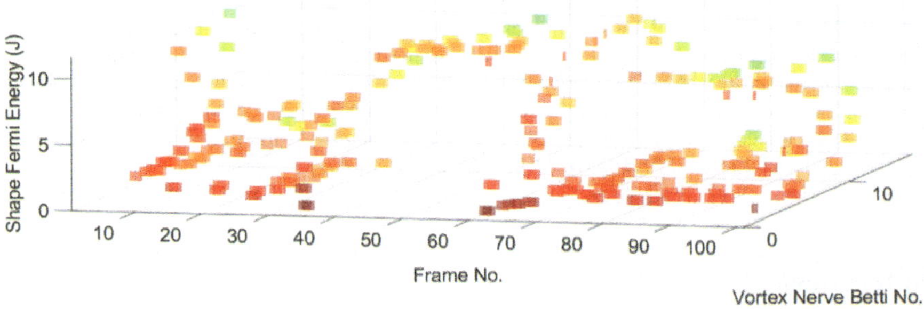

Fig. 6 Example 3D barcode for traffic video

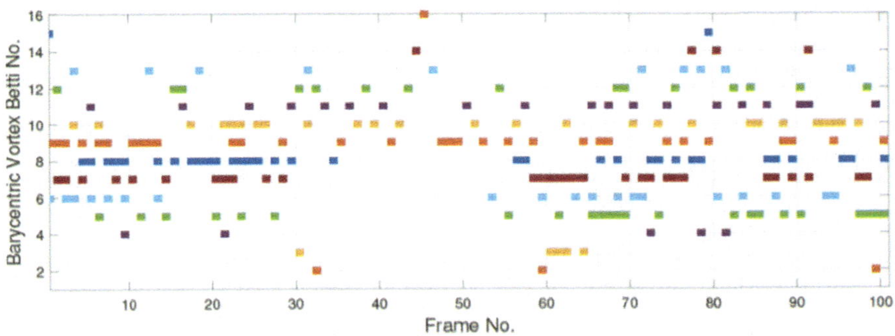

Fig. 7 Example barcode for traffic video

Example 7 In the 2D persistence barcode given in Fig. 7, the vortex cycles Betti number β_1 ranges from 2 to 16 over a sequence of traffic video frames. The mid-range β_1 counts (from 7 to 9) have frequent extended horizontal bars spanning a sequence of frames with extended bars ranging over 2–7 consecutive frames. In other words, shapes with mid-range β_1 counts tend to be more persistent. ■

5 Conclusion

This paper introduces the construction of both 2D and 3D persistence-based barcoded videos using a topology of data approach in the search for appropriate niches for different videos. A major contribution of this work is a combination of the triangulation of moving shapes recorded in video frames and a form of descriptive proximity formulated in terms of geometric Betti numbers and Fermi energy. A major gap in this paper is the absence of a comparison with comparable video barcoding methods. This comparison will be given in a future follow-up to this paper.

Acknowledgements This research has been supported by the University of Manitoba Graduate Fellowship and Gorden P. Osler Graduate Scholarship, the Natural Sciences and Engineering Research Council of Canada (NSERC) discovery grants 185986 and 194376, Instituto Nazionale di Alta Matematica (INdAM) Francesco Severi, Gruppo Nazionale per le Strutture Algebriche, Geometriche e Loro Applicazioni grant 9 920160 000362, n.prot U 2016/000036 and Scientific and Technological Research Council of Turkey (TÜBİTAK) Scientific Human Resources Development (BIDEB) under grant no: 2221-1059B211301223.

References

1. S.A. Naimpally, J.F. Peters, *Topology with Applications. Topological Spaces via Near and Far* (World Scientific, Singapore, 2013), xv + 277pp. American Mathematical Society MR3075111
2. P. Alexandroff, H. Hopf, *Topologie. Band I* (Springer, Berlin, 1935), Zbl 13, 79, iii+637pp. Reprinted Chelsea Publishing Co., Bronx, N. Y., 1972, MR0345087
3. J.H.C. Whitehead, Simplicial spaces, nuclei and m-groups. Proc. Lond. Math. Soc. **45**, 243–327 (1939)
4. J.H.C. Whitehead, Combinatorial homotopy I. Bull. Am. Math. Soc. **55**(3), 213–245 (1949). Part 1
5. M.Z. Ahmad, J.F. Peters, Maximal centroidal vortices in triangulations. a descriptive proximity framework in analyzing object shapes. Theory Appl. Math. Comput. Sci. **8**(1), 38–59 (2018). ISSN 2067-6202
6. J.F. Peters, S. Ramanna, Shape descriptions and classes of shapes. A proximal physical geometry approach, in *Advances in Feature Selection for Data and Pattern Recognition, ed. by B. Zielosko U. Stanczyk, L.C. Jain* (Springer, 2018), pp. 203–225. MR3811252
7. J.F. Peters, Proximal vortex cycles and vortex nerve structures. non-concentric, nesting, possibly overlapping homology cell complexes. J. Math. Sci. Model. **1**(2), 56–72 (2018), www.dergipark.gov.tr/jmsm, See, also, arXiv:1805.03998. ISSN 2636-8692
8. J.F. Peters, Proximal planar shapes. Correspondence between triangulated shapes and nerve complexes. Bull. Allahabad Math. Soc. **33**, 113–137. *MR3793556, Zbl 06937935* (Review by D. Leseberg (Berlin), 2018)
9. J.F. Peters, Proximal planar shape signatures. Homology nerves and descriptive proximity. Adv. Math.: Sci. J. **6**(2), 71–85 (2017). Zbl 06855051
10. J.F. Peters, *Computational Geometry, Topology and Physics of Digital Images with Applications. Shape Complexes, Optical Vortex Nerves and Proximities* (Springer Nature, Cham, Switzerland, 2020), xxv+440pp. ISBN 978-3-030-22191-1/hbk; 978-3-030-22192-8/ebook, Zbl07098311
11. R. Ghrist, Barcodes: the persistent topology of data. Bull. Am. Math. Soc. (N.S.) **45**(1), 61–75 (2008). MR2358377
12. R.W. Ghrist, *Elementary Applied Topology*. University of Pennsylvania (2014), vi+269pp. ISBN: 978-1-5028-8085-7
13. H. Edelsbrunner, D. Letscher, and A. Zomorodian. Topological persistence and simplification, in *41st Annual Symposium on Foundations of Computer Science*. (IEEE Comput. Soc. Press, Los Alamitos, California, 2000), pp. 454–463. MR1931842
14. H. Edelsbrunner, D. Letscher, A. Zomorodian, Topological persistence and simplification. Discret. Comput. Geom. **28**(4), 511–533 (2001). MR1949898, reviewed by H.W. Guggenheimer
15. J.A. Perea, A brief history of persistence, 1–11 (2018), arXiv:1809(036249)
16. J.R. Munkres, Elements of Algebraic Topology, 2nd edn. (Perseus Publishing, Cambridge, MA, 1984), ix + 484pp. ISBN: 0-201-04586-9, MR0755006
17. A.J. Zomorodian, Computing and comprehending topology persistence and hierarchical morse complexes. Ph.D. thesis, University of Illinois at Urbana-Champaign, Graduate College (2001), 199pp. Supervisor: H. Edelsbrunner

18. J.F. Peters, S. Ramanna, Shape descriptions and classes of shapes. a proximal physical geometry approach, in *Advances in Feature Selection for Data and Pattern Recognition, ISRL 138* (Springer, 2018), pp. 203–225. https://doi.org/10.1007/978-3-319-67588-6, MR3811252
19. R.E. Hummel, *Electronic Properties of Materials*, 4th edn. (Springer, New York, U.S.A., 2011), viii+488pp., 539 pp. ISBN: ISBN 978-1-4419-8163-9
20. S. Pradikar, J. Sil, A.D. Das, Region identification of infected rice images using the concept of fermi energy, in *Advances in Computing & Information Technology*, ed. by N. Meghanathan et al. (Springer, Berlin, 2013), pp. 805–811
21. J.F. Peters, Planar ribbon complexes and their approximate proximities. Ribbon nerves, betti numbers and planar divisions. Bull. Allahabad Math. Soc. 1–14 (2020). To appear. See, also, arXiv:1911.09014
22. A.P.H. Don, J.F. Peters, Ghrist barcoded video frames. Application in detecting persistent visual scene surface shapes captured in videos. Quaestiones Mathematicae **37**(2), 249–263 (2014). Zbl 1397.54037, Zbl 1397.54038
23. A.P.H. Don, J.F. Peters, S. Ramanna, A. Tozzi, Topological inference from spontaneous activity structures in FMRI videos with peristence barcodes. *bioRxiv* 1101(809293), 1–11 (2019). https://doi.org/10.1101/809293

Heat Transfer of MHD Channel Flow of Viscoelastic (PTT) Fluid

B. K. Swain, M. Das, and G. C. Dash

Abstract The study presents the heat transfer aspect of MHD channel flow of Phan-Thien-Tanner (PTT) conducting flow accounting for the viscous dissipation. The role of Deborah number substantiates the dual behavior of Newtonian and non-Newtonian aspects of the flow model. The inclusion of two body forces due to magnetic field (force act at a distance) and porosity of the medium enrich the analysis. The important findings are the role of magnetic parameter is to enhance the temperature across the flow domain, whereas Deborah number and other parameters act adversely. Thus, the simulation of the flow parameters provides ample scopes to meet the design requirements in cooling/heating. The most interesting observation is that contribution of viscous dissipative heat seems to be insignificant due to linear variation across the temperature field in the present PTT model indicating the preservation of thermal energy loss.

Keywords MHD · PTT · Homotopy perturbation method · Porous medium · Heat transfer

Nomenclature

$\vec{B}$	Magnetic flux
B_0	Constant flux density
D/D_t	Material time derivative
D_e	Deborah number

B. K. Swain (✉)
Department of Mathematics, IGIT, Sarang, Dhenknal, Odisha, India
e-mail: bharatkeshari1@gmail.com

M. Das · G. C. Dash
Department of Mathematics, S.O.A, Deemed to Be University, Bhubaneswar, Odisha, India
e-mail: manjuladas@soa.ac.in

G. C. Dash
e-mail: gcdash45@gmail.com

S. K. Paikray et al. (eds.), *New Trends in Applied Analysis and Computational Mathematics*, Advances in Intelligent Systems and Computing 1356,
https://doi.org/10.1007/978-981-16-1402-6_4

$\vec{J}$	Electric current density
I	Identity tensor
K_p	Porosity parameter
L	Characteristic length
M	Magnetic parameter
p	Pressure
$\overrightarrow{T^*}$	Cauchy stress tensor
$\vec{V}$	Velocity vector

Greek Symbol

$\vec{\tau}$	Extra stress tensor
λ	Relaxation time
μ	Constant viscosity coefficient
σ	electrical conductivity
ρ	fluid density
ε	elongation parameter
∇	Gradient operator

1 Introduction

The Navier–Stokes equations governing the motion of the viscous fluid are nonlinear due to convected forms and their exact solutions are possible for simple flow problems. The non-Newtonian fluids exhibit many physical structures; therefore, it is impossible to interpret their mechanical behavior with a single constitutive equation. For better exploration, many varieties of constitutive equations have been proposed. The constitutive equations proposed by Phan-Thien and Tanner [1, 2], and Tanner [3] have been an interesting subject in recent years for researchers. Oliveira and Pinho [4] derived the analytical expressions for velocity field and stress component of the flow of PTT fluid through the fully developed channel and pipe. Some other researches regarding PTT fluid have been carried out in [5–7], [12, 13]. Letelier and Siginer [8] gave the solution for the problem of pipe flow of a class of nonlinear viscoelastic fluids considering PTT and Johnson–Segalman models as special cases. Some other works using PTT fluid have been carried out by Siddiqui et al. [9] and Hayat et al. [14–16].

Newtonian fluid mechanics deals with the interplay of inertial and viscous forces. As eluded to many common or industrial fluids exhibit complex behavior due to their intricate molecular structure, for example, molten plastics paint, blood and egg white, etc. The non-linear relationships between stress and rate of strain are exhibited by many real fluids. In the present study, the second-order nonlinear differential equations appearing in viscoelastic Phan-Thein-Tanner (PTT) flows in a channel has been obtained and the behavior of these solutions has been elaborately discussed. This class of fluids exhibits memory property termed viscoelastic fluids which act as elastic material when deformed. The flow considered herein is subjected to an external magnetic field of low intensity, to avoid the effects of induced magnetic field as well as external electric field. Hence, the magnetic Reynolds number is assumed to be small. From literature survey, it is found that not much work has been reported dealing with flow and heat transfer simultaneously. The heat transfer phenomenon is an important phenomena rarely attempted, as the analysis of PTT flow requires the solution of momentum equation as well as rheological equations. More importantly, we have considered the effect of Deborah number (D_e) which behaves as Newtonian viscous fluid for small D_e ($D_e < 1$) and elastic fluid for large $D_e(D_e > 1)$. Due to numerous applications of PTT fluid, the work of Nasseria et al. [10] is particularly interesting as it predicts the behavior of soft tissue (pig liver) and bread dough under viscous compressions usingf the PTT model. The present investigation encompasses the domain of viscous and elastic properties of the fluid with special importance to flow characteristics.

2 Mathematical Formulation

The constitutive equation of an incompressible, PTT fluid [1, 2] is of the form

$$\overrightarrow{T^*} = -pI + \overrightarrow{\tau}, \tag{1}$$

$$f(tr(\vec{\tau}))\vec{\tau} + \lambda \overset{\nabla}{\tau} = 2\mu A_1, \tag{2}$$

where A_1, the first Rivlin–Ericksen tensor and $\overset{\nabla}{\tau}$, the Oldroyd's upper convected derivative defined as

$$A_1 = \nabla \overrightarrow{V} + (\nabla \overrightarrow{V})^T,$$

$$\overset{\nabla}{\tau} = \frac{D\overrightarrow{V}}{Dt} - \overrightarrow{\tau}.\nabla\overrightarrow{V} - \nabla\overrightarrow{V}^T.\overrightarrow{\tau}. \tag{3}$$

$\vec{V}$ and $\nabla\vec{V}$ are defined in two dimensions as follows:

$$\overrightarrow{V} = \begin{pmatrix} u(x, y) \\ v(x, y) \end{pmatrix}, \nabla\overrightarrow{V} = \begin{pmatrix} \frac{\partial u}{\partial x} & \frac{\partial u}{\partial y} \\ \frac{\partial v}{\partial x} & \frac{\partial v}{\partial y} \end{pmatrix}.$$

In the present case,

$$\vec{V} = \begin{pmatrix} u(y) \\ 0 \end{pmatrix} \text{ and } \nabla\vec{V} = \begin{pmatrix} 0 & \frac{\partial u}{\partial y} \\ 0 & 0 \end{pmatrix}$$

Two forms of the PTT models are in common use, where the function f is defined by

$$f(tr(\vec{\tau})) = 1 + \frac{\varepsilon\lambda}{\mu}tr(\vec{\tau}), \text{ linear form (Phan-Thein[1])} \tag{4}$$

$$f(tr(\vec{\tau})) = \exp\left(\frac{\varepsilon\lambda}{\mu}tr(\vec{\tau})\right), \text{ exponential form (Phan-Thein[2]).} \tag{5}$$

When ε tends to zero and the trace of the stress tensor is small, then (4) and (5) become upper convected Maxwell (UCM) model.

The field equation of MHD flow becomes

$$\rho\frac{D\vec{V}}{Dt} = \nabla.\vec{T^*} + \vec{J} \times \vec{B}. \tag{6}$$

$$\vec{T^*} = -pI + \vec{\tau},$$

where $\vec{\tau} = \begin{pmatrix} \tau_{xx} & \tau_{xy} \\ \tau_{yx} & \tau_{yy} \end{pmatrix}$, the extra stress tensor and $\vec{T^*}$ is the total stress tensor.

The continuity equation is given by

$$trA_1 = 0 \tag{7}$$

For the present problem, the stress tensor and velocity field turn into the form

$$\vec{V} = u(y)\hat{i}, \vec{\tau} = \vec{\tau}(y)\hat{i}, \tag{8}$$

where $\hat{i}$, the unit vectors and $u(y)$, the velocity in the x-direction, respectively.

Suppose the external electric field is negligible and the magnetic Reynolds number is very small as in [17]. Therefore, the MHD body force can be considered as

$$\vec{J} \times \vec{B} = -\sigma B_0^2\vec{V} \tag{9}$$

The continuity equation is satisfied by the assumptions in (8).

The equation of motion (6) gives the following equations:

$$-\frac{\partial p}{\partial x} + \frac{\partial \tau_{xy}}{\partial y} - \sigma B_0^2 u - \frac{\rho\nu u}{K_p} = 0, \tag{10}$$

$$-\frac{\partial p}{\partial y}+\frac{\partial \tau_{yy}}{\partial y}=0 \tag{11}$$

$$\frac{-\partial p}{\partial z}=0 \tag{12}$$

Substituting (8) in (1) and (2), we get

$$f(tr(\tau))\tau_{xx}-2\lambda\tau_{xy}\frac{du}{dy}=0, \tag{13}$$

$$f(tr(\tau))\tau_{xy}-\lambda\tau_{yy}\frac{du}{dy}=\mu\frac{du}{dy}, \tag{14}$$

$$f(tr(\tau))\tau_{yy}=0. \tag{15}$$

Applying the linear form of f, we get

$$\tau_{yy}=0,\ \tau_{xx}=2\frac{\lambda}{\mu}\tau_{xy}^2. \tag{16}$$

Substituting following nondimensional variables and parameters:

$$x^*=\frac{x}{L},\ y^*=\frac{y}{L},\ u^*=\frac{u}{U},\ \tau^*=\frac{\tau L}{\mu U},\ p^*=\frac{pL}{\mu U},$$
$$M^*=\frac{\sigma B_0^2 L^2}{\mu},\ K_p^*=\frac{K_p}{L^2},\ D_e=\frac{\lambda U}{L} \tag{17}$$

in (10), we get (after dropping the asterisks)

$$-\frac{\partial p}{\partial x}+\frac{\partial \tau_{xy}}{\partial y}-Nu=0,\ \text{where} N=M+\frac{1}{K_p} \tag{18}$$

Now putting (16) in (4), we have

$$\left(1+2\varepsilon D_e^2\tau_{xy}^2\right)\tau_{xy}=\frac{du}{dy}. \tag{19}$$

Since $\frac{dp}{dy}=0$, from (18) and (19), we get

$$\frac{d^2\tau_{xy}}{dy^2}-\left(M+\frac{1}{K_p}\right)\left(\tau_{xy}+2\varepsilon D_e^2 N\tau_{xy}^3\right)=0. \tag{20}$$

The suitable no-slip boundary conditions are as follows:

$$u(0) = 0, u(1) = 0 \Rightarrow \frac{d}{dy}\tau_{xy}(0) = \frac{d}{dy}\tau_{xy}(1) = P, \tag{21}$$

where P is constant pressure gradient.

For brevity, we introduce $\psi = \tau_{xy}$, $N = M + \frac{1}{K_p}$, $\beta = 2\varepsilon D_e^2\left(M + \frac{1}{K_p}\right)$ and we get

$$\frac{d^2\psi}{dy^2} = N\psi + \beta\psi^3 \tag{22}$$

$$\psi'(0) = \psi'(1) = P, \tag{23}$$

where $'$ denotes derivative with respect to y.

3 Solution of the Problem

We set (22) into the following form by introducing 'q' as the homotopy perturbation parameter [11] and get

$$(1-q)\{\psi'' - N\psi\} + q\{\psi'' - N\psi - \beta\psi^3\} = 0, \tag{24}$$

$$\psi'(0) = \psi'(1) = P \tag{25}$$

Now putting

$\psi = \psi_0 + \psi_1 q + \psi_2 q^2 + \cdots$ in (24) and (25) and collecting the coefficients of q^0 and q^1, we get

$$(i)\ \psi_0'' - N\psi_0 = 0 \tag{26}$$

$$\psi_0'(0) = P, \psi_0'(1) = P, \tag{27}$$

$$(ii)\ \psi_1'' - N\psi_1 = \beta\psi_0^3, \tag{28}$$

$$\psi_1'(0) = 0, \psi_1'(1) = 0. \tag{29}$$

Solving (26) and (28) with the boundary conditions (27) and (29), respectively, we get

$$\psi_0 = c_1 e^{\sqrt{N}y} + c_2 e^{-\sqrt{N}y} \tag{30}$$

$$\psi_1 = c_3 e^{\sqrt{N}y} + c_4 e^{-\sqrt{N}y} + \beta\left\{\frac{c_1^3 e^{3\sqrt{N}y}}{8N} + \frac{c_2^3 e^{-3\sqrt{N}y}}{8N} + \frac{3c_1^2 c_2 e^{\sqrt{N}y}}{2\sqrt{N}}y - \frac{3c_1 c_2^2 e^{-\sqrt{N}y}}{2\sqrt{N}}y\right\} \tag{31}$$

The HPM iterative process is considered up to first order. Hence,

$$\psi = \lim_{q\to 1}(\psi_0 + \psi_1 q). \tag{32}$$

The constant of coefficients can be calculated using boundary conditions.

$$c_2 = c_1 - \frac{P}{\sqrt{N}},\ c_3 = c_4 + c_6,\ c_4 = \frac{c_6 e^{\sqrt{N}} - c_8}{e^{-\sqrt{N}} - e^{\sqrt{N}}},$$

$$c_5 = \beta\left\{\frac{3c_1^3}{8\sqrt{N}} - \frac{3c_2^3}{8\sqrt{N}} + \frac{3c_1^2 c_2}{2\sqrt{N}} - \frac{3c_1 c_2^2}{2\sqrt{N}}\right\},\ c_6 = -\frac{c_5}{\sqrt{N}},\ c_8 = -\frac{c_7}{\sqrt{N}},$$

$$c_7 = \beta\left\{\begin{array}{l}\dfrac{3c_1^3 e^{3\sqrt{N}}}{8\sqrt{N}} - \dfrac{3c_2^3 e^{-3\sqrt{N}}}{8\sqrt{N}} + \dfrac{3c_1^2 c_2}{2\sqrt{N}}\left(e^{\sqrt{N}} + \sqrt{N}e^{\sqrt{N}}\right)\\ -\dfrac{3c_1 c_2^2}{2\sqrt{N}}\left(e^{-\sqrt{N}} - \sqrt{N}e^{-\sqrt{N}}\right)\end{array}\right\}$$

4 Heat Transfer Analysis

Now, considering the heat equation for the PTT fluid model which takes care of heat conduction phenomena and energy loss due to viscous dissipation. However, we have neglected the Joule heating as the study is confined to a low magnetic Reynolds number. Further, we have not considered Darcy dissipation term because our present flow field is not of extreme size or at extremely low temperature or high gravity field. In such a situation, the Darcy dissipation term can be neglected while studying the flow through porous media.

Using boundary layer approximations, the heat transfer in the steady one-dimensional flow of viscoelastic PTT fluid with boundary condition can be expressed as follows:

$$k\frac{\partial^2 T}{\partial y^2} + \tau_{xy}\frac{\partial u}{\partial y} = 0 \tag{33}$$

$$T(0) = T_0,\ T(L) = T_1 \tag{34}$$

Introducing the following nondimensional parameters in Eq. (33)

$$\theta = \frac{T - T_0}{T_1 - T_0}, \Pr = \frac{\mu C_p}{k}, Ec = \frac{U^2}{C_p(T_1 - T_0)}, y^* = \frac{y}{L}, u^* = \frac{u}{U}, \tau^* = \frac{\tau L}{\mu U}$$

we get (after dropping asterisk),

$$\frac{\partial^2 \theta}{\partial y^2} + P_r E_c \tau_{xy} \frac{\partial u}{\partial y} = 0 \tag{35}$$

$$\theta(0) = 0, \theta(1) = 1 \tag{36}$$

Substituting the values of τ_{xy} and u from (32) and integrating twice, we get the expression for θ.

5 Results and Discussion

The following discussion on temperature distribution reveals the effects of Deborah number and Eckert number which accounts for both momentum and thermal energy transport, resulting in heating/cooling of the bounding surface besides other parameters on temperature field.

Figures 1, 2, 3, 4, and 5 depict the effects of pertinent parameters on temperature distribution. The temperature distribution exhibit monotonically increasing behavior across the flow domain. The effects of important parameters are

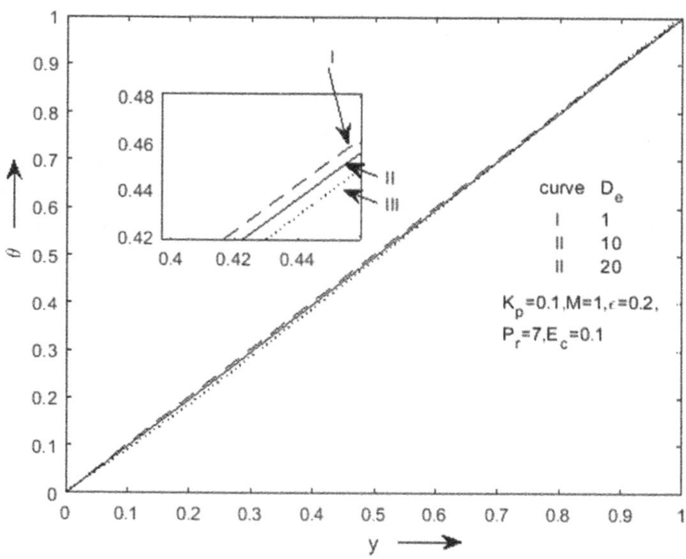

Fig. 1 Temperature profiles for different values of D_e

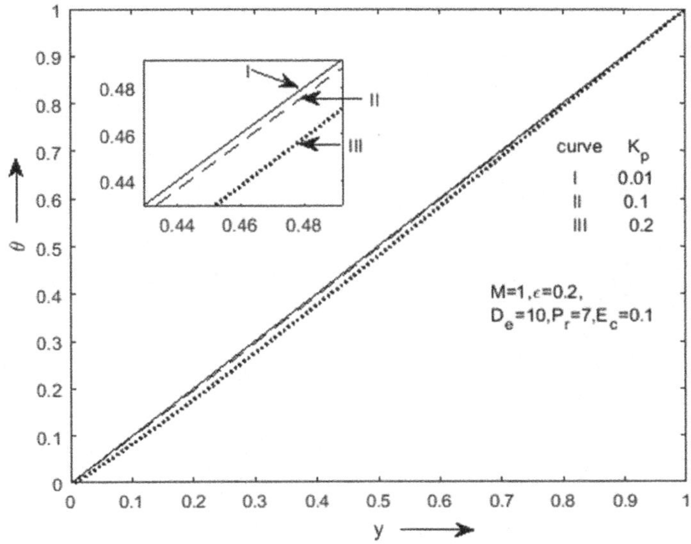

Fig. 2 Temperature profiles for different values of K_p

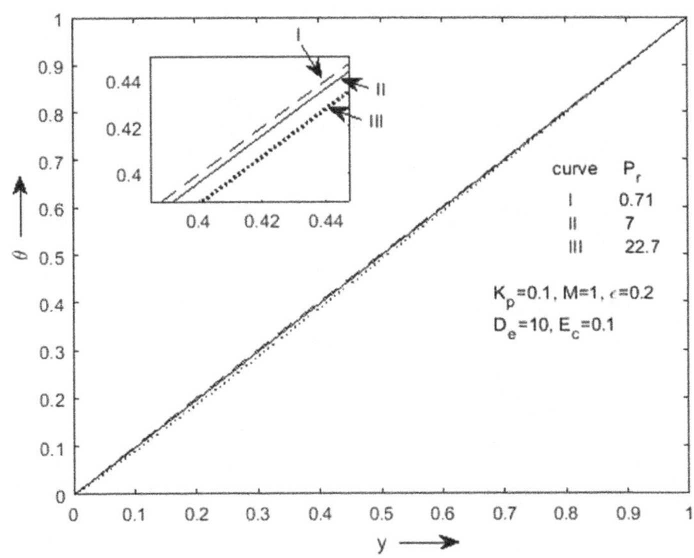

Fig. 3 Temperature profiles for different values of P_r

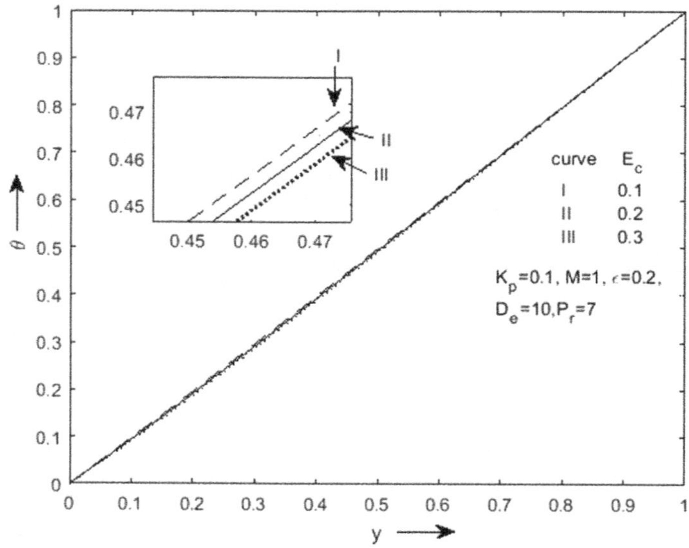

Fig. 4 Temperature profiles for different values of E_c

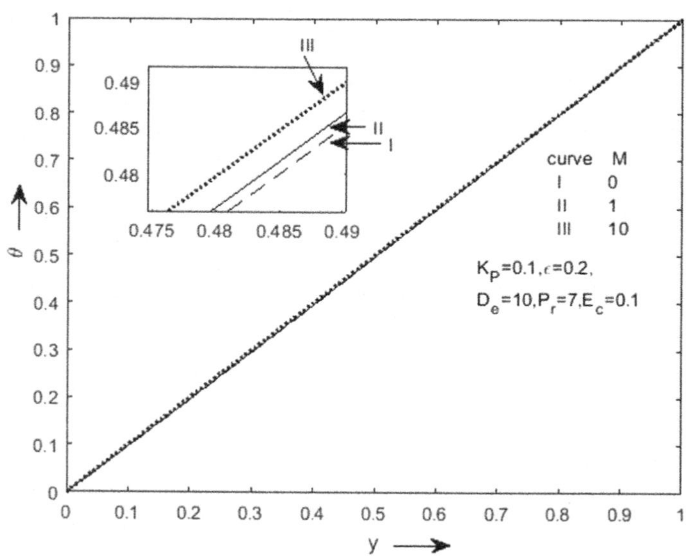

Fig. 5 Temperature profiles for different values of M

The effect of an increase in the magnetic parameter is to increase the temperature throughout the flow domain but reverse effect is observed for all other parameters such as Deborah number D_e, porosity parameter K_p, Prandtl number P_r, and Eckert number E_c. In the absence of viscous dissipation, Ohmic dissipation and strain deformation energy or strain energy in the flow ($E_c = 0$), the temperature distribution is linear which is evident from energy Eq. (35). The reason for the decrease in temperature with the increase in Pr is akin to the material property as Pr signifies the ratio of momentum diffusivity to thermal diffusivity. Fluid with higher $\mathrm{P_r}$ will possess low conductivity and hence temperature decreases as Pr increases. On the other hand, an increase in magnetic force density parameter M generates force which is proportional to negative of the velocity of the moving medium and acts as a viscous breaking force [12] and hence resists the motion and generates heat energy. Further, the effect of Deborah number which characterizes both Newtonian ($D_e < 1$) and non-Newtonian elastic fluid ($D_e > 1$) is to decrease the temperature in the present study for both low and moderate values of D_e. This may be attributed to elastic property of the fluid for which some strain energy is stored up in the fluid mass, decreasing the temperature in the flow domain.

Figure 6 illustrates the impact of electromagnetic force, a resistive electromagnetic force, created due to the interaction of transverse magnetic field with the conducting-flowing PTT fluid. Due to the resistive force generated and acted upon in the main direction of flow, the velocity decreases. On careful observation, it is revealed that for low magnetic number vis-à-vis for low intensity of the applied magnetic field, the significant increase in velocity is marked. Therefore, during clinical/mechanical

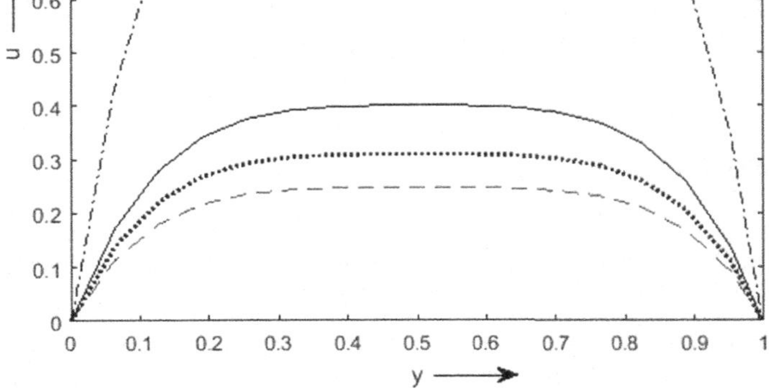

Fig. 6 Velocity profiles for different values of M

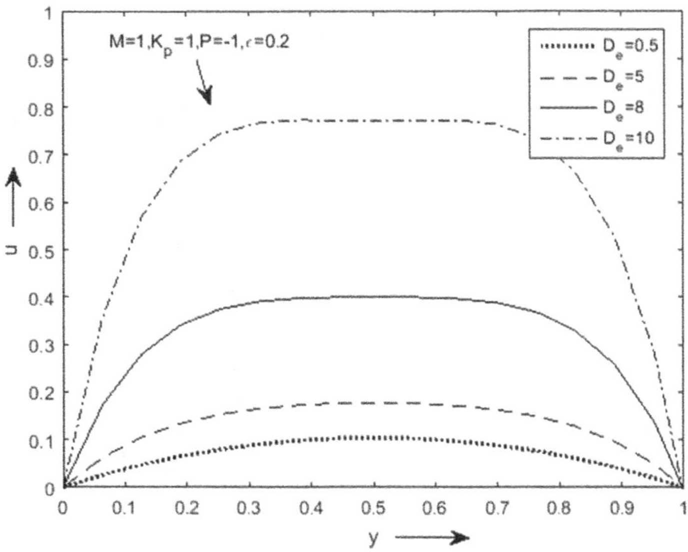

Fig. 7 Velocity profiles for different values of D_e

necessity to control the flow of biological or industrial fluid flow, one can regulate the intensity of the external magnetic field to obtain the desired flow rate. As regard to the effect of magnetic parameter M, it is mentioned that the effect of magnetic parameter is to increase the velocity profile, i.e., the magnetic field acts as an aiding force; it seems to be unusual.

Figure 7 shows that the higher values of D_e enhance the velocity of flow. On careful observation, it is revealed that the velocity profiles are more noticeable in the viscoelastic case than that in the Newtonian case, i.e., for higher value of $D_e(D_e = 10)$ correspond to non-Newtonian viscoelastic case and lower D_e ($D_e = 0.5$) to Newtonian case. The same observation was made by Akyildiz and Vajravelu [18] (2008). Deborah number accelerates the velocity that means that elastic property of the fluid accelerates the momentum transport process.

6 Conclusion

The above discussion presents a flexible means to simulate the heat transfer parameters to make use of PTT fluid as coolant or otherwise.

- Fluid with higher Pr possesses low conductivity and hence temperature decreases.
- The temperature is decreased for both low and moderate values of Deborah Number.
- Increase in magnetic force intensity parameter leads to an increase in temperature.

- Temperature decreases for increasing values of both Eckert number and porosity parameter.
- The linear variation of temperature distribution is the significant revelation of heat transfer for the property of PTT flow, ignoring the contribution of viscous dissipative heat (Eq. 35).
- Increase in magnetic force intensity increase the fluid temperature may result in cooling of the bounding surface.

References

1. N. Phan-Thien, R.I. Tanner, A new constitutive equation derived from network theory. J. Nonnewton. Fluid Mech. **2**, 353–365 (1977)
2. N. Phan-Thein, J. Rheol. **22**, 259–283 (1978)
3. R.I. Tanner, *Engineering Rheology* (Clarendon Press, Oxford, 2000)
4. P.J. Oliveira, F.T. Pinho, Analytical solution for the fully-developed channel and pipe flow of Phan-Thien-Tanner fluids. J. Fluid Mech. **387**, 271–280 (1999)
5. F.T. Pinho, P.J. Oliveira, Analysis of forced convection in pipes and channels with simplified Phan-Thien-Tanner fluid. Int. J. Heat Mass Transf. **43**, 2273–2287 (2000)
6. F.T. Pinho, P.J. Oliveria, Axial annular flow of a nonlinear viscoelastic fluid: an analytical solution. J. Nonnewton. Fluid Mech. **93**, 325–337 (2000)
7. M.A. Alves, F.T. Pinho, P.J. Oliveira, Study of steady pipe and channel flows of single-mode Phan-Thien-Tanner fluid. J. Nonnewton. Fluid Mech. **101**, 55–76 (2001)
8. M.F. Letelier, D.A. Siginer, On the fully developed tube flow of a class of non-linear viscoelastic fluids. Int. J. Non-Linear Mech. **40**, 485–493 (2005)
9. A.M. Siddiqui, R. Mahmood, Q.K. Ghori, Some exact solutions for the thin film flow of a PTT fluid. Phys. Lett. A **356**, 353–356 (2006)
10. S. Nasseri, L. Bilston, B. Fasheun, R. Tanner, Modelling the biaxial elongational deformation of soft solids. Rheol. Acta **43**, 68–79 (2004)
11. J.H. He, Int. J. Nonlinear Sci. Numer. Simul. **6**(2), 207 (2005)
12. P. Loraih, *Magneto Fluid Dynamics*, Second edn. (Springer), p. 85
13. R.B. Bird, R.C. Armstrong, O. Hassager, *Dynamics of Polymeric Liquids.Vol. 1 Fluid Mechanics*, 2nd edn. (John Wiley and Sons, Inc., Hoboken, New Jersey, 1987)
14. T. Hayat, S. Noreen, A. Hendi, Peristaltic motion of Phan-Thien-Tanner fluid in the presence of slip condition. J. Biorheol. **25**(1), 8–17 (2011)
15. T. Hayat, S. Noreen, N. Ali, S. Abbasbanday, Peristaltic motion of Phan-Thien-Tanner fluid ina planar channel. Numer. Methods Partial Differ. Equ. **28**(3), 737–748 (2012)
16. T. Hayat, S. Noreen, S. Asghar, A. Hendi, Influence of an induced magnetic field on peristaltic transport of a Phan-Thien-Tanner fluid in an asymmetric channel. Chem. Eng. Commun. **198**(5), 609– 628 (2011)
17. N. Faraz, Y. Khan, D.S. Shankar, Decomposition-transform method for fractional differential equation. Int. J. Nonlinear Sci. Sim. **11**, 305–310 (2010)
18. F. Talay Akyildiz, K. Vajravelu, Magnetohydrodynamic flow of a viscoelastic fluid. Physics Letters A. **372**, 3380–3384 (2008)

Properties of Characteristic Polynomials of Oriented Graphs

Musa Demirci, Ugur Ana, and Ismail Naci Cangul

Abstract Corresponding to a graph, one can define several matrices, and finding the eigenvalues of such matrices gives us the spectrum. Graph energy is defined as the sum of the absolute values of the eigenvalues and it has important applications related to molecular graphs. In this paper, we introduce some spectral properties of oriented graphs. We determine characteristic polynomials of several oriented graph classes, the effect of edge addition to characteristic polynomial, and also give several recursive results for the characteristic polynomial of an oriented graph by means of cut vertices, bridges, paths, and pendant edges.

Keywords Oriented graph · Characteristic polynomial · Spectrum · Energy · Spectral polynomial

MSC 2000 Numbers 05C20 · 05C31 · 05C50

1 Introduction

Studying a graph by means of matrices gives information about the properties of graph. There are many types of matrices defined to this aim. The most popular ones are adjacency, incidency, and Laplacian matrices. There are hundreds of matrices defined in literature having different applications.

One of the most popular applications of the matrices is the notion of energy which was defined as the sum of absolute values of all eigenvalues of the adjacency matrix.

M. Demirci (✉) · U. Ana · I. N. Cangul
Department of Mathematics, Bursa Uludag University, 16059 Gorukle, Bursa, Turkey
e-mail: mdemirci@uludag.edu.tr

U. Ana
e-mail: ugurana2@gmail.com

I. N. Cangul
e-mail: cangul@uludag.edu.tr

S. K. Paikray et al. (eds.), *New Trends in Applied Analysis and Computational Mathematics*, Advances in Intelligent Systems and Computing 1356,
https://doi.org/10.1007/978-981-16-1402-6_5

The study with graph energy is an important sub-area of graph theory named as the spectral graph theory, see [1].

For a given graph G, the set of all eigenvalues is called the spectrum of G. As the eigenvalues of cycles and paths are trigonometric algebraic numbers and they are different than other classes of graphs, these two classes need special attention. In [2], the spectral polynomials and some recurrence relations for the spectral polynomials of these two graph classes were studied. It was shown that we can obtain the spectrum of the cycle graph C_{2n} and path graph P_{2n+1} without detailed calculations just in terms of the spectrums of C_n and P_n, respectively. See e.g. [3, 6–8].

Let G be a graph. We say that two vertices v_i and v_j of G are adjacent to each other, whenever these two vertices are connected by an edge $e = v_i v_j$. The most popular matrix corresponding to graphs, the adjacency matrix of G denoted by $A = A(G) = [a_{ij}]$ is defined by

$$a_{ij} = \begin{cases} 1 \text{ if } v_i \text{ is adjacent to } v_j \\ 0 \text{ if not.} \end{cases}$$

It is an $n \times n$ matrix. If $e = v_i v_j$, we say that e is incident to v_i and v_j.

A directed graph is a graph that is a set of vertices connected by edges where each edge has a direction. The direction of an edge is shown by an arrow on the edge. The main difference between a graph and a directed graph is that the edges in the latter one have an order.

Recall that we denote an edge connecting two vertices u and v in a graph by uv and this means u is adjacent to v and v is adjacent to u. If we denote an edge connecting two vertices u and v in a directed graph, we have to choose either uv or vu and the former one reads as u is adjacent to v and v is adjacent from u, where the latter one reads as v is adjacent to u and u is adjacent from v. Although very rare, there are situations that allow an edge to be bidirected in a directed graph. That is, both uv and vu may be the edges of the directed graph. See [5] for a spectral study of directed graphs.

A directed graph is called an oriented graph if it has no bidirected edges, that is, at most one of uv and vu for every vertex pair u and v may be an edge of the graph. Therefore, an oriented graph in this work will be a directed graph with no loops nor multiple edges. In [4], the characteristic polynomials of oriented graphs have been studied for the first time. Although the directed and oriented graphs have more applications compared to the classical graphs, there is not much done about these two graph types. Throughout this paper, we take G to be a simple, connected, and oriented graph unless stated otherwise.

It is known that the total number of non-isomorphic oriented graphs having $n = 1, 2, 3, \ldots$ vertices are

$$1, 2, 7, 42, 582, 21480, 21422888, \ldots \ (OEIS\ A001174)$$

and the number of connected oriented graphs with $n = 1, 2, 3, \ldots$ vertices are

$$1,\ 1,\ 5,\ 34,\ 535,\ 20848,\ 2120098,\ \ldots\ (OEIS\ A086345).$$

2 Some Properties of the Characteristic Polynomials of Oriented Graphs

We now obtain some new properties of the characteristic polynomials of graphs by means of some smaller graphs. In graph theory, and in general in mathematics, it is usually a clever idea to study properties of smaller pieces to obtain the properties of the main part. The classical graph parts which are used for this aim are cut vertices and bridges. So, we shall use them here to calculate the characteristic polynomials of several graph structures. First, we have

Theorem 1 *Let G be a graph formed by identifying only one vertex, say v, of two cycles having lengths r and s each. Then*

$$P(G) = \begin{cases} \dfrac{P(C_r)P(C_s) - 1}{\lambda} & \textit{if all orientations are the same} \\ \dfrac{P(C_r)P(C_s)}{\lambda} & \textit{otherwise.} \end{cases}$$

Proof Note that G is as in Fig. 1.

First, let all the orientations be the same in G. In this case, note that $P(G)$ has the form

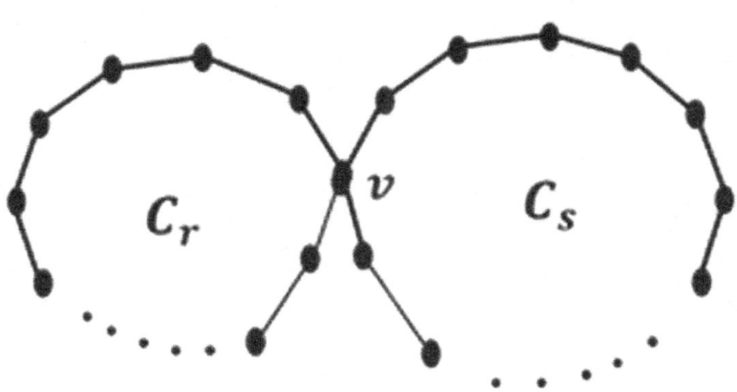

Fig. 1 Two cycles joined with a cut vertex v

$$P(G) = \begin{array}{c|cccccccccc|}
 & v_1 & v_2 & v_3 & \cdots & v_{r-1} & v_r & v_{r+1} & \cdots & v_{r+s-1} \\
v_1 & \lambda & -1 & 0 & \cdots & 0 & 0 & 0 & \cdots & 0 \\
v_2 & 0 & \lambda & -1 & \cdots & 0 & 0 & 0 & \cdots & 0 \\
v_3 & 0 & 0 & \lambda & \cdots & 0 & 0 & 0 & \cdots & 0 \\
\vdots & \vdots & \vdots & \vdots & \vdots & \vdots & \vdots & \vdots & \vdots & \vdots \\
v_{r-1} & 0 & 0 & 0 & \cdots & \lambda & -1 & 0 & \cdots & 0 \\
v_r & -1 & 0 & 0 & \cdots & 0 & \lambda & -1 & \cdots & 0 \\
v_{r+1} & 0 & 0 & 0 & \cdots & 0 & 0 & \lambda & \cdots & 0 \\
\vdots & \vdots & \vdots & \vdots & \vdots & \vdots & \vdots & \vdots & \vdots & \vdots \\
v_{r+s-1} & 0 & 0 & 0 & \cdots & 0 & -1 & 0 & \cdots & \lambda
\end{array}.$$

Calculating this determinant, we get the required formula.

Second, let at least one orientation be different than the others. Let the different orientation be $v_k v_{k+1}$. In this case, we have

$$\begin{array}{c|ccccccccccccc|}
 & v_1 & v_2 & v_3 & \cdots & v_k & v_{k+1} & v_{k+2} & \cdots & v_{r-1} & v_r & v_{r+1} & \cdots & v_{r+s-1} \\
v_1 & \lambda & -1 & 0 & \cdots & 0 & 0 & 0 & \cdots & 0 & 0 & 0 & \cdots & 0 \\
v_2 & 0 & \lambda & -1 & \cdots & 0 & 0 & 0 & \cdots & 0 & 0 & 0 & \cdots & 0 \\
v_3 & 0 & 0 & \lambda & \cdots & 0 & 0 & 0 & \cdots & 0 & 0 & 0 & \cdots & 0 \\
\vdots & \vdots & \vdots & \vdots & \vdots & \vdots & \vdots & \vdots & \vdots & \vdots & \vdots & \vdots & \vdots & \vdots \\
v_k & 0 & 0 & 0 & \cdots & 0 & \lambda & 0 & \cdots & 0 & 0 & 0 & \cdots & 0 \\
v_{k+1} & 0 & 0 & 0 & \cdots & 0 & -1 & \lambda & \cdots & 0 & 0 & 0 & \cdots & 0 \\
v_{k+2} & 0 & 0 & 0 & \cdots & 0 & 0 & \lambda & \cdots & 0 & 0 & 0 & \cdots & 0 \\
\vdots & \vdots & \vdots & \vdots & \vdots & \vdots & \vdots & \vdots & \vdots & \vdots & \vdots & \vdots & \vdots & \vdots \\
v_{r-1} & 0 & 0 & 0 & \cdots & 0 & 0 & 0 & \cdots & \lambda & -1 & 0 & \cdots & 0 \\
v_r & -1 & 0 & 0 & \cdots & 0 & 0 & 0 & \cdots & 0 & \lambda & -1 & \cdots & 0 \\
v_{r+1} & 0 & 0 & 0 & \cdots & 0 & 0 & 0 & \cdots & 0 & 0 & \lambda & \cdots & 0 \\
\vdots & \vdots & \vdots & \vdots & \vdots & \vdots & \vdots & \vdots & \vdots & \vdots & \vdots & \vdots & \vdots & \vdots \\
v_{r+s-1} & 0 & 0 & 0 & \cdots & 0 & 0 & 0 & \cdots & 0 & -1 & 0 & \cdots & \lambda
\end{array}.$$

Hence, we get the required result after some row and column operations. ■

Using the previous result we can easily generalize Theorem 1 as follows:

Corollary 2 *Let G be a bicyclic graph and let v be a cut vertex of G connecting two cycles C_r and C_s in G (Fig. 2).*

Then

$$P(G) = \begin{cases} \lambda^{r+s-1}(P(C_r)P(C_s) - 1) & \text{if all orientations are the same} \\ \lambda^{r+s-1}P(C_r)P(C_s) & \text{otherwise.} \end{cases}$$

Theorem 3 *The orientation of a bridge between two unicyclic components G_1 and G_2 of a bicyclic oriented graph G does not effect $P(G)$ and*

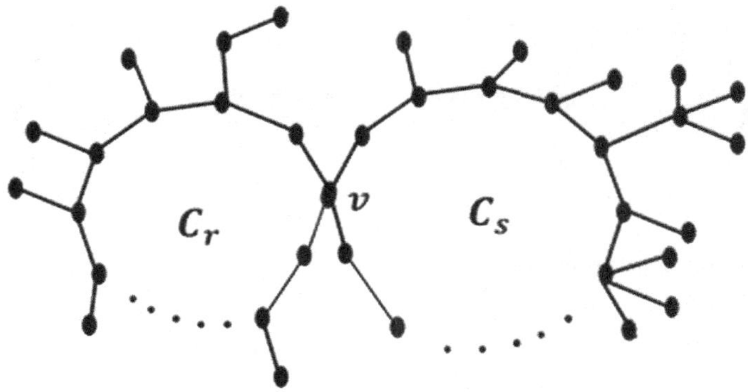

Fig. 2 A bicyclic graph with cut vertex v and k pendant edges

$$P(G) = P(G_1) \cdot P(G_2).$$

Proof Let $e = u_k u_{k+1}$ be an oriented bridge of G from u_k to u_{k+1}. Let G_1 and G_2 be two unicyclic oriented components of $G - e$. Let r and s be the lengths of the cycles in G_1 and G_2, respectively, See Fig. 3.

Then the characteristic polynomial of G is obtained by the determinant

$$P(G) = \begin{vmatrix} \lambda I - AG_1 & B \\ C & \lambda I - AG_2 \end{vmatrix}$$

where B is a matrix of dimension $r \times s$ with all the entries being 0 except for the entry $a_{r(r+1)}$ which is -1, and C is a matrix of dimension $s \times r$ with all the entries being 0. This determinant is equivalent to $|\lambda I - AG_1|\,|\lambda I - AG_2|$. Therefore

$$P(G) = P(G_1) \cdot P(G_2).$$

Now let us reverse the orientation of e. In this case, the characteristic polynomial of G is obtained by means of the determinant

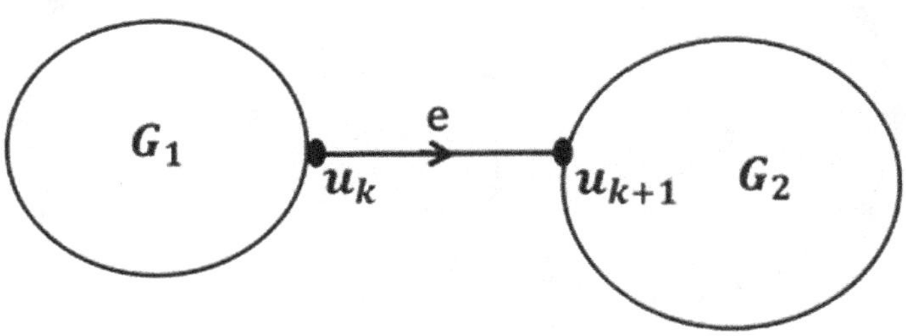

Fig. 3 Two unicyclic graphs joined with a bridge

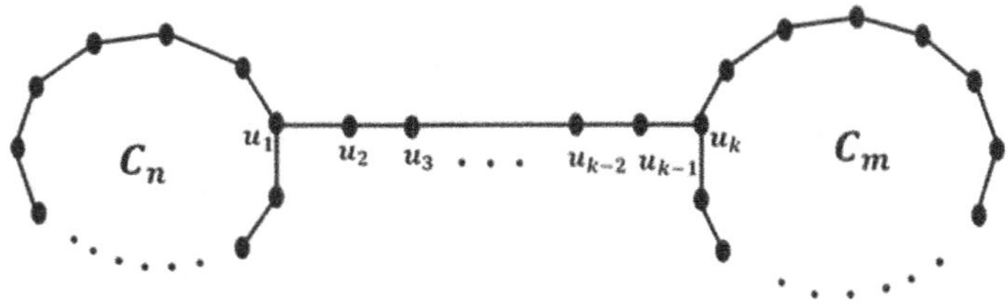

Fig. 4 Two cycles joined with a path

$$P(G) = \begin{vmatrix} \lambda I - AG_1 & B \\ C & \lambda I - AG_2 \end{vmatrix}$$

where B is a matrix of dimension $r \times s$ with all the entries being 0, and C is a matrix of dimension $s \times r$ with all the entries being 0 except for the entry $a_{(r+1)r}$ which is -1. This gives the same result. ■

Theorem 4 *Let G be the oriented graph by joining two cycle graphs C_m and C_n by a path P_k as in Fig. 4.*

Then the adjacency characteristic polynomial of the oriented graph G is given by

$$P(G) = P(C_m) \cdot P(C_n) \cdot P(P_{k-2}) = (\lambda^m - 1)(\lambda^n - 1)\lambda^{k-2}.$$

Proof It follows by induction. ■

Next, we calculate the characteristic polynomial of an oriented bicyclic graph where two cycles of the graph have a common edge. The proof can be done using row and column operations.

Theorem 5 *Let G be a bicyclic graph obtained by joining the two cycles C_r and C_s along a common edge e, see Fig. 5.*

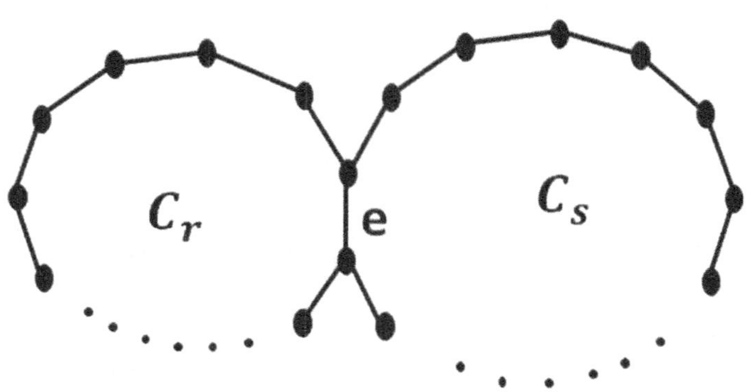

Fig. 5 A bicyclic graph where two cycles have a common edge

Then

$$P(G)=\begin{cases}\dfrac{P(C_r)P(C_s)}{\lambda^2} & \textit{if either there is at least one edge in each of } C_r \textit{and } C_s \textit{with different orientation or one has all the orientations the same and the other has at least one edge different frome with different orientation}\\[2ex] \dfrac{P(C_r)P(C_s)-1}{\lambda^2} & \textit{if all edges of both cycles have the same orientations}\\[2ex] \dfrac{P(C_r)P(C_s)-\lambda^2}{\lambda^2} & \textit{if all the edges of one of } C_r \textit{and } C_s \textit{are the same and only} e \textit{has different orientation in the other cycle.}\end{cases}$$

References

1. A.E. Brouwer, W.H. Haemers, Spectra of graphs (2012)
2. F. Celik, I.N. Cangul, Formulae and recurrence relations on spectral polynomials of some graphs. Adv. Stud. Contemp. Math. **27**(3), 325–332 (2017)
3. K.C. Das, P. Kumar, Some new bounds on the spectral radius of graphs. Discret. Math. **281**, 149–161 (2004)
4. M. Demirci, U. Ana, I.N. Cangul, Characteristic polynomials of oriented graphs (communicated)
5. N. Jahanbakht, Energy of graphs and digraphs (2010)
6. M.K. Kumar, Characteristic polynomial and domination energy of some special class of graphs. Int. J. Math. Comb. **1**, 37–48 (2014)
7. J. Shu, Y. Wu, Sharp upper bounds on the spectral radius of graphs. Linear Algebr. Appl. **377**, 241–248 (2004)
8. A. Yu, M. Lu, F. Tian, On the spectral radius of graphs. Linear Algebr. Appl. **387**, 41–49 (2004)

An Inventory Model for Linear Deteriorating Item with Shortages Under Partial-Backlogged Condition

S. K. Indrajitsingha, A. K. Sahu, and U. K. Misra

Abstract The objective of this paper is to develop an inventory model for deteriorating items with price-dependent linear demand. Allowable shortage is considered which is partial backlogged. Since selling value assumes a significant job in stock framework, we use price-dependent demand and holding up time subordinate build-up rates in a general structure to develop the model. The solution method of this optimization model is adorned by a numerical example. A convexity checkup of the total average cost function is executed by plotting a 3D graph. Finally, sensitivity analysis and managerial insights are completed to examine the impact of various system parameters of this model.

Keywords Demand · Deterioration · Shortages · Partial-backlogged · Lost-sale cost

AMS classification no 90B05

1 Introduction

In recent times, organizations are highly competitive in the market to promote the reduction of inventory cost in terms of physical stock which provides flexibility in operating business affairs. Moreover, inventory plays an important role drastically. Inventory management is a necessary part for getting profit in business.

S. K. Indrajitsingha (✉)
Department of Mathematics, Saraswati Degree Vidya Mandir, Neelakantha Nagar, Neelakantha Nagar, Berhampurr 760002, Odisha, India
e-mail: skismath86@gmail.com

A. K. Sahu
Department of Mathematics, GIET University, Gunupur, Rayagada 765002, Odisha, India
e-mail: anitaksahu92@gmail.com

U. K. Misra
NIST, Palur Hills, Berhampur 761008, Odisha, India
e-mail: umakanta_misra@yahoo.com

S. K. Paikray et al. (eds.), *New Trends in Applied Analysis and Computational Mathematics*, Advances in Intelligent Systems and Computing 1356,
https://doi.org/10.1007/978-981-16-1402-6_6

Now, investors or business persons are facing many threats in decisive the methods to accomplish the consistently increasing consumer's requirement and to maintain stability in the competitive market while managing the total costs conveniently. About 80% of customers in India eagerly waiting for the decrease in the price of products for purchasing. So, selling price with dependent demand plays a vital role to attract the customers toward the low-price products and hence increases the sales which ultimately give profit. There are many factors which affect the total cost of the products like deterioration, shortages, selling price, lost sale cost, time, etc. Deterioration occurs in most of the product during its storage period or at the end of the season. Therefore, high deterioration rate may decrease the profit. Basing upon these above parameters, many research papers have been published since the twentieth century.

At the beginning of twentieth century, a new concept of inventory was introduced by Harris [11] considering very few parameters. In our day-to-day life, there are various parameters which directly influence the profit of inventory. In 1957, Whitin [25] had developed a theory on inventory management. Later, these models were adorned and incorporated extra parameters. This is found in Inventory Control: Theory and practice established by Pal [21]. Ghare and Schrader [9] had developed an inventory problem on exponentially deteriorating goods. Covert and Phillip [6] worked on an economic ordered quantity model for commodities with fluctuating deterioration rate and assuming Weibull distribution. In 1957, Mishra [20] developed a deteriorating inventory model where he focused on optimizing the production of lot-size. During 1980–2000, several researchers like Giri and Chaudhuri [10], Elsayed and Terasi [8], Dave [7], Pal [21], Benkherouf [2], Chakrabarti et al. [4], Bhuina and Maity [3], Balkhi [1], Chang and Dye [5], etc., had been carried out on inventory modeling by taking different parameters time to time.

Huang [12] established an inventory model using two types of trade credit policy. Jaggi and Verma [15] proposed an inventory model where jointly optimized the price and ordering quantity for a two-warehouse system. Later in 2010, Jaggi et al. [13] published an inventory model with deteriorating items and time-proportional backlogging rate. Again in 2017, Jaggi et al. [14] established a two-storage model, i.e., one is owned store and the other is rented store with deteriorating products. Since the quality of goods is not good, they give duration to the dealer for payment. Kumar et al. [17] developed a cost optimization model for items having fuzzy demand and deterioration with two-warehouse facility under the trade credit financing Mandal and Giri [19] developed a two-storage mixed stock model with stock-dependent demand under imperfect production process through quantity discount offer. Mahapatra et al. [18] established a mathematical inventory model on focusing the demand which is assumed as depending upon time and reliability and the unsatisfied demand is backlogged. Tiwari et al. [24] proposed a deteriorating inventory model in supply chain system in which they focused on optimizing the price and lot-size.

In 2019, Sahoo et al. [22] established an inventory model to allow shortages. Taleizadeh et al. [23] represent a problem on payment delayed model in which, the order is reached late at retailers and also considered for lifetime decaying products. Kumar [16] proposed an inventory decision problem for time fluctuate linear demand and parabolic holding cost with salvage value.

In this proposed model, to obtain vigorous and general outcomes, we try to impose a deterministic model where demand is price-dependent and allowable shortages. The shortage is considered as partial backlogging. In addition, the constant deterioration rate is taken into the model. The aim is then to find the optimal total inventory cost. The remainder of the paper is organized as cited below: Sect. 2, presents all the notations and explains the assumptions which are considered followed by the Sect. 3, where we formulate a mathematical model. In Sect. 4, a numerical example is provided for validation of the developed model. In Sect. 5, sensitivity analysis is carried out by using Mathematica 11 software with graphical representation. In Sect. 6, we conclude and give some future research scope.

2 Model Notation and Assumptions

2.1 Notations

m	The purchase cost per item.
p	The selling price per unit, where $p > m$
$D(t)$	The demand rate per unit time and taken as $a - bp$. Where a and b are positive constant and $a > bp$.
θ	The rate of deterioration.
t_1	Time at which shortage start.
T	Duration of a cycle.
O	Ordering cost.
Q_1	The inventory level at which each cycle starts.
Q_2	Backorder quantity.
h	Holding cost per unit per unit time.
μ	The shortage cost per unit.The shortage cost per unit.
δ	Cost sale per unit.
ρ	Backlogging rate.
$TAC(t_{1,} p)$	Total profit for one replenishment cycle.

2.2 Assumptions

In establishing the mathematical model, the following assumptions are considered:

i. Single item is used throughout the system.
ii. The sales prices ρ and back order δ are predetermined.
iii. The selling price is greater than that of purchasing price of a unit.
iv. Demand is deterministic and is a function of selling price where stock out period is partially lost due to impatient customers.
v. There is neither replaced nor repair of decaying goods during a cycle.

vi. Shortage is allowed, and is a mixture of partially backlogged and lost sales
vii. Lead time is zero.

3 Mathematical Model Development

Given the assumptions referenced previously, the stock level over time is shown in Fig. 1. The exhaustion of stock level occurs because of the combined impacts of the demand and decay in the interval $[0, t_1)$ and backlogged during the shortage period $[t1, T)$ with a rate of ρ. Let Q_1 be the initial inventory at time $t = 0$.

Thus, the variation of inventory level, $I(t)$, w.r.t. time can be governed by the accompanying differential equations:

$$\frac{dI_1(t)}{dt} = -\theta I_1(t) - (a - bp), 0 \le t \le t_1 \tag{1}$$

$$\frac{dI_2(t)}{dt} = -(a - bp), t_1 \le t \le T \tag{2}$$

$$\text{with } I_1(t_1) = I_2(t_1) = 0 \tag{3}$$

Solution of (1) and (2) yields,

$$I_1(t) = (a - bp)\left[(t_1 - t) + \frac{\theta}{6}\left(t_1^3 - t^3\right)\right]e^{-\frac{\theta t^2}{2}}, 0 \le t \le t_1 \tag{4}$$

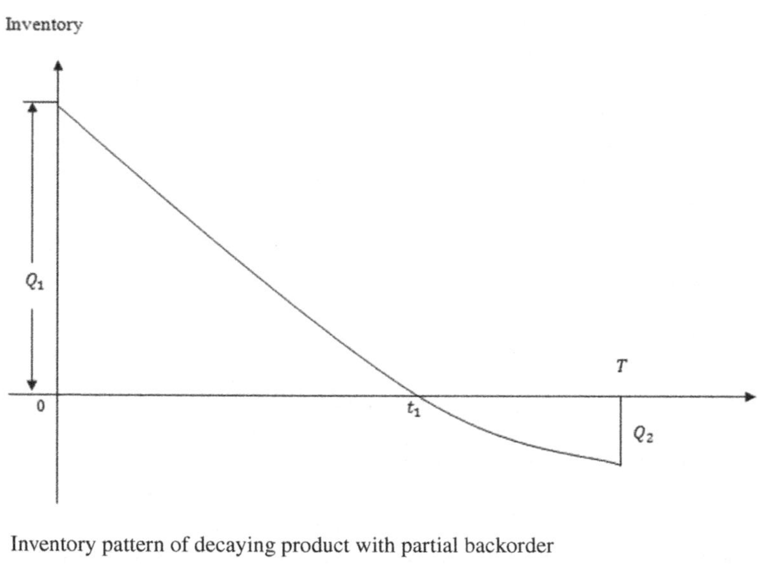

Fig. 1 Inventory pattern of decaying product with partial backorder

$$I_2(t) = (a - bp)(t_1 - t), t_1 \leq t \leq T \tag{5}$$

Based on the above equations, the total average cost consists of the following elements:

1. Ordering cost (O.C) per cycle is O.
2. Sales revenue (S.R) per cycle is

$$\begin{aligned} S.R &= (Q_1 + Q_2)p \\ &= (a - bp)\left[t_1 + \frac{\theta t_1^3}{6} + \rho(T - t_1)\right]p \end{aligned} \tag{6}$$

where

$$Q_1 = (a - bp)\left[t_1 + \frac{\theta t_1^3}{6}\right]$$

$$Q_2 = \int_{t_1}^{T} (a - bp)\rho dt$$

3. Purchasing cost (P.C) per cycle is

$$\begin{aligned} P.C &= (Q_1 + Q_2)m \\ &= (a - bp)\left[t_1 + \frac{\theta t_1^3}{6} + \rho(T - t_1)\right]m \end{aligned} \tag{7}$$

4. Deterioration cost (D.C) per cycle is

$$\begin{aligned} D.C &= \left(I_1(0) - \int_0^{t_1} (a - bp)dt\right)m \\ &= \left[(a - bp)\frac{\theta t_1^3}{6}\right]m \end{aligned} \tag{8}$$

5. Holding cost (H.C) per cycle is

$$H.C = h\int_0^{t_1}(a-bp)\left[(t_1-t)+\frac{\theta}{6}\left(t_1^3-t^3\right)\right]e^{-\frac{\theta t^2}{2}}dt$$
$$= h(a-bp)\left(\frac{t_1^2}{2}+\frac{\theta t_1^4}{12}\right) \tag{9}$$

6. Shortage cost (S.C) per cycle is

$$S.C = \mu\int_{t_1}^{T}(a-bp)dt$$
$$= \mu(a-bp)(T-t_1) \tag{10}$$

7. Lost sale cost (L.C) per cycle is

$$L.C = \delta\int_{t_1}^{T}(1-\rho)(a-bp)dt$$
$$= \delta(1-\rho)(a-bp)(T-t_1) \tag{11}$$

8. Unit time profit (TAC) is

$$TAC(t\_1, p) = 1/T[S.R - P.C - D.C - H.C - S.C - L.C - O.C]$$
$$= \frac{1}{T}\begin{bmatrix}\left((a-bp)\left(t_1+\frac{\theta t_1^3}{6}+\rho(T-t_1)\right)(p-m)\right)-\left((a-bp)\frac{\theta t_1^3}{6}\right)m\\ -h(a-bp)\left(\frac{t_1^2}{2}+\frac{\theta t_1^4}{12}\right)-\mu(a-bp)(T-t_1)\\ -\delta(1-\rho)(a-bp)(T-t_1)-O\end{bmatrix} \tag{12}$$

For minimization of the total average cost function $TAC(t_1, p)$ per unit time, the values of t_1 and p can be obtained by solving the equations

$$\frac{\partial TAC(t_1,p)}{\partial t_1} = 0 \text{ and } \frac{\partial TAC(t_1,p)}{\partial p} = 0 \tag{13}$$

Equation (13) is equivalent to

$$\frac{1}{\mathrm{T}}\begin{bmatrix}-(a-bp)\mathrm{t}_1^5\theta - h(a-bp)\left(t_1+\frac{t_1^3\theta}{3}\right)+(a-bp)\mu\\ +(a-bp)\delta(1-\rho)+(-m+p)(a-bp)\left(1+\frac{t_1^2\theta}{2}-\rho\right)\end{bmatrix} = 0$$

and

$$\frac{1}{\mathrm{T}}\left[\begin{array}{l} \frac{1}{6}bt_1^6\theta + bh\left(\frac{t_1^2}{2} + \frac{t_1^4\theta}{12}\right) + b(T - t_1)\mu + b(T - t_1)\delta(1 - \rho) \\ -b(-m + p)(t_1 + \frac{t_1^3\theta}{6} + (T - t_1)\rho + (a - bp)(t_1 + \frac{t_1^3\theta}{6} + (T - t_1)\rho \end{array}\right] = 0$$

Provided it satisfies the equations

$$\frac{\partial^2 TAC(t_1, p)}{\partial t_1^2} > 0, \quad \frac{\partial^2 TAC(t_1, p)}{\partial \mathrm{p}^2} > 0$$

and

$$\left(\frac{\partial^2 TAC(t_1, p)}{\partial t_1^2}\right)\left(\frac{\partial^2 TAC(t_1, p)}{\partial \mathrm{p}^2}\right) - \left(\frac{\partial^2 TAC(t_1, p)}{\partial t_1 \partial \mathrm{p}}\right)^2 > 0 \quad (14)$$

4 Numerical Example

In order to endorse the proposed model, we considered an inventory system with the following parametric data:

$T = 12$ days, $\theta = 0.001$, $a = 250$ units, $b = 10$, $\rho = \$0.6$/unit/day, m = \$10/unit, $h = \$0.01$/unit/day, $\mu = 2$, $\delta = \$15$/unit/day. The values of different system parameters considered here are realistic in nature, in spite of the fact that these are not taken from any contextual analysis. Corresponding to these input values, the optimum value of $t_1 = 6.57779$ days, $p = \$20.3899$ and the minimum value of $TAC(t_1, p) = \$174.543$. To show the convexity of cost function $TAC(t_1, p)$, we plot a 3D graph. A 3D graph is shown in Fig. 2.

5 Sensitivity Analysis

Sensitivity analysis is carefully analyzed in Table 1 and observed the optimal solution by a change in values of different parameters (Figs. 3, 4, 5, 6 and 7).

Effect due to change in

i. **Demand parameter (*a*)**: The shortage starting time t_1, selling price p, and the total average cost TAC increases with the increase of a.
ii. **Demand parameter (*b*)**: The shortage starting time t_1 increases with the increase of b, while selling price p and the total average cost TAC decreases.
iii. **Purchasing cost (*m*)**: The shortage starting time t_1 increases slowly and selling price p increases with the increase of m. However, the total average cost decreases rapidly.

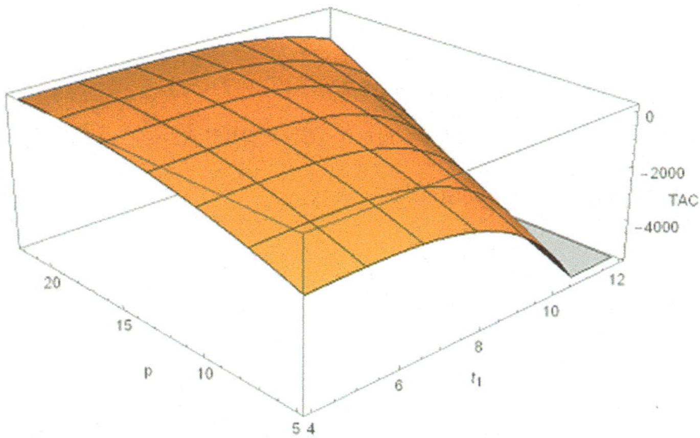

Fig. 2 Convexity of the given function

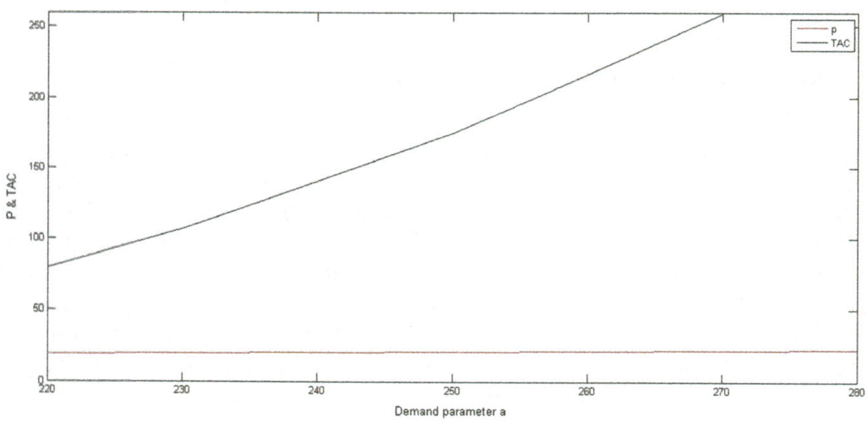

Fig. 3 Graphical representation of *a* w.r.t. *p* and *TAC*

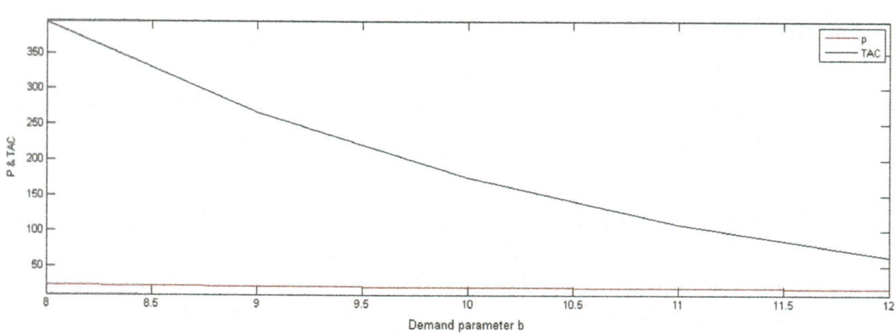

Fig. 4 Graphical representation of *b* w.r.t. *p* and *TAC*

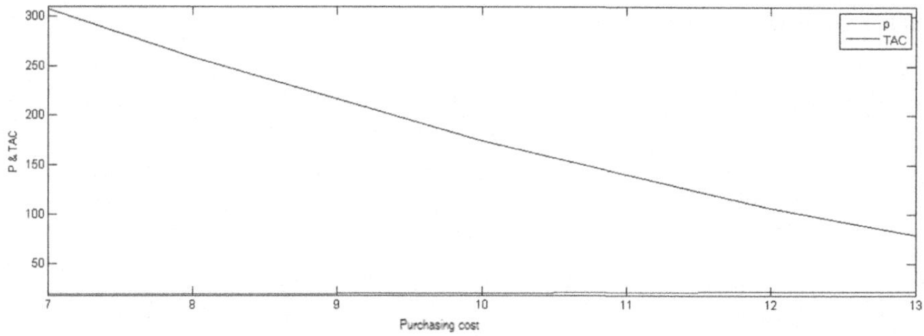

Fig. 5 Graphical representation of m w.r.t. p and TAC

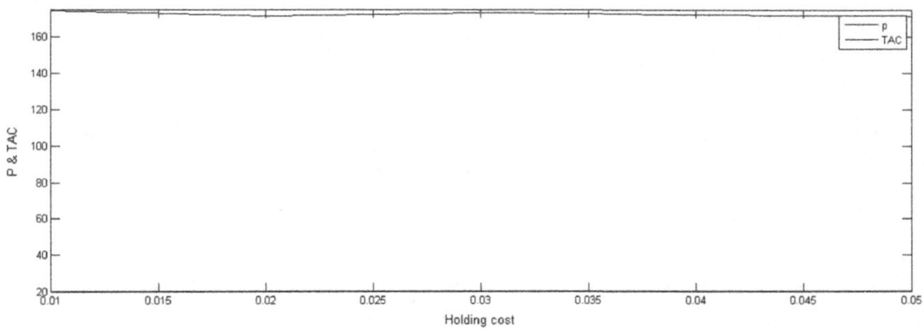

Fig. 6 Graphical representation of h w.r.t. p and TAC

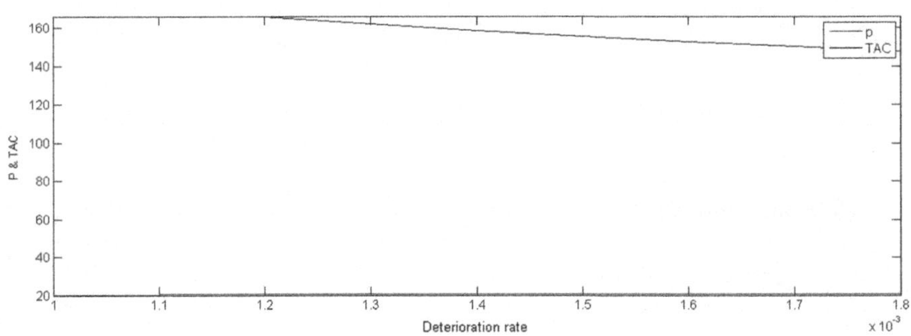

Fig. 7 Graphical representation of θ w.r.t. p and TAC

iv. **Holding cost (*h*)**: The shortage starting time t_1 and total average cost TAC decreases slowly with the increase of holding cost h, while the selling price p increases.

Table 1 Sensitivity test, for example

Parameter	Parameter value	t_1	p	TAC
a	220	6.50815	18.8841	79.2687
	230	6.53167	19.3858	106.902
	250	6.57779	20.3899	174.543
	270	6.62274	21.3948	258.722
	280	6.64481	21.8975	307.028
b	8	23.5328	23.5328	394.092
	9	21.7858	21.7858	266.299
	10	6.57779	20.3899	174.543
	11	6.52528	19.249	108.895
	12	6.48031	18.2991	62.7623
m	7	6.64481	18.8975	307.028
	8	6.62274	19.3948	258.722
	10	6.57779	20.3899	174.543
	12	6.53167	21.3858	106.902
	13	6.50815	21.8841	79.2687
h	0.01	6.57779	20.3899	174.543
	0.02	6.57108	20.4002	171.207
	0.03	6.56435	20.4106	172.876
	0.04	6.55762	20.4209	172.048
	0.05	6.55087	20.4312	171.224
θ	0.001	6.57779	20.3899	174.543
	0.0012	6.34949	20.4881	165.681
	0.0014	6.16285	20.5697	158.565
	0.0016	6.00576	20.6392	152.669
	0.0018	5.87066	20.6996	147.668

v. **Deterioration ($\boldsymbol{\theta}$)**: The shortage starting time t_1 and the total average cost TAC decreases slowly with the increase of deterioration. However, selling price p also increases slowly as θ increases.

6 Conclusion

In this paper, a single deteriorating item in a deterministic approach is developed where demand is depending upon selling price. Shortage is allowed and, due to some impatient customers, it is taken as partially backlogged and incorporates some lost sales. This model is solved analytically by optimizing the total average cost. Moreover, the mathematical model is illustrated with a numerical example. Convexity of total cost function concerning to t_1 and selling price p are studied. The robustness of this proposed model is examined through sensitivity analysis by altering the value of different related parameters.

There are a few expansions of this work which can be done by considering realistic values of different system parameters. Also one can extend this model to the deteriorating problems in supply chains. The model is suggested to the retailers and big shopkeepers like malls to determine the more accurate total average cost.

References

1. Z.T. Balkhi, On the global optimal solution to an integrated inventory system with general time-varying demand, production and deterioration rates. Eur. J. Oper. Res. **114**, 29–37 (1999)
2. L. Benkherouf, On the optimality of a replenishment policy for an inventory model with deteriorating items and time-varying demand and shortages. Arab J. Math. Sci. **3**, 59–67 (1997)
3. A.K. Bhunia, M. Maity, Deterministic inventory model for deteriorating items with finite rate of replenishment dependent on inventory level. Comput. Oper. Res. **25**(11), 997–1006 (1998)
4. T. Chakrabarti, B.C. Giri, K.S. Chaudhury, A heuristic for replenishment of deteriorating items with time varying demand and shortages in all cycles. Int. J. Syst. Sci. **29**(6), 551–556 (1998)
5. H.J. Chang, C.Y. Dye, An EOQ model for deteriorating items with time varying demand and partial backlogging. J. Oper. Res. Soc. **50**, 1176–1182 (1999)
6. R.P. Covert, G.C. Philip, An EOQ model for items with Weibull distribution deterioration. AIIE Trans. **5**, 323–326 (1973)
7. U. Dave, On the EOQ models with two levels of storage. Opsearch **25**(3), 190–196 (1988)
8. E.A. Elsayed, C. Terasi, Analysis of inventory systems with deteriorating items. Int. J. Prod. Res. **21**, 449–460 (1983)
9. P.M. Ghare, S.F. Schrader, A model for exponentially decaying inventory. J. Ind. Eng. **14**, 238–243 (1963)
10. B.C. Giri, K.S. Chaudhuri, Heuristic models for deteriorating items with shortages and time varying demand and costs. Int. J. Syst. Sci. **28**(2), 153–159 (1997)
11. F. Harris, *Operations and Cost* (AW Shaw Co., Chicago, 1915)
12. Y.F. Huang, An inventory model under two levels of trade credit and limited storage space derived without derivatives. Appl. Math. Model. **30**(5), 418–436 (2006)
13. C.K. Jaggi, K.K. Aggarwal, P. Verma, Inventory and pricing strategies for deteriorating items with limited capacity and time proportional backlogging rate. Int. J. Oper. Res. **8**(3), 331–354 (2010)
14. C.K. Jaggi, L.E. Cardenas-Barron, S. Tiwari, A.A. Shafi, Two-warehouse inventory model for deteriorating items with imperfect quality under the conditions of permissible delay in payments. Sci. Iran. **157**(2), 344–356 (2017)
15. C.K. Jaggi, P. Verma, Joint optimization of price and order quantity with shortages for a two-warehouse system. Top (Spain) **16**(1), 195–213 (2008)
16. P. Kumar, An inventory planning problem for time-varying linear demand and parabolic cost with salvage value. Croatian Oper. Res. Rev. **10**, 187–199
17. B.A. Kumar, S.K. Paikray, H. Dutta, Cost optimization model for items having fuzzy demand and deterioration with two-warehouse facility under the trade credit financing. AIMS Math. **5**, 1603–1620 (2020)
18. G.S. Mahapatra, S. Adak, T.K. Mandal, S. Pal, Inventory model for deteriorating item with time and reliability dependent demand and partial backorder. Int. J. Oper. Res. **9**(3), 344–359 (2017)
19. P. Mandal, B.C. Giri, A two-warehouse integrated inventory model with imperfect production process under stock dependent demand quantity discount offer. Int. J. Syst. Sci.: Oper. Logist. **4**(4), 1–12 (2017)

20. R.B. Mishra, Optimum production lot-size model for a system with deteriorating inventory. Int. J. Prod. Res. **13**, 495–505 (1975)
21. M. Pal, An inventory model for deteriorating items when demand is random. Calcutta Stat. Assoc. Bull. **39**, 201–207 (1990)
22. A.K. Sahoo, S.K. Indrajitsingha, P.N. Samanta, U.K. Misra, Selling price dependent demand with allowable shortages model under partially backlogged- deteriorating items. Int. J. Appl. Comp. Math. **5**, 104. https://doi.org/10.1007/s40819-019-0670-7
23. A.A. Taleizadeh, P. Nadia, I. Konstantras, Partial linked-to-order delayed payment and life time effects on decaying items ordering. Oper. Res. (2019)
24. S. Tiwari, C.K. Jaggi, M. Gupta, L.E. Cardensan-Barron, Optimal pricing and lot-sizing policy for supply chain system with deteriorating items under limited storage capacity. Int. J. Prod. Econ. **200**, 278–290 (2018)
25. T.M. Whitin, *The Theory of Inventory Management*, 2nd edn. (Princeton University Press, Princeton, 1957)

On the Local Convergence of a Sixth-Order Iterative Scheme in Banach Spaces

Sanjaya Kumar Parhi and Debasis Sharma

Abstract Applying the Lipschitz continuity of the first-order Fréchet derivative, we describe the local convergence of a sixth-order nonlinear system solver in Banach spaces. This study eliminates the standard practice of Taylor expansion in local analysis and enhances the algorithm applicability through the use of a set of conditions on the first-order derivative. Also, our analysis offers the radii of convergence balls and computable error distances together with the unique result. Various numerical experiments are conducted to demonstrate that our technique is beneficial when prior studies fail to solve problems.

Keywords Nonlinear equation · Banach space · Iterative schemes · Local convergence · Fréchet derivative · Lipschitz condition

AMS Codes (2010) 47H99 · 49M15 · 65J15 · 65D99 · 65G99

1 Introduction

The main objective of the analysis discussed in this study is to find a solution x^* of the equation

$$\mathcal{A}(t) = 0, \tag{1}$$

where $\mathcal{A} : \Omega \subseteq X_1 \rightarrow X_2$ is a Fréchet differentiable operator and Ω is a convex subset of X_1. X_1 and X_2 are Banach spaces. Taking the reality into consideration

S. K. Parhi (✉) · D. Sharma
Department of Mathematics, International Institute of Information Technology Bhubaneswar, Bhubaneswar, Odisha, India
e-mail: sanjaya@iiit-bh.ac.in

D. Sharma
e-mail: iiit.debasis@gmail.com

S. K. Paikray et al. (eds.), *New Trends in Applied Analysis and Computational Mathematics*, Advances in Intelligent Systems and Computing 1356,
https://doi.org/10.1007/978-981-16-1402-6_7

that a large number of problems arise in diverse disciplines of applied sciences and engineering can be solved by obtaining the solutions of nonlinear equations in the form (1), a lot of successful algorithms have been constructed. "The solutions for these nonlinear equations are generally not obtained in closed form" [1]. Iterative algorithms are therefore often used to avoid these issues.

The second-order convergent Newton's scheme is extensively used as the solver of (1), which can be expressed as

$$t_{i+1} = t_i - [\mathcal{A}'(t_i)]^{-1}\mathcal{A}(t_i),\ i = 0,\ 1,\ 2\ldots \tag{2}$$

Chebyshev's, Halley's, and Super-Halley's schemes can be produced by substituting $\alpha = 0$, $\alpha = \frac{1}{2}$ and $\alpha = 1$, respectively, in

$$t_{i+1} = t_i - \left(1 + \frac{1}{2}(1 - \alpha H_{\mathcal{A}}(t_i))^{-1} H_{\mathcal{A}}(t_i)\right)[\mathcal{A}'(t_i)]^{-1}\mathcal{A}(t_i), \tag{3}$$

where $H_{\mathcal{A}}(t_i) = \mathcal{A}'(t_i)^{-1}\mathcal{A}''(t_i)\mathcal{A}'(t_i)^{-1}\mathcal{A}(t_i)$.

Several researchers have been designing Newton-like methods [2–16] and other variants of Chebyshev–Halley type methods to remove the computation of high-order derivatives found in conventional third-order schemes.

Local and semi-local analysis of iterative approaches have been studied by numerous researchers [1, 17–30], and a lot of essential findings have been obtained. "The semi-local convergence analysis, which is based on the information around an initial guess gives us the necessary condition to ensure the convergence and the local convergence analysis, which is based on the information around a solution provides the radius of convergence ball" [1]. Analysis of local convergence for several modifications of the schemes described in (3) has been discussed in [1, 18–20]. Also, the local study of efficient algorithms such as Weerakoon-type, Jarratt-type, and Newton-like schemes is studied in Banach spaces by Argyros, Cordero, Martínez and others [21, 23–27].

In this work, we employ Lipschitz condition only on $\mathcal{A}'$ to boost the performance of a sixth-order convergent iterative algorithm discussed in [7]. The authors modified the fourth-order formula proposed by Cordero et al. [8] to obtain a sixth-order convergent scheme for systems of nonlinear equations, which is given by

$$\begin{aligned} y_i &= t_i - \mathcal{A}'(t_i)^{-1}\mathcal{A}(t_i) \\ z_i &= y_i - \mathcal{A}'(t_i)^{-1}[2I - \mathcal{A}'(y_i)\mathcal{A}'(t_i)^{-1}]\mathcal{A}(y_i) \\ t_{i+1} &= z_i - \mathcal{A}'(y_i)^{-1}\mathcal{A}(z_i) \end{aligned} \tag{4}$$

Evaluation of only $\mathcal{A}'$ is necessary for the application of this algorithm. But the convergence analysis was shown in [7] using sixth-order Fréchet derivative. The utility of the scheme for problems where the higher order derivatives are undefined or unobtainable is therefore limited. Consider, for example,

$$\mathcal{A}(t) = \begin{cases} t^3 \log(t^2) + t^5 - t^4, & \text{if } t \neq 0 \\ 0, & \text{if } t = 0 \end{cases}$$

defined on $\Omega = [-\frac{1}{2}, \frac{5}{2}]$. Since $\mathcal{A}'''$ is unbounded on Ω, the earlier analysis given in [7] is unable to establish the convergence of (4). In this study, the local analysis of the method (4) is provided considering the hypotheses only on $\mathcal{A}'$ to stay away from the evaluation of high-order derivatives. In specific, Lipschitz condition is applied on $\mathcal{A}'$ for extending the usage of the algorithm.

The other parts of this manuscript are designed as follows: Sect. 2 deals with the local study of the scheme (4). Section 3 describes the applicability of our analytical findings on standard numerical. The last section contains conclusions of this work.

2 Local Convergence Analysis

Let $B(c, \rho) = \{t \in X_1 : ||c - t|| < \rho\}$, $\bar{B}(c, \rho) = \{t \in X_1 : ||c - t|| \leq \rho\}$, $BL(X_2, X_1) = \{K : X_2 \to X_1 \textit{ is linear and bounded}\}$ and IVT stands for the intermediate value theorem. Suppose $w_0 > 0$ and $w_1 > 0$ are real parameters with $w_0 \leq w_1$. We construct the real function Θ_1 on $[0, \frac{1}{w_0})$ by

$$\Theta_1(v) = \frac{w_1 v}{2(1 - w_0 v)} \tag{5}$$

and the parameter

$$\sigma_1 = \frac{2}{2w_0 + w_1} < \frac{1}{w_0}.$$

Now, $\Theta_1(\sigma_1) = 1$. Again, we introduce real functions Θ_2 and Φ_2 on $[0, \frac{1}{w_0})$ by

$$\Theta_2(v) = \left(1 + \frac{1 + w_0\Theta_1(v)v}{1 - w_0 v} + \frac{(1 + w_0\Theta_1(v)v)^2}{(1 - w_0 v)^2}\right)\Theta_1(v) \tag{6}$$

and

$$\Phi_2(v) = \Theta_2(v) - 1.$$

Now, $\Phi_2(0) = -1 < 0$ and $\lim_{v \to \frac{1}{w_0}^-} \Phi_2(v) = +\infty$. The function $\Phi_2(v)$ has zeros in $(0, \frac{1}{w_0})$ due to IVT. The notation of the smallest zero of $\Phi_2(v)$ in $(0, \frac{1}{w_0})$ is σ_2. Also, we introduce Θ_3 and Φ_3 on $[0, \frac{1}{w_0})$ by

$$\Theta_3(v) = w_0\Theta_1(v)v \tag{7}$$

and

$$\Phi_3(v) = \Theta_3(v) - 1.$$

Now, $\Phi_3(0) = -1 < 0$ and $\lim_{v \to \frac{1}{w_0}^-} \Phi_3(v) = +\infty$. So, the smallest zero of the function $\Phi_3(v)$ is in the interval $(0, \frac{1}{w_0})$ and denoted by σ_3. Finally, we define Θ_4 and Φ_4 on $[0, \sigma_3)$ by

$$\Theta_4(v) = w_1 \left(\frac{[2\Theta_1(v) + \Theta_2(v)]\, v}{2(1 - \Theta_3(v))} \right) \Theta_2(v) \tag{8}$$

and

$$\Phi_4(v) = \Theta_4(v) - 1.$$

Now, $\Phi_4(0) = -1 < 0$ and $\lim_{v \to \sigma_3^-} \Phi_4(v) = +\infty$. Hence, the interval $(0, \sigma_3)$ holds the smallest zero σ_4 of $\Phi_4(v)$.

Let us choose

$$R = min\{\sigma_1, \sigma_2, \sigma_4\}. \tag{9}$$

Now, we have

$$0 \leq \Theta_1(v) < 1, \tag{10}$$

$$0 \leq \Theta_2(v) < 1, \tag{11}$$

$$0 \leq \Theta_3(v) < 1 \tag{12}$$

and

$$0 \leq \Theta_4(v) < 1, \tag{13}$$

for each $v \in [0, R)$. In addition, we assume that the the Fréchet differentiable operator $\mathcal{A} : \Omega \subseteq X_1 \to X_2$ obeys the following conditions:

$$\mathcal{A}(t^*) = 0, \ \mathcal{A}'(t^*)^{-1} \in BL(X_2, X_1), \tag{14}$$

$$||\mathcal{A}'(t^*)^{-1}(\mathcal{A}'(t) - \mathcal{A}'(t^*))|| \leq w_0||t - t^*||, \ \forall t \in \Omega \tag{15}$$

and

$$||\mathcal{A}'(t^*)^{-1}(\mathcal{A}'(t) - \mathcal{A}'(y))|| \leq w_1||t - y||, \ \forall t, y \in \Omega, \tag{16}$$

where $BL(X_2, X_1)$ is the set of all bounded linear operators from X_2 to X_1.

An extra condition

$$||\mathcal{A}'(t^*)^{-1}\mathcal{A}'(t)|| \leq M, \ \forall t \in B\left(t^*, \frac{1}{w_0}\right) \tag{17}$$

is used in earlier studies [1, 17, 18, 22, 26]. However, we avoid this extra condition by the following results.

Lemma 1 *If $\mathcal{A}$ obeys (15) and $\bar{B}(t^*, R) \subseteq \Omega$, then $\forall t \in B(t^*, R)$, we get*

$$||\mathcal{A}'(t^*)^{-1}\mathcal{A}'(t)|| \le 1 + w_0||t - t^*|| \tag{18}$$

and

$$||\mathcal{A}'(t^*)^{-1}\mathcal{A}(t)|| \le (1 + w_0||t - t^*||)||t - t^*|| \tag{19}$$

Proof Applying (15), we obtain

$$||\mathcal{A}'(t^*)^{-1}\mathcal{A}'(t)|| \le 1 + ||\mathcal{A}'(t^*)^{-1}(\mathcal{A}'(t) - \mathcal{A}'(t^*))|| \le 1 + w_0||t - t^*||.$$

For $\theta \in [0, 1]$,

$$||\mathcal{A}'(t^*)^{-1}\mathcal{A}'(t^* + \theta(t - t^*))|| \le 1 + w_0\theta||t - t^*|| \le 1 + w_0||t - t^*||$$

and

$$\begin{aligned} ||\mathcal{A}'(t^*)^{-1}\mathcal{A}(t)|| &= ||\mathcal{A}'(t^*)^{-1}(\mathcal{A}(t) - \mathcal{A}(t^*))|| \\ &\le (1 + w_0||t - t^*||)||t - t^*||. \end{aligned}$$

Next, the local analysis of the iterative scheme (4) is derived in the following result.

Theorem 2 *Let $t^* \in \Omega$, $\mathcal{A}$ obeys (14)–(16) and*

$$\bar{B}(t^*, R) \subseteq \Omega, \tag{20}$$

where R is provided in (9). With the starter $t_0 \in B(t^*, R)$ the method (4) generates the sequence of iterates $\{t_i\}$ which is well defined, $\{t_i\}_{i\ge 0} \in B(t^*, R)$ and converges to the solution t^* of (1). Moreover, the following estimations hold $\forall i \ge 0$:

$$||y_i - t^*|| \le \Theta_1(||t_i - t^*||)||t_i - t^*|| < ||t_i - t^*|| < R, \tag{21}$$

$$||z_i - t^*|| \le \Theta_2(||t_i - t^*||)||t_i - t^*|| < ||t_i - t^*|| < R \tag{22}$$

and

$$||t_{i+1} - t^*|| \le \Theta_4(||t_i - t^*||)||t_i - t^*|| < ||t_i - t^*|| < R, \tag{23}$$

where Θ_1, Θ_2 and Θ_4 are introduced in (5), (6) and (8) respectively. In addition to this, t^* is unique in $\bar{B}(t^*, \xi) \cap \Omega$, where $\xi \in [R, \frac{2}{w_0})$.

Proof As $t_0 \in B(t^*, R)$, we find from (9) and (15) that

$$||\mathcal{A}'(t^*)^{-1}(\mathcal{A}'(t_0) - \mathcal{A}'(t^*))|| \le w_0||t_0 - t^*|| < w_0 R < 1.$$

Now, Banach lemma on invertible operators [2, 4, 5, 12, 14] confirms that $\mathcal{A}'(t_0)^{-1} \in BL(X_2, X_1)$ and

$$||\mathcal{A}'(t_0)^{-1}\mathcal{A}'(t^*)|| \le \frac{1}{1 - w_0||t_0 - t^*||} < \frac{1}{1 - w_0 R}. \tag{24}$$

Hence, y_0 is well defined. Again,

$$\begin{aligned} y_0 - t^* &= t_0 - t^* - \mathcal{A}'(t_0)^{-1}\mathcal{A}(t_0) \\ &= -\left[\mathcal{A}'(t_0)^{-1}\mathcal{A}'(t^*)\right]\left[\int_0^1 \mathcal{A}'(t^*)^{-1}(\mathcal{A}'(t^* + \theta(t_0 - t^*)) - \mathcal{A}'(t_0))(t_0 - t^*)\, d\theta\right]. \end{aligned} \tag{25}$$

Using (5), (9), (10), (16), (24) and (25) we find

$$\begin{aligned} ||y_0 - t^*|| &\le \left[||\mathcal{A}'(t_0)^{-1}\mathcal{A}'(t^*)||\right]\left[\left|\left|\int_0^1 \mathcal{A}'(t^*)^{-1}(\mathcal{A}'(t^* + \theta(t_0 - t^*)) - \mathcal{A}'(t_0))(t_0 - t^*)\, d\theta\right|\right|\right] \\ &\le \frac{w_1||t_0 - t^*||}{2(1 - w_0||t_0 - t^*||)}||t_0 - t^*|| \\ &= \Theta_1(||t_0 - t^*||)||t_0 - t^*|| < ||t_0 - t^*|| < R \end{aligned} \tag{26}$$

and this shows (21) for $i = 0$. Since $\mathcal{A}'(t_0)^{-1} \in BL(X_2, X_1)$ so, z_0 is well defined. From the definition of R, (4), (6), (11), (18), (19) and (26), we have

$$\begin{aligned} ||z_0 - t^*|| &\le ||y_0 - t^*|| + 2||\mathcal{A}'(t_0)^{-1}\mathcal{A}(y_0)|| + ||\mathcal{A}'(t_0)^{-1}\mathcal{A}'(y_0)||\; ||\mathcal{A}'(t_0)^{-1}\mathcal{A}(y_0)|| \\ &\le ||y_0 - t^*|| + ||\mathcal{A}'(t_0)^{-1}\mathcal{A}'(t^*)||\; ||\mathcal{A}'(t^*)^{-1}\mathcal{A}(y_0)|| \\ &+ ||\mathcal{A}'(t_0)^{-1}\mathcal{A}'(t^*)||\; ||\mathcal{A}'(t^*)^{-1}\mathcal{A}'(y_0)||\; ||\mathcal{A}'(t_0)^{-1}\mathcal{A}'(t^*)||\; ||\mathcal{A}'(t^*)^{-1}\mathcal{A}(y_0)|| \\ &\le ||y_0 - t^*|| + \frac{(1 + w_0||y_0 - t^*||)||y_0 - t^*||}{1 - w_0||t_0 - t^*||} \\ &+ \frac{(1 + w_0||y_0 - t^*||)}{1 - w_0||t_0 - t^*||}\frac{(1 + w_0||y_0 - t^*||)||y_0 - t^*||}{1 - w_0||t_0 - t^*||} \\ &\le \left(1 + \frac{(1 + w_0||y_0 - t^*||)}{1 - w_0||t_0 - t^*||} + \frac{(1 + w_0||y_0 - t^*||)}{1 - w_0||t_0 - t^*||}\frac{(1 + w_0||y_0 - t^*||)}{1 - w_0||t_0 - t^*||}\right)||y_0 - t^*|| \\ &\le \left(1 + \frac{(1 + w_0\Theta_1(||t_0 - t^*||)||t_0 - t^*||)}{1 - w_0||t_0 - t^*||}\right)\Theta_1(||t_0 - t^*||)||t_0 - t^*|| \\ &+ \left(\frac{(1 + w_0\Theta_1(||t_0 - t^*||)||t_0 - t^*||)^2}{(1 - w_0||t_0 - t^*||)^2}\right)\Theta_1(||t_0 - t^*||)||t_0 - t^*|| \\ &= \Theta_2(||t_0 - t^*||)||t_0 - t^*|| < ||t_0 - t^*|| < R. \end{aligned} \tag{27}$$

Hence, we establish (22) for $n = 0$. Again,

$$\begin{aligned} ||\mathcal{A}'(t^*)^{-1}(\mathcal{A}'(y_0) - \mathcal{A}'(t^*))|| &\le w_0||y_0 - t^*|| < w_0\Theta_1(||t_0 - t^*||)||t_0 - t^*|| \\ &= \Theta_3(||t_0 - t^*||) < 1. \end{aligned} \tag{28}$$

So, $\mathcal{A}'(y_0)^{-1} \in BL(X_2, X_1)$ with

$$||\mathcal{A}'(y_0)^{-1}\mathcal{A}'(t^*)|| \le \frac{1}{1 - \Theta_3(||t_0 - t^*||)}. \tag{29}$$

Now, t_1 exists and it is well defined. Also,

$$\begin{aligned} t_1 - t^* &= z_0 - \mathcal{A}'(y_0)^{-1}\mathcal{A}(z_0) - t^* \\ &= z_0 - t^* - \mathcal{A}'(y_0)^{-1}\mathcal{A}(z_0) \\ &= -\left[\mathcal{A}'(y_0)^{-1}\mathcal{A}'(t^*)\right]\left[\int_0^1 \mathcal{A}'(t^*)^{-1}(\mathcal{A}'(t^* + \theta(z_0 - t^*)) \right. \\ &\qquad \left. - \mathcal{A}'(y_0))(z_0 - t^*)\, d\theta\right] \end{aligned} \tag{30}$$

Finally, we use (8), (9), (13), (19), (27), (29) and (30) to get

$$\begin{aligned} \|t_1 - t^*\| &\le \left[\|\mathcal{A}'(y_0)^{-1}\mathcal{A}'(t^*)\|\right]\left[\left\|\int_0^1 \mathcal{A}'(t^*)^{-1}(\mathcal{A}'(t^* + \theta(z_0 - t^*)) - \mathcal{A}'(y_0))(z_0 - t^*)\, d\theta\right\|\right] \\ &\le \frac{w_1 \int_0^1 \|t^* + \theta(z_0 - t^*) - y_0\|\, d\theta\, \|z_0 - t^*\|}{1 - \Theta_3(\|t_0 - t^*\|)} \\ &\le w_1 \left(\frac{\|y_0 - t^*\| + \frac{\|z_0 - t^*\|}{2}}{1 - \Theta_3(\|t_0 - t^*\|)}\right) \|z_0 - t^*\| \\ &\le w_1 \left(\frac{\Theta_1(\|t_0 - t^*\|)\|t_0 - t^*\| + \frac{\Theta_2(\|t_0 - t^*\|)\|t_0 - t^*\|}{2}}{1 - \Theta_3(\|t_0 - t^*\|)}\right) \|z_0 - t^*\| \\ &= w_1 \left(\frac{2\Theta_1(\|t_0 - t^*\|)\|t_0 - t^*\| + \Theta_2(\|t_0 - t^*\|)\|t_0 - t^*\|}{2(1 - \Theta_3(\|t_0 - t^*\|))}\right) \|z_0 - t^*\| \\ &\le w_1 \left(\frac{2\Theta_1(\|t_0 - t^*\|)\|t_0 - t^*\| + \Theta_2(\|t_0 - t^*\|)\|t_0 - t^*\|}{2(1 - \Theta_3(\|t_0 - t^*\|))}\right) \Theta_2(\|t_0 - t^*\|)\|t_0 - t^*\| \\ &= w_1 \left(\frac{\left[2\Theta_1(\|t_0 - t^*\|) + \Theta_2(\|t_0 - t^*\|)\right]\|t_0 - t^*\|}{2(1 - \Theta_3(\|t_0 - t^*\|))}\right) \Theta_2(\|t_0 - t^*\|)\|t_0 - t^*\| \\ &= \Theta_4(\|t_0 - t^*\|)\|t_0 - t^*\| < \|t_0 - t^*\| < R. \end{aligned} \tag{31}$$

Thus, (23) is true for $i = 0$. We derive (21)–(23) applying the substitution of t_i, y_i, z_i and t_{i+1} in place of t_0, y_0, z_0 and t_1, respectively, in the prior results. We deduce from $\|t_{i+1} - t^*\| \le \Theta_4(R)\|t_i - t^*\| < R$ that $t_{i+1} \in B(t^*, R)$ and $\lim_{i\to\infty} t_i = t^*$. Now, we need to prove that t^* is unique. Suppose y^* ($\neq t^*$) is another solution of $\mathcal{A}(t) = 0$ in $B(t^*, \xi)$. Consider $K = \int_0^1 \mathcal{A}'(\theta t^* + (1 - \theta)y^*)\, d\theta$. From Eq. (15), we get

$$\begin{aligned} \|\mathcal{A}'(t^*)^{-1}(K - \mathcal{A}'(t^*))\| &\le \int_0^1 w_0\|y^* + \theta(t^* - y^*) - t^*\|\, d\theta \\ &\le \frac{w_0}{2}\|t^* - y^*\| \\ &\le \frac{w_0 \xi}{2} < 1. \end{aligned}$$

We find $K^{-1} \in BL(X_2, X_1)$ employing Banach lemma. Now, the identity $0 = \mathcal{A}(t^*) - \mathcal{A}(y^*) = K(t^* - y^*)$ confirms that $t^* = y^*$.

Table 1 Radius of convergence ball for Example 1

Method (4)
$\sigma_1 = 0.324947$
$\sigma_2 = 0.133649$
$\sigma_4 = 0.150710$
$R = 0.133649$

Table 2 Radius of convergence ball for Example 2

Method (4)
$\sigma_1 = 0.066667$
$\sigma_2 = 0.026527$
$\sigma_4 = 0.034069$
$R = 0.026527$

3 Numerical Examples

Numerical examples for the validation of theoretical findings are given in this section.

Example 1 ([23]) Let $\mathcal{A}$ is defined on $\bar{B}(0, 1)$ for $(t_1, t_2, t_3)^T$ by

$$\mathcal{A}(t) = \left(e^{t_1} - 1, \frac{e-1}{2}t_2^2 + t_2, t_3\right)^T$$

We have $t^* = (0, 0, 0)^T$. Also, we have $w_0 = e - 1$ and $w_1 = e$. R is derived using "Θ" functions (Table 1).

Example 2 ([25]) Consider the following integral equation defined by

$$\mathcal{A}(t)(s) = t(s) - 5\int_0^1 s\, x\, t(x)^3\, dx,$$

where $t(s) \in C[0, 1]$. Also, $t^* = 0$, $w_0 = 7.5$ and $w_1 = 15$. From "Θ" functions the value of R is obtained (Table 2).

Example 3 ([23]) Define $\mathcal{A}$ on $\Omega = [-\frac{1}{2}, \frac{5}{2}]$ by

$$\mathcal{A}(t) = \begin{cases} t^3 \log(t^2) + t^5 - t^4, & \text{if} t \neq 0 \\ 0, & \text{if } t = 0 \end{cases}.$$

We have $t^* = 1$. Also, $w_0 = w_1 = 96.6628$. We find R using "Θ" functions (Table 3).

Thus, the convergence of the scheme (4) is guaranteed with the radius of the ball of convergence $R = 0.003054$.

Table 3 Radius of convergence ball for Example 3

Method (4)
$\sigma_1 = 0.006897$
$\sigma_2 = 0.003054$
$\sigma_4 = 0.003532$
$R = 0.003054$

4 Conclusions

We studied the local analysis of a sixth-order convergent scheme (4) to approximate a solution of a nonlinear equation in Banach spaces. This analysis is suggested using the assumption that the first derivative belongs to the Lipschitz class. Also, our study is useful when high-order derivative-based previous studies can not be applied. At last, standard examples like systems of nonlinear equations and Hammerstein integral equation is solved to establish the convergence of this algorithm.

Conflict of interest The authors declare that they do not have conflict of interests.

Acknowledgements This research is financially supported by the University Grants Commission of India (Ref. No.: 988/(CSIR-UGC NET JUNE 2017); ID: NOV2017-402662).

References

1. I.K. Argyros, Á.A. Magreñán, A study on the local convergence and the dynamics of Chebyshev-Halley-type methods free from second derivative. Numer. Algorithm **71**(1), 1–23 (2015)
2. I.K. Argyros, *Convergence and Application of Newton-type Iterations* (Springer, Berlin, 2008)
3. I.K. Argyros, Y.J. Cho, S. Hilout, *Numerical Methods for Equations and its Applications* (Taylor & Francis, CRC Press, New York, 2012)
4. I.K. Argyros, S. Hilout, *Computational Methods in Nonlinear Analysis* (World Scientific Publishing House, New Jersey, 2013)
5. M.S. Petković, B. Neta, L. Petković, D. Džunić, Multipoint Methods for Solving Nonlinear Equations (Elsevier, 2013)
6. F.A. Potra, V. Ptak, *Nondiscrete Induction and Iterative Processes*. Research Notes in Mathematics (Pitman Publ, Boston, 1984), p. 103
7. A. Cordero, J.L. Hueso, E. Martínez, J.R. Toregrossa, Increasing the convergence order of an iterative method for nonlinear systems. Appl. Math. Lett. **25**, 2369–2374 (2012)
8. A. Cordero, E. Martínez, J.R. Toregrossa, Iterative methods of order four and five for systems of nonlinear equations. J. Comput. Appl. Math. **231**, 541–551 (2012)
9. M.V. Kanwar, V.K. Kukreja, S. Singh, On some third-order iterative methods for solving nonlinear equations. Appl. Math. Comput. **171**(1), 272–280 (2005)
10. J. Kou, Y. Li, X. Wang, A composite fourth-order iterative method for solving non-linear equations. Appl. Math. Comput. **184**, 71–475 (2007)
11. A.Y. Özban, Some new variants of Newton's method. Appl. Math. Lett. **17**(6), 677–682 (2004)
12. L.B. Rall, *Computational Solution of Nonlinear Operator Equations* (Robert E, Krieger, New York, 1979)

13. H. Ren, Q. Wu, W. Bi, New variants of Jarratt method with sixth-order convergence. Numer. Algorithm **52**(4), 585–603 (2009)
14. J.F. Traub, *Iterative Methods for Solution of Equations* (Prentice-Hall, Englewood Cliffs, 1964)
15. S. Weerakoon, T.G.I. Fernando, A variant of Newton's method with accelerated third-order convergence. Appl. Math. Lett. **13**(8), 87–93 (2000)
16. H.P.S. Nishani, S. Weerakoon, T.G.I. Fernando, M. Liyanag, Weerakoon-Fernando Method with accelerated third-order convergence for systems of nonlinear equations. Int. J. Math. Model. Numer. Optim. **8**(3), 287–304 (2018)
17. I.K. Argyros, Y.J. Cho, S. George, Local convergence for some third order iterative methods under weak conditions. J. Korean Math. Soc. **53**(4), 781–793 (2016)
18. I.K. Argyros, S. George, Local convergence of deformed Halley method in Banach space under Hölder continuity conditions. J. Nonlinear Sci. Appl. **8**, 246–254 (2015)
19. I.K. Argyros, S. George, Á.A. Magreñán, Local convergence for multi-point-parametric Chebyshev-Halley-type methods of higher convergence order. J. Comput. Appl. Math. **282**, 215–224 (2015)
20. I.K. Argyros, S. George, Local convergence of modified Halley-like methods with less computation of inversion. NOVI SAD J. Math. **45**, 47–58 (2015)
21. D. Sharma, S.K. Parhi, On the local convergence of a third-order iterative scheme in Banach spaces. Rendiconti del Circolo Matematico di Palermo, II. Series (2020). https://doi.org/10.1007/s12215-020-00500-x
22. I.K. Argyros, D. González, Local convergence for an improved Jarratt-type method in Banach space. Int. J. Interact. Multimed. Artif. Intell. **3**(Special Issue on Teaching Mathematics Using New and Classic Tools), 20–25 (2015)
23. S. Singh, D.K. Gupta, R.P. Badoni, E. Martínez, J.L. Hueso, Local convergence of a parameter based iteration with Hölder continuous derivative in Banach spaces. Calcolo **54**(2), 527–539 (2017)
24. A. Cordero, J.A. Ezquerro, M.A. Hernandez-Veron, On the local convergence of a fifth-order iterative method in Banach spaces. J. Math. **46**, 53–62 (2014)
25. E. Martínez, S. Singh, J.L. Hueso, D.K. Gupta, Enlarging the convergence domain in local convergence studies for iterative methods in Banach spaces. Appl. Math. Comput. **281**, 252–265 (2016)
26. I.K. Argyros, S. Hilout, On the local convergence of fast two-step Newton-like methods for solving nonlinear equations. J. Comput. Appl. Math. **245**, 1–9 (2013)
27. I.K. Argyros, S.K. Khattri, Local convergence for a family of third order methods in Banach spaces. Appl. Math. Comput. **251**, 396–403 (2015)
28. S.K. Parhi, D.K. Gupta, Semilocal convergence of a Stirling-like method in Banach spaces. Int. J. Comput. Methods **7**(2), 215–228 (2010)
29. I.K. Argyros, Y.J. Cho, S.K. Khattri, On a new semilocal convergence analysis for the Jarratt method. J. Inequal. Appl., **194**, 16pp. (2013)
30. I.K. Argyros, Y.J. Cho, H.M. Ren, Convergence of Halley's method for operators with the bounded second derivative in Banach spaces. J. Inequal. Appl. **260**, 12pp. (2013)

New Results on Chromatic Polynomials

Utkum Sanli, Susanta Kumar Paikray, and Ismail Naci Cangul

Abstract Colouring graphs is one of the oldest problems of graph theory. As it is a kind of labelling, colouring has always been one of the main study areas in graph theory and related fields, especially when combinatorial calculations are needed. The most well-known and cited result on colouring is the Birkhoff–Lewis Theorem. It gives a step-by-step reduction method to calculate the chromatic polynomial of any given graph as the difference of the chromatic polynomials of two smaller graphs, one is edge deleted and the other is edge contracted. Here, we give some short-cut results which enables us to calculate the chromatic polynomial of a relatively large graph by dividing it into smaller graphs.

Keywords Chromatic polynomial · Chromatic number · Birkhoff–Lewis theorem · Graph colouring

MSC 2000 Numbers 05C10 · 05C30 · 68R10 · 68Uxx

1 Introduction

One of the well-known applications of graph theory is the four-colour problem. There are many notions related to colouring of graphs. A (vertex) colouring of a graph $G(V, E)$ is a mapping $f : V \to C$, where C is the set of colours, with $f(u) \neq f(v)$ for $uv \in E$.

U. Sanli · I. N. Cangul (✉)
Department of Mathematics, Bursa Uludag University, 16059 Gorukle, Bursa, Turkey
e-mail: cangul@uludag.edu.tr

U. Sanli
e-mail: utkumsanli@hotmail.com

S. K. Paikray
Department of Mathematics, VSS University of Technology, Burla, Sambalpur 768018, India
e-mail: skpaikray_math@vssut.ac.in

S. K. Paikray et al. (eds.), *New Trends in Applied Analysis and Computational Mathematics*, Advances in Intelligent Systems and Computing 1356,
https://doi.org/10.1007/978-981-16-1402-6_8

If there is a colouring of G with n colours, then G is said to be n-colourable. The smallest number n for which G is n-colourable is called the chromatic number and denoted by $\chi(G)$. Clearly, a graph with at least one loop cannot have any colourings. As we shall be considering simple graphs, this case will be out of the question. When we say colouring, we normally understand vertex colouring. But there are also other types of colouring using other graph pieces like edge colourings or face colourings.

Here, we shall study another aspect related to colourings, the chromatic polynomial of a graph. The chromatic polynomial of G is defined to be a function $C_G(k)$ which expresses the number of distinct k-colourings possible for the graph G for each integer $k > 0$. This number was first used by Birkhoff [3], in 1912. Chromatic polynomials are widely used in determining several properties of graphs. Although there is no known formula for the chromatic polynomial of any given graph, there are algorithms to do that. The most well-known algorithm is the Birkhoff–Lewis theorem stated as below:

Theorem 1 (Birkhoff–Lewis 1946) *The chromatic polynomial of a graph G can be found by the formula*

$$C_G(k) = C_{G-e}(k) - C_{G/e}(k),$$

where G-e is the graph obtained by deleting the edge e from G and G/e is the graph obtained from G by removing e and identifying the end vertices of e, called contracting, and leaving only one copy of any resulting multiple edges (see Figs. 1 and 2).

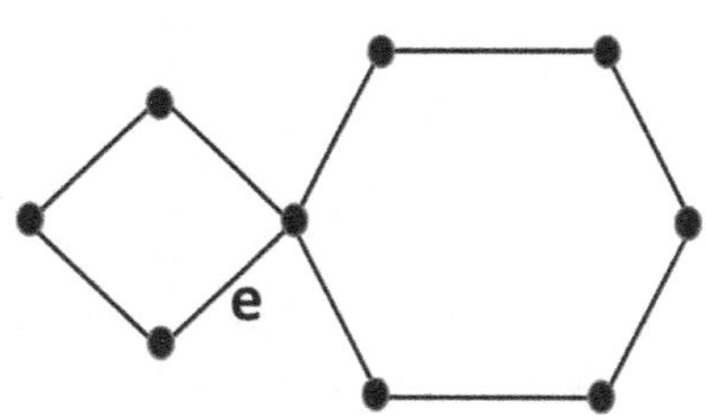

Fig. 1 A graph G

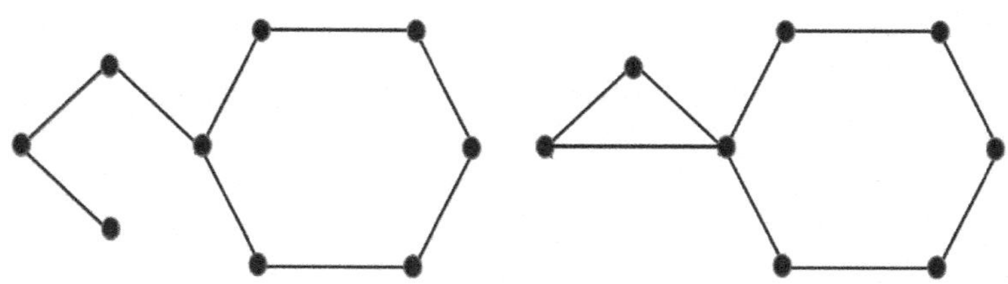

Fig. 2 The graphs $G - e$ and G/e

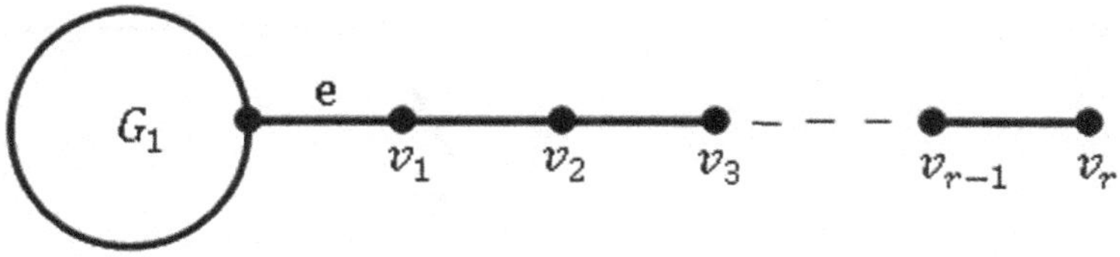

Fig. 3 The graph $G = G_1 \cup P_r$

Fig. 4 A graph G splitted through an edge

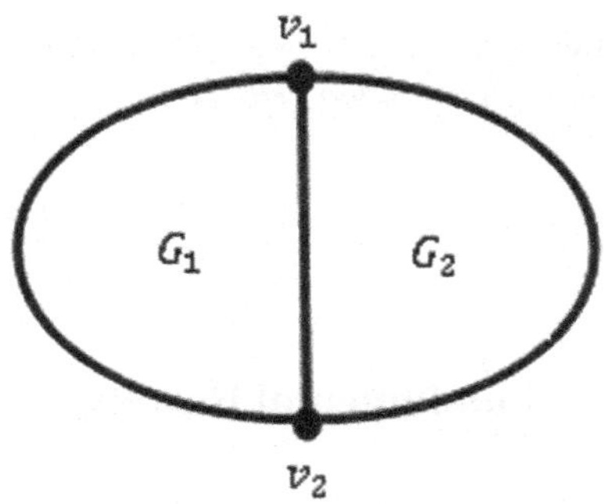

As usual, we denote path, cycle, star, complete, complete bipartite and tadpole graphs by P_n, C_n, S_n, K_n, $K_{r,s}$ and $T_{r,s}$, respectively.

A bridge of a connected graph is an edge whose removal disconnects the graph. For this and other definitions, see e.g. [1, 2, 4–7, 9]. In other words, a bridge is an edge of a graph G whose removal increases the number of components of G. Also, it is well known that an edge of a connected graph is a bridge iff it does not lie on any cycle.

The following few results which will be used in getting our results are given in [8]:

Lemma 2 *Let G_1 be a connected graph. Let G be a graph obtained by joining a path to G_1, see Fig. 3. Then*

$$C_G(k) = \frac{C_{G_1}(k) \cdot C_{P_r}(k)}{k}$$

$$= C_{G_1(k)} \cdot (k-1)^r.$$

These two last results show that we can get rid of each edge which is not on a cycle in G and at the end of each such deletion, the chromatic polynomial of G is reduced by $k - 1$.

Now we deal with another case where we split a graph into two subgraphs through a common edge as in Fig. 4.

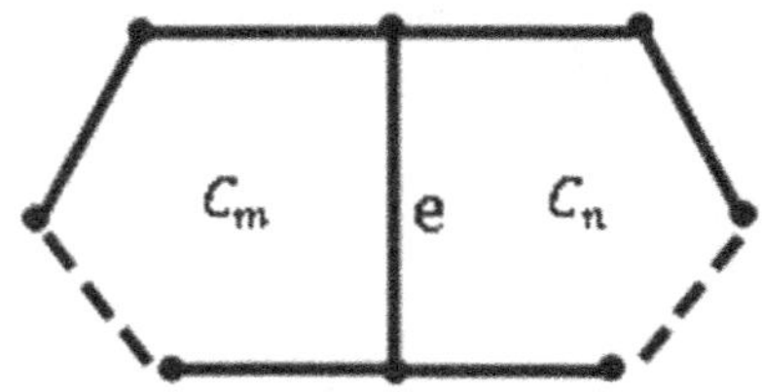

Fig. 5 A graph G

Lemma 3 *Let G be a graph which can be splitted into two subgraphs G_1 and G_2 through a common edge $e = v_1v_2$. Then*

$$C_G(k) = \frac{C_{G_1}(k) \cdot C_{G_2}(k)}{k \cdot (k-1)}.$$

2 Fundamental Results

In this section, we want to obtain some shortcut moves to find the chromatic polynomial of any given graph. The idea behind this will be splitting the given graph into smaller subgraphs and apply Birkhoff–Lewis Theorem when necessary. We shall state results for the chromatic polynomial of a given graph G in terms of two smaller graphs G_1 and G_2. When suitable, these can be generalized to a finite number of subgraphs. We shall consider four different cases where a graph is splitted into two parts at a cut vertex or at a bridge.

In this paper, we use $G_1 \cup G_2$ to denote the union of two graphs G_1 and G_2 which have one common vertex.

One can think of starting by any graph and applying Theorem 1 successively until getting to a point eventually. But this process might be time-consuming for many complicated graphs. It is therefore wise to have handy some short-cut methods for some special situations in complex graphs. We first have the following useful result:

Theorem 4 *Let G be a bicyclic graph constructed by joining two cycle graphs C_n and C_m along an edge e, see Fig. 5. Then*

$$C_G(k) = C_{C_{n+m-2}}(k) - \frac{C_{C_{n-1}}(k) \cdot C_{C_{m-1}}(k)}{k}.$$

Proof The graph G is seen in Fig. 5.

By Lemma 3, we can easily reduce that

$$C_G(k) = \frac{C_{C_n}(k) \cdot C_{C_m}(k)}{k(k-1)}.$$

If n and m are both even, then by [8], we can write

$$\begin{aligned} C_{C_n}(k) &= (k-1)^n + (k-1) \\ &= (k-1)[(k-1)^{n-1} + 1] \end{aligned}$$

and

$$\begin{aligned} C_{C_m}(k) &= (k-1)^m + (k-1) \\ &= (k-1)[(k-1)^{m-1} + 1] \end{aligned}.$$

Then

$$\begin{aligned} C_G(k) &= \frac{(k-1)^2[(k-1)^{n-1}+1][(k-1)^{m-1}+1]}{k(k-1)} \\ &= \tfrac{k-1}{k}[(k-1)^{n-1}+1][(k-1)^{m-1}+1]. \end{aligned}$$

Also as $n+m$ is even, we have

$$\begin{aligned} C_{C_{n+m-2}}(k) &= (k-1)^{n+m-2} + (k-1) \\ &= (k-1)[(k-1)^{n+m-3}+1]. \end{aligned}$$

Hence

$$\begin{aligned} C_G(k) &= C_{C_{n+m-2}}(k) + \tfrac{k-1}{k}[(k-1)^{n-1}+1][(k-1)^{m-1}+1] \\ &\quad -(k-1)[(k-1)^{n+m-3}+1] \\ &= C_{C_{n+m-2}}(k) + \tfrac{k-1}{k}[-(k-1)^{n+m-3} + (k-1)^{n-1} + (k-1)^{m-1} - k + 1] \\ &= C_{C_{n+m-2}}(k) + \frac{(k-1)^2}{k}[-(k-1)^{n+m-4} + (k-1)^{n-2} + (k-1)^{m-2} - 1] \\ &= C_{C_{n+m-2}}(k) + \frac{-(k-1)^2}{k}[(k-1)^{n-2} - 1][(k-1)^{m-2} - 1] \\ &= C_{C_{n+m-2}}(k) + \tfrac{-1}{k} C_{C_{n-1}}(k) C_{C_{m-1}}(k) \end{aligned}$$

giving the required result. The other cases where n and m are both odd or n and m have opposite parities can be proven similarly. ■

Fig. 6 The graph $A_{n,m,r}$

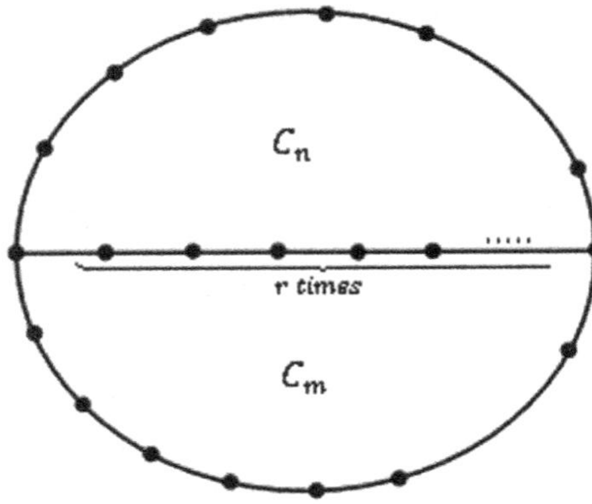

This result can be generalized to a more general class of graphs where the two adjacent cycles have a path in common instead of a single edge:

Theorem 5 *Let $A_{n,m,r}$ be the graph consisting of two cycles of lengths n and m having a path of length $r+1$ in common as in Fig. 6. Then*

$$C_{A_{n,m,r}}(k) = \frac{C_{C_{n+m-2}}(k) \cdot C_{C_{r+2}}(k)}{k(k-1)} + (-1)^{r+1} \cdot \frac{C_{C_{n-1}}(k) \cdot C_{C_{m-1}}(k)}{k}.$$

Proof By induction. First, we prove that the result is true for $r = 1$. By Theorem 1, we can write

$$\begin{aligned} C_{A_{n,m,1}}(k) &= \frac{C_{C_{n+m-2}}(k)k(k-1)}{k} - \left[C_{C_{n+m-2}}(k) - \frac{C_{C_{n-1}}(k) \cdot C_{C_{m-1}}(k)}{k} \right] \\ &= C_{C_{n+m-2}}(k)(k-2) + \frac{C_{C_{n-1}}(k) \cdot C_{C_{m-1}}(k)}{k} \\ &= \frac{C_{C_{n+m-2}}(k) C_{C_3}(k)}{k(k-1)} + \frac{C_{C_{n-1}}(k) \cdot C_{C_{m-1}}(k)}{k}. \end{aligned}$$

Hence, we showed that it is true for $r = 1$. Now let us assume that the result is true for $r = f$. That is, we assume that

$$C_{A_{n,m,f}}(k) = \frac{C_{C_{n+m-2}}(k) \cdot C_{C_{f+2}}(k)}{k(k-1)} + (-1)^{f+1} \cdot \frac{C_{C_{n-1}}(k) \cdot C_{C_{m-1}}(k)}{k}.$$

We must show that it is true for $r = f + 1$. By Theorem 1, we see that

$$
\begin{aligned}
C_{A_{n,m,f+1}}(k) &= \frac{C_{C_{n+m-2}}(k)\cdot C_{P_{f+2}}(k)}{k} - C_{A_{n,m,f}}(k)\\
&= \frac{C_{C_{n+m-2}}(k)\cdot C_{P_{f+2}}(k)}{k}\\
&\quad -[\frac{C_{C_{n+m-2}}(k)\cdot C_{C_{f+2}}(k)}{k(k-1)} + (-1)^{f+1}\cdot\frac{C_{C_{n-1}}(k)\cdot C_{C_{m-1}}(k)}{k}]\\
&= \frac{C_{C_{n+m-2}}(k)}{k}\cdot[P_{f+2}(k) - \frac{C_{C_{f+2}}(k)}{k-1}]\\
&\quad -(-1)^{f+1}\cdot\frac{C_{C_{n-1}}(k)\cdot C_{C_{m-1}}(k)}{k}\\
&= \frac{C_{C_{n+m-2}}(k)}{k}[k(k-1)^{f+1} - \frac{(k-1)^{f+2}+(-1)^{f+2}\cdot(k-1)}{k-1}]\\
&\quad -(-1)^{f+1}\cdot\frac{C_{C_{n-1}}(k)\cdot C_{C_{m-1}}(k)}{k}\\
&= \frac{C_{C_{n+m-2}}(k)}{k}[k(k-1)^{f+1} - [(k-1)^{f+1} - (-1)^{f+1}]]\\
&\quad -(-1)^{f+1}\cdot\frac{C_{C_{n-1}}(k)\cdot C_{C_{m-1}}(k)}{k} \qquad (1)\\
&= \frac{C_{C_{n+m-2}}(k)}{k}[(k-1)^{f+1}\cdot(k-1) + (-1)^{f+1}]\\
&\quad -(-1)^{f+1}\cdot\frac{C_{C_{n-1}}(k)\cdot C_{C_{m-1}}(k)}{k}\\
&= \frac{C_{C_{n+m-2}}(k)}{k}[(k-1)^{f+2}\cdot(k-1) + (-1)^{f+1}]\\
&\quad -(-1)^{f+1}\cdot\frac{C_{C_{n-1}}(k)\cdot C_{C_{m-1}}(k)}{k}\\
&= \frac{C_{C_{n+m-2}}(k)}{k}\left[\frac{(k-1)^{f+3}+(-1)^{f+3}\cdot(k-1)}{k-1}\right]\\
&\quad -(-1)^{f+1}\cdot\frac{C_{C_{n-1}}(k)\cdot C_{C_{m-1}}(k)}{k}\\
&= \frac{C_{C_{n+m-2}}(k)C_{C_{f+3}}(k)}{k(k-1)} + (-1)^{f+2}\cdot\frac{C_{C_{n-1}}(k)\cdot C_{C_{m-1}}(k)}{k}.
\end{aligned}
$$

■

The following immediate result of this is frequently used:

Corollary 6 *When $n = m = 3$, that is, when the neighbour faces are triangles, we have*

$$C_{A_{3,3,r}}(k) = C_{A_{3,3,0}}(k)\cdot(k-1)^r + C_{C_{r+1}}(k)\cdot(2k-3).$$

We now study a more complicated graph:

Theorem 7 *Let $G_{n,m,r}$ be the graph consisting of three cycles of lengths n, m and $r+3$ which pairwise have a common edge and having a single vertex in common as in Fig. 7. Then*

$$C_{G_{3,3,r}}(k) = (k-1)^{r+2}(k-2)^2 - 2(k-1)(k-2).$$

Proof First note that by Theorem 1, we can write

$$C_{G_{3,3,r}}(k) = (k-1)^{r+2}(k-2)^2 + 2(-1)^n(k-1)^2(k-2) + (-1)^{n+1}2k(k-1)(k-2).$$

Fig. 7 The graph $G_{n,m,r}$

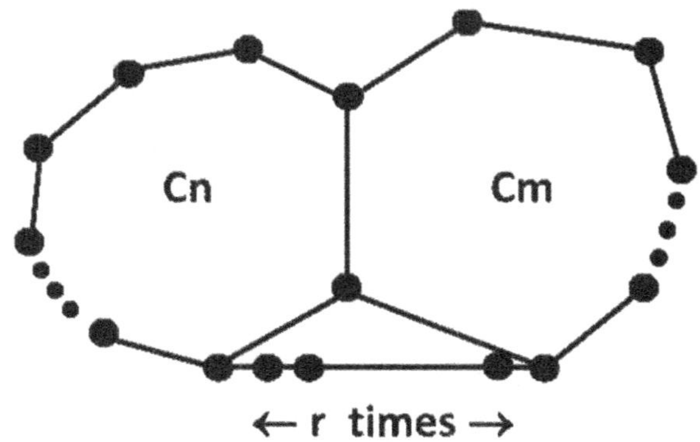

Therefore, if n is even, then this formula becomes

$$\begin{aligned} C_{G_{3,3,r}}(k) &= (k-1)^{r+2}(k-2)^2 + 2(k-1)^2(k-2) - 2k(k-1)(k-2) \\ &= (k-1)^{r+2}(k-2)^2 + 2(k-1)(k-2)[k-1-k] \\ &= (k-1)^{r+2}(k-2)^2 - 2(k-1)(k-2) \end{aligned}$$

and if n is odd, then we have

$$\begin{aligned} C_{G_{3,3,r}}(k) &= (k-1)^{r+2}(k-2)^2 - 2(k-1)^2(k-2) + 2k(k-1)(k-2) \\ &= (k-1)^{r+2}(k-2)^2 - 2(k-1)(k-2)[k-1-k] \\ &= (k-1)^{r+2}(k-2)^2 + 2(k-1)(k-2). \end{aligned}$$

That is, in all cases, the result is true. ∎

Corollary 8 *The following recurrence relation holds:*

$$C_{G_{3,3,r}}(k) = C_{G_{3,3,r-1}}(k) \cdot (k-1) + (-1)^{r+1} 2C_{C_3}(k).$$

In the following result, a much more useful and direct recurrence relation is given:

Theorem 9

$$C_{G_{3,3,r}}(k) = C_{G_{3,3,0}}(k) \cdot (k-1)^r + 2C_{C_{r+1}}(k) \cdot (k-2).$$

Proof By induction. First, we check whether the statement is true for $r = 2$ or not. That is,

$$C_{G_{3,3,2}}(k) = C_{G_{3,3,0}}(k) \cdot (k-1)^2 + 2C_{C_3}(k) \cdot (k-2)?$$

As both sides are $k(k-1)(k-2)(k^3 - 5k^2 + 9k - 7)$, we conclude that it is true for $r = 2$. Next we assume that the statement is true for $r = t$. That is, let

$$C_{G_{3,3,t}}(k) = C_{G_{3,3,0}}(k) \cdot (k-1)^t + 2C_{C_{t+1}}(k) \cdot (k-2).$$

We need to show that it is also true for $r = t + 1$? That is,

$$C_{G_{3,3,t+1}}(k) = C_{G_{3,3,0}}(k) \cdot (k-1)^{t+1} + 2C_{C_{t+2}}(k) \cdot (k-2)?$$

Now by Theorem 1 and Lemma 3, we have

$$\begin{aligned} C_{G_{3,3,t+1}}(k) &= C_{A_{3,3,t+1}}(k) - C_{C_{t+4}}(k) \\ &= (k-1)[C_{G_{3,3,0}}(k) \cdot (k-1)^t + 2C_{C_{t+1}}(k) \cdot (k-2)] - 2k(k-1)(k-2) \\ &= C_{G_{3,3,0}}(k) \cdot (k-1)^{t+1} + 2C_{C_{t+1}}(k) \cdot (k-1)(k-2) - 2k(k-1)(k-2). \\ &= C_{G_{3,3,0}}(k) \cdot (k-1)^{t+1} + 2C_{C_{t+2}}(k) \cdot (k-2). \end{aligned}$$

Therefore, the proof follows. ■

The following result states $C_{G_{n,m,r}}(k)$ in terms of $C_{G_{n,m,0}}(k)$ recursively:

Theorem 10 *The following recurrence relation is satisfied:*

$$C_{G_{n,m,r}}(k) = (-1)^{n+1} C_{G_{n,m,0}}(k) + \frac{C_{C_n}(k) C_{C_m}(k)}{k} \sum (k-1)^n (-1)^{n+1}.$$

Proof By Theorem 1, we can write

$$\begin{aligned} C_{G_{n,m,r}}(k) + C_{G_{n,m,r-1}}(k) &= \frac{C_{C_n}(k) C_{C_m}(k) C_{P_{r+1}}(k)}{k^2(k-1)} \\ &= \frac{C_{C_n}(k) C_{C_m}(k) \cdot (k-1)^{r-1}}{k} \\ &= \frac{C_{C_n}(k) C_{C_m}(k) \cdot (k-1)^{r-2}}{k} \\ &\dots \\ &= \frac{C_{C_n}(k) C_{C_m}(k)(k-1)}{k} \\ &= \frac{C_{C_n}(k) C_{C_m}(k)}{k}. \end{aligned}$$

If r is even, then we obtain

$$\begin{aligned} C_{G_{n,m,r}}(k) - C_{G_{n,m,0}}(k) &= \frac{C_{C_n}(k) C_{C_m}(k)}{k} [(k-1)^{r-1} - (k-1)^{r-2} + \cdots + (k-1) - 1] \\ &= \frac{C_{C_n}(k) C_{C_m}(k)}{k} \sum (k-1)^n \cdot (-1)^{n+1} \end{aligned}$$

and if r is odd, then we write

$$\begin{aligned} C_{G_{n,m,r}}(k) + C_{G_{n,m,0}}(k) &= \frac{C_{C_n}(k) C_{C_m}(k)}{k} [(k-1)^{r-1} - (k-1)^{r-2} + \cdots - (k-1) + 1] \\ &= \frac{C_{C_n}(k) C_{C_m}(k)}{k} \sum (k-1)^n . (-1)^n . \end{aligned}$$

giving the result. ■

Theorem 11 *The following recurrence relation is satisfied:*

$$C_{G_{n,m,r}}(k) = \frac{C_{C_n}(k)C_{C_m}(k)C_{C_{r+1}}(k)}{k^2(k-1)} + (-1)^r C_{G_{n,m,0}}(k).$$

Proof By successively applying Lemma 2, we obtain

$$\begin{aligned}
C_{G_{n,m,r}}(k) &= \frac{C_{C_n}(k)C_{C_m}(k)C_{P_{r+1}}(k)}{k^2(k-1)} - C_{G_{n,m,r-1}}(k) \\
&= \frac{C_{C_n}(k)C_{C_m}(k)C_{P_{r+1}}(k)}{k^2(k-1)} - [\frac{C_{C_n}(k)C_{C_m}(k)C_{P_r}(k)}{k^2(k-1)} - C_{G_{n,m,r-2}}(k)] \\
&= \frac{C_{C_n}(k)C_{C_m}(k)C_{P_r}(k)}{k^2(k-1)}(k-2) + C_{G_{n,m,r-2}}(k) \\
&= \frac{C_{C_n}(k)C_{C_m}(k)C_{P_{r-1}}(k)}{k^2(k-1)}(k^2-3k+3) - C_{G_{n,m,r-3}}(k) \\
&= \frac{C_{C_n}(k)C_{C_m}(k)C_{P_{r-2}}(k)}{k^2(k-1)}(k^3-4k^2+6k-4) + C_{G_{n,m,r-4}}(k) \\
&\dots \\
&= \frac{C_{C_n}(k)C_{C_m}(k)C_{P_2}(k)}{k^2(k-1)} \frac{C_{C_{r+1}}(k)}{k(k-1)} + (-1)^r C_{G_{n,m,0}}(k),
\end{aligned}$$

so the claimed result is obtained. ■

References

1. J.M. Aldous, R.J. Wilson, *Graphs and Applications* (The Open University, UK, 2004)
2. C. Berge, *The Theory of Graphs* (Dover, 2001)
3. G.D. Birkhoff, A determinant formula for coloring a map. Ann. Math. **14**, 42–46 (1912)
4. B. Bollobas, *Modern Graph Theory* (Springer, New York, 1998)
5. W. Chen, *Applied Graph Theory* (North-Holland Publishing Company, New York, 1976)
6. W. Dicks, M.J. Dunwoody, *Groups Acting on Graphs* (Cambridge UK, 1989)
7. F. Harary, *Graph Theory* (Addison-Wesley, US, 1994)
8. U. Sanli, I.N. Cangul, A new method for calculating the chromatic polynomial. Appl. Sci. **19**, 110–121 (2017)
9. D.B. West, *Introduction to Graph Theory* (Prentice Hall, Upper Saadle River, 1996)

Recurrence Relations of Edge-Zagreb and Sum-Edge Characteristic Polynomials of Some Graphs

Mert Sinan Oz and Ismail Naci Cangul

Abstract Spectral graph theory is a rising area of graph theory in the last decades due to its applications related to molecular energy in mathematical chemistry. Classical molecular energy was defined by means of the adjacency matrix of the graph modeling the given molecular structure. In the last decade, many various variants of classical energy have been introduced and studied. Usually, as the spectrum of a graph does not have a reductive form, that is, as we cannot relate the spectrum of a graph to some other graph obtained by deleting or adding a vertex, an edge, or applying some operation, one cannot obtain general formulae for most graph classes. In this paper, we recall two of such graph matrices and obtain recurrence relations for the graph energies corresponding to these matrices making it easy to calculate the energy of large graphs by means of smaller graphs.

Keywords Graphs · Adjacency · Edge-Zagreb · Sum-edge · Second Zagreb index · Graph matrix · Characteristic polynomial

MSC 2000 Numbers 05C07 · 05C31 · 05C38 · 05C50

1 Introduction

Let $G = (V, E)$ be a connected simple graph, that is, an undirected graph without any loops nor multiple edges having vertex set $V = \{v_1, v_2, \ldots, v_n\}$ and edge set $E = \{e_1, e_2, \ldots, e_m\}$, [1]. Two vertices v_i and v_j of G are called adjacent vertices if

M. S. Oz (✉)
Department of Mathematics, Faculty of Engineering and Natural Sciences, Bursa Technical University, 16320 Bursa, Turkey
e-mail: sinan.oz@btu.edu.tr

I. N. Cangul
Department of Mathematics, Bursa Uludag University, 16059 Gorukle, Bursa, Turkey
e-mail: cangul@uludag.edu.tr

S. K. Paikray et al. (eds.), *New Trends in Applied Analysis and Computational Mathematics*, Advances in Intelligent Systems and Computing 1356,
https://doi.org/10.1007/978-981-16-1402-6_9

there is an edge that connects v_i to v_j and it is denoted by $e = v_i v_j$. Let us denote the degree of a vertex v by $d(v)$. A component of G is a maximal connected subgraph of G. If the removal of an edge e increases the number of components in G, then e is called a bridge in G. A path graph of order n is a straight line with vertices where the degrees of the two terminal vertices are one and degrees of other (internal) vertices are two. We denote a path graph with n vertices by P_n. A trail is a walk without any repeated edges. Cycle graph is a closed trail where two end vertices are the same. We denote a cycle graph with n vertices by C_n. A star graph with n vertices, denoted by S_n, is a graph, in which the degree of the central vertex is $n-1$ and the degree of each adjacent remaining vertex is one. Tadpole graph $T_{r,s}$ is a graph that consists of a combination of a cycle graph C_r and P_s joined by means of a bridge. A complete bipartite graph $K_{r,s}$ is a graph whose vertices are separated into two subsets V_i and V_j so that there is no edge between the vertices in the same subset and all edges having one vertex in V_i and the other in V_j are drawn.

Spectral graph theory deals with the properties of several types of matrices of a graph G, and also with the eigenvalues and eigenvectors of these matrices. The most useful one is the adjacency matrix. There are lots of other special types of adjacency matrices corresponding to a graph. For years, lots of mathematicians and chemists have used these matrices to get spectral properties of given graphs by means of linear algebraic methods. Therefore, the study of adjacency with the aid of corresponding matrices gains a prominent seat in molecular chemistry and it is counted as one of the most applicable sub-areas of graph theory. Energy of a graph is defined as the sum of absolute values of eigenvalues of the corresponding adjacency matrix and it has a huge importance in molecular chemistry. Similar to vertex adjacency matrix of G, also edge-Zagreb, sum-edge, Laplacian, incidency, etc., matrices were defined and used in graph theory for several applications. Moreover, some topological graph invariants were defined by using adjacency notion. One of the most important of them is the second Zagreb index and it is denoted by $M_2(G)$. This invariant were defined in [2] as follows:

$$M_2(G) = \sum_{v_i v_j \in E} d(v_i) d_G(v_j).$$

In 2017, Celik and Cangul gave some formulae and recurrence relations on spectral polynomials of some special graph types, [3], and Yamac et al. gave the exact formulae for the edge-Zagreb characteristic polynomials of some graph types in [4]. In this paper, we recall the edge-Zagreb adjacency matrix and we obtain some recurrence formulae for the edge-Zagreb characteristic polynomials of the most-widely used graph classes P_n, C_n, S_n, $T_{r,s}$ and $K_{r,s}$.

2 Edge-Zagreb Adjacency Matrix and Recurrence Relations of Edge-Zagreb Characteristic Polynomials of Some Graph Types

Let us recall the edge-Zagreb adjacency matrix of a graph $G = (V, E)$ with n vertices. The edge-Zagreb adjacency matrix $Z(G)$ of G is a symmetric matrix and $Z(G) = [z_{ij}]_{n\times n}$ is specified by using adjacency of vertices as follows, [5]:

$$z_{ij} = \begin{cases} d(v_i)d(v_j), & \text{if the vertices } v_i \text{ and } v_j \text{ are adjacent vertices} \\ 0, & \text{otherwise.} \end{cases}$$

Now, we are ready to start with the edge-Zagreb matrix and characteristic polynomials of path graphs P_n:

Theorem 1 *The recurrence formula for the edge-Zagreb characteristic polynomial of the path graph P_n obtained by means of the edge-Zagreb matrix is*

$$P_{P_n}(x) = xP_{P_{n-1}}(x) - 16P_{P_{n-2}}(x),$$

where $n \geq 4$.

Proof Let us form the edge-Zagreb characteristic polynomial of P_n by using the definition of characteristic polynomial as $P_{P_n}(x) = |xI - Z|$, where I is the identity matrix of order n and Z is the edge-Zagreb adjacency matrix of P_n.

$$P_{P_n}(x) = \begin{vmatrix} x & -2 & 0 & 0 & 0 & \cdots & 0 & 0 & 0 \\ -2 & x & -4 & 0 & 0 & \cdots & 0 & 0 & 0 \\ 0 & -4 & x & -4 & 0 & \cdots & 0 & 0 & 0 \\ 0 & 0 & -4 & x & -4 & \cdots & 0 & 0 & 0 \\ . & . & . & . & . & \cdots & . & . & . \\ 0 & 0 & 0 & 0 & 0 & \cdots & x & -4 & 0 \\ 0 & 0 & 0 & 0 & 0 & \cdots & -4 & x & -2 \\ 0 & 0 & 0 & 0 & 0 & \cdots & 0 & -2 & x \end{vmatrix}.$$

We calculate this $n \times n$ determinant with respect to the first row to get two $(n-1) \times (n-1)$ determinants

$$P_{P_n}(x) = x \begin{vmatrix} x & -4 & 0 & \cdots & 0 & 0 & 0 \\ -4 & x & -4 & \cdots & 0 & 0 & 0 \\ 0 & -4 & x & \cdots & 0 & 0 & 0 \\ . & . & . & & . & . & . \\ 0 & 0 & 0 & \cdots & x & -4 & 0 \\ 0 & 0 & 0 & \cdots & -4 & x & -2 \\ 0 & 0 & 0 & \cdots & 0 & -2 & x \end{vmatrix} - (-2) \begin{vmatrix} -2 & -4 & 0 & \cdots & 0 & 0 & 0 \\ 0 & x & -4 & \cdots & 0 & 0 & 0 \\ 0 & -4 & x & \cdots & 0 & 0 & 0 \\ . & . & . & & . & . & . \\ 0 & 0 & 0 & \cdots & x & -4 & 0 \\ 0 & 0 & 0 & \cdots & -4 & x & -2 \\ 0 & 0 & 0 & \cdots & 0 & -2 & x \end{vmatrix}.$$

Here, we calculate the second determinant with respect to the first column to get, respectively, the following $(n-1) \times (n-1)$ and $(n-2) \times (n-2)$ determinants:

$$P_{P_n}(x) = x\begin{vmatrix} x & -4 & 0 & \cdots & 0 & 0 & 0 \\ -4 & x & -4 & \cdots & 0 & 0 & 0 \\ 0 & -4 & x & \cdots & 0 & 0 & 0 \\ \vdots & \vdots & \vdots & & \vdots & \vdots & \vdots \\ 0 & 0 & 0 & \cdots & x & -4 & 0 \\ 0 & 0 & 0 & \cdots & -4 & x & -2 \\ 0 & 0 & 0 & \cdots & 0 & -2 & x \end{vmatrix} - (-2)(-2)\begin{vmatrix} x & -4 & 0 & \cdots & 0 & 0 & 0 \\ -4 & x & -4 & \cdots & 0 & 0 & 0 \\ 0 & -4 & x & \cdots & 0 & 0 & 0 \\ \vdots & \vdots & \vdots & & \vdots & \vdots & \vdots \\ 0 & 0 & 0 & \cdots & x & -4 & 0 \\ 0 & 0 & 0 & \cdots & -4 & x & -2 \\ 0 & 0 & 0 & \cdots & 0 & -2 & x \end{vmatrix}.$$

Now we expand both determinants according to the first rows, we get $P_{P_n}(x)$ as

$$x\left(x\begin{vmatrix} x & -4 & 0 & \cdots & 0 & 0 & 0 \\ -4 & x & -4 & \cdots & 0 & 0 & 0 \\ 0 & -4 & x & \cdots & 0 & 0 & 0 \\ \vdots & \vdots & \vdots & & \vdots & \vdots & \vdots \\ 0 & 0 & 0 & \cdots & x & -4 & 0 \\ 0 & 0 & 0 & \cdots & -4 & x & -2 \\ 0 & 0 & 0 & \cdots & 0 & -2 & x \end{vmatrix} - (-4)\begin{vmatrix} -4 & -4 & 0 & \cdots & 0 & 0 & 0 \\ 0 & x & -4 & \cdots & 0 & 0 & 0 \\ 0 & -4 & x & \cdots & 0 & 0 & 0 \\ \vdots & \vdots & \vdots & & \vdots & \vdots & \vdots \\ 0 & 0 & 0 & \cdots & x & -4 & 0 \\ 0 & 0 & 0 & \cdots & -4 & x & -2 \\ 0 & 0 & 0 & \cdots & 0 & -2 & x \end{vmatrix}\right)$$
$$-4\left(x\begin{vmatrix} x & -4 & 0 & \cdots & 0 & 0 & 0 \\ -4 & x & -4 & \cdots & 0 & 0 & 0 \\ 0 & -4 & x & \cdots & 0 & 0 & 0 \\ \vdots & \vdots & \vdots & & \vdots & \vdots & \vdots \\ 0 & 0 & 0 & \cdots & x & -4 & 0 \\ 0 & 0 & 0 & \cdots & -4 & x & -2 \\ 0 & 0 & 0 & \cdots & 0 & -2 & x \end{vmatrix} - (-4)\begin{vmatrix} -4 & -4 & 0 & \cdots & 0 & 0 & 0 \\ 0 & x & -4 & \cdots & 0 & 0 & 0 \\ 0 & -4 & x & \cdots & 0 & 0 & 0 \\ \vdots & \vdots & \vdots & & \vdots & \vdots & \vdots \\ 0 & 0 & 0 & \cdots & x & -4 & 0 \\ 0 & 0 & 0 & \cdots & -4 & x & -2 \\ 0 & 0 & 0 & \cdots & 0 & -2 & x \end{vmatrix}\right).$$

If we calculate the second and fourth determinants according to the first columns, we get

$$x\left(x\begin{vmatrix} x & -4 & 0 & \cdots & 0 & 0 & 0 \\ -4 & x & -4 & \cdots & 0 & 0 & 0 \\ 0 & -4 & x & \cdots & 0 & 0 & 0 \\ \vdots & \vdots & \vdots & & \vdots & \vdots & \vdots \\ 0 & 0 & 0 & \cdots & x & -4 & 0 \\ 0 & 0 & 0 & \cdots & -4 & x & -2 \\ 0 & 0 & 0 & \cdots & 0 & -2 & x \end{vmatrix} - 16\begin{vmatrix} x & -4 & 0 & \cdots & 0 & 0 & 0 \\ -4 & x & -4 & \cdots & 0 & 0 & 0 \\ 0 & -4 & x & \cdots & 0 & 0 & 0 \\ \vdots & \vdots & \vdots & & \vdots & \vdots & \vdots \\ 0 & 0 & 0 & \cdots & x & -4 & 0 \\ 0 & 0 & 0 & \cdots & -4 & x & -2 \\ 0 & 0 & 0 & \cdots & 0 & -2 & x \end{vmatrix}\right)$$
$$-4\left(x\begin{vmatrix} x & -4 & 0 & \cdots & 0 & 0 & 0 \\ -4 & x & -4 & \cdots & 0 & 0 & 0 \\ 0 & -4 & x & \cdots & 0 & 0 & 0 \\ \vdots & \vdots & \vdots & & \vdots & \vdots & \vdots \\ 0 & 0 & 0 & \cdots & x & -4 & 0 \\ 0 & 0 & 0 & \cdots & -4 & x & -2 \\ 0 & 0 & 0 & \cdots & 0 & -2 & x \end{vmatrix} - 16\begin{vmatrix} x & -4 & 0 & \cdots & 0 & 0 & 0 \\ -4 & x & -4 & \cdots & 0 & 0 & 0 \\ 0 & -4 & x & \cdots & 0 & 0 & 0 \\ \vdots & \vdots & \vdots & & \vdots & \vdots & \vdots \\ 0 & 0 & 0 & \cdots & x & -4 & 0 \\ 0 & 0 & 0 & \cdots & -4 & x & -2 \\ 0 & 0 & 0 & \cdots & 0 & -2 & x \end{vmatrix}\right).$$

After rearrangements, we get the $P_{P_n}(x)$ as follows:

$$xP_{P_{n-1}}(x) - 16\left(x\begin{vmatrix} x & -4 & 0 & \cdots & 0 & 0 & 0 \\ -4 & x & -4 & \cdots & 0 & 0 & 0 \\ 0 & -4 & x & \cdots & 0 & 0 & 0 \\ \vdots & \vdots & \vdots & & \vdots & \vdots & \vdots \\ 0 & 0 & 0 & \cdots & x & -4 & 0 \\ 0 & 0 & 0 & \cdots & -4 & x & -2 \\ 0 & 0 & 0 & \cdots & 0 & -2 & x \end{vmatrix} - 4\begin{vmatrix} x & -4 & 0 & \cdots & 0 & 0 & 0 \\ -4 & x & -4 & \cdots & 0 & 0 & 0 \\ 0 & -4 & x & \cdots & 0 & 0 & 0 \\ \vdots & \vdots & \vdots & & \vdots & \vdots & \vdots \\ 0 & 0 & 0 & \cdots & x & -4 & 0 \\ 0 & 0 & 0 & \cdots & -4 & x & -2 \\ 0 & 0 & 0 & \cdots & 0 & -2 & x \end{vmatrix}\right).$$

As the result, we get the recurrence relation as $P_{P_n}(x) = xP_{P_{n-1}}(x) - 16P_{P_{n-2}}(x)$, where $n \geq 4$. ■

Next, we consider the cycle graphs:

Theorem 2 *The recurrence formula for the edge-Zagreb characteristic polynomial of the cycle graph C_n obtained by means of the edge-Zagreb matrix is*

$$P_{C_n}(x) = xP_{C_{n-1}}(x) - 16P_{C_{n-2}}(x) + 2^{2n-1}x - 2^{2n+2}$$

where $n \geq 5$.

Proof We know that $P_{C_n}(x) = |xI - Z|$, where I and Z are the identity matrix of order n and the edge-Zagreb adjacency matrix of C_n, respectively. So, we have

$$P_{C_n}(x) = \begin{vmatrix} x & -4 & 0 & 0 & 0 & \cdots & 0 & 0 & -4 \\ -4 & x & -4 & 0 & 0 & \cdots & 0 & 0 & 0 \\ 0 & -4 & x & -4 & 0 & \cdots & 0 & 0 & 0 \\ 0 & 0 & -4 & x & -4 & \cdots & 0 & 0 & 0 \\ . & . & . & . & . & \cdots & . & . & . \\ 0 & 0 & 0 & 0 & 0 & \cdots & x & -4 & 0 \\ 0 & 0 & 0 & 0 & 0 & \cdots & -4 & x & -4 \\ -4 & 0 & 0 & 0 & 0 & \cdots & 0 & -4 & x \end{vmatrix}.$$

Let us calculate the $P_{C_n}(x)$ by calculating the determinant with respect to first row.

$$P_{C_n}(x) = x\begin{vmatrix} x & -4 & 0 & \cdots & 0 & 0 & 0 \\ -4 & x & -4 & \cdots & 0 & 0 & 0 \\ 0 & -4 & x & \cdots & 0 & 0 & 0 \\ . & . & . & & . & . & . \\ 0 & 0 & 0 & \cdots & x & -4 & 0 \\ 0 & 0 & 0 & \cdots & -4 & x & -4 \\ 0 & 0 & 0 & \cdots & 0 & -4 & x \end{vmatrix} + 4\begin{vmatrix} -4 & -4 & 0 & \cdots & 0 & 0 & 0 \\ 0 & x & -4 & \cdots & 0 & 0 & 0 \\ 0 & -4 & x & \cdots & 0 & 0 & 0 \\ . & . & . & . & . & . & . \\ 0 & 0 & 0 & \cdots & x & -4 & 0 \\ 0 & 0 & 0 & \cdots & -4 & x & -4 \\ -4 & 0 & 0 & \cdots & 0 & -4 & x \end{vmatrix}$$

$$(-1)^{n+1}(-4)\begin{vmatrix} -4 & x & -4 & \cdots & 0 & 0 & 0 \\ 0 & -4 & x & \cdots & 0 & 0 & 0 \\ 0 & 0 & -4 & \cdots & 0 & 0 & 0 \\ . & . & . & . & . & . & \\ 0 & 0 & 0 & \cdots & x & -4 & 0 \\ 0 & 0 & 0 & \cdots & -4 & x & -4 \\ -4 & 0 & 0 & \cdots & 0 & -4 & x \end{vmatrix}$$

Hence, after some more row and column operations, we finally have the following:

$$P_{C_n}(x) = xP_{C_{n-1}}(x) - 16P_{C_{n-2}}(x) + 2^{2n-1}x - 2^{2n+2} + 16x\begin{vmatrix} x & -4 & \cdots & 0 & 0 & 0 \\ -4 & x & \cdots & 0 & 0 & 0 \\ . & . & & . & . & . \\ 0 & 0 & \cdots & x & -4 & 0 \\ 0 & 0 & \cdots & -4 & x & -4 \\ 0 & 0 & \cdots & 0 & -4 & x \end{vmatrix}$$

$$-16\left(x\begin{vmatrix} x & -4 & \cdots & 0 & 0 & 0 \\ -4 & x & \cdots & 0 & 0 & 0 \\ . & . & & . & . & . \\ 0 & 0 & \cdots & x & -4 & 0 \\ 0 & 0 & \cdots & -4 & x & -4 \\ 0 & 0 & \cdots & 0 & -4 & x \end{vmatrix} - 16\begin{vmatrix} x & -4 & \cdots & 0 & 0 & 0 \\ -4 & x & \cdots & 0 & 0 & 0 \\ . & . & & . & . & . \\ 0 & 0 & \cdots & x & -4 & 0 \\ 0 & 0 & \cdots & -4 & x & -4 \\ 0 & 0 & \cdots & 0 & -4 & x \end{vmatrix}\right)$$

$$-16^2\begin{vmatrix} x & -4 & \cdots & 0 & 0 & 0 \\ -4 & x & \cdots & 0 & 0 & 0 \\ . & . & & . & . & . \\ 0 & 0 & \cdots & x & -4 & 0 \\ 0 & 0 & \cdots & -4 & x & -4 \\ 0 & 0 & \cdots & 0 & -4 & x \end{vmatrix}$$

After rearranging the last equation, we get the desired result. ■

Next, we consider the star graphs:

Theorem 3 *The formula for the edge-Zagreb characteristic polynomial of the star graph S_n obtained by means of the edge-Zagreb matrix is*

$$P_{S_n}(x) = xP_{S_{n-1}}(x) - 6\binom{n-1}{2}x^{n-2} - x^{n-2}$$

where $n \geq 3$.

Proof Let us form the $P_{S_n}(x)$ as follows:

$$P_{S_n}(x) = \begin{vmatrix} x & 0 & 0 & 0 & 0 & \cdots & 0 & 0 & -(n-1) \\ 0 & x & 0 & 0 & 0 & \cdots & 0 & 0 & -(n-1) \\ 0 & 0 & x & 0 & 0 & \cdots & 0 & 0 & -(n-1) \\ 0 & 0 & 0 & x & 0 & \cdots & 0 & 0 & -(n-1) \\ \cdot & \cdot & \cdot & \cdot & \cdot & \cdots & \cdot & \cdot & \cdot \\ 0 & 0 & 0 & 0 & 0 & \cdots & x & 0 & -(n-1) \\ 0 & 0 & 0 & 0 & 0 & \cdots & 0 & x & -(n-1) \\ -(n-1) & -(n-1) & -(n-1) & -(n-1) & -(n-1) & \cdots & -(n-1) & -(n-1) & x \end{vmatrix}.$$

Hence, it can be observed that $P_{S_n}(x) = x^n - (n-1)^2\binom{n-1}{1}x^{n-2}$ by using elementary row and column operations. Therefore, since $P_{S_{n-1}}(x) = x^{n-1} - (n-2)^2\binom{n-2}{1}x^{n-3}$, we have $P_{S_n}(x) = xP_{S_{n-1}}(x) + \left[(n-2)^2\binom{n-2}{1} - (n-1)^2\binom{n-1}{1}\right]x^{n-2}$. Also, since $(n-2)^3 - (n-1)^3 = -3(n-1)(n-2) = -6\binom{n-1}{2}x^{n-2} - x^{n-2}$, we get the desired result. ■

Theorem 4 *The recurrence formula for the edge-Zagreb characteristic polynomial of the tadpole graph $T_{r,s}$ obtained by means of the edge-Zagreb matrix is*

$$P_{T_{r,s}}(x) = xP_{T_{r,s-1}}(x) - 16P_{T_{r,s-2}}(x)$$

where $s > 3$.

Proof Let s be greater than 3. Let us express the $P_{T_{r,s}}(x)$:

$$P_{T_{r,s}}(x) = \begin{vmatrix} x & -2 & 0 & 0 & \cdots & 0 & 0 & 0 & 0 & 0 & 0 & 0 & 0 & 0 \\ -2 & x & -4 & 0 & \cdots & 0 & 0 & 0 & 0 & 0 & 0 & 0 & 0 & 0 \\ 0 & -4 & x & 0 & \cdots & 0 & 0 & 0 & 0 & 0 & 0 & 0 & 0 & 0 \\ \cdot & \cdots & \cdot & \cdot & \cdot & \cdot & \cdot & \cdot & \cdot & \cdot & \cdot & \cdot & \cdot & \cdot \\ 0 & \cdots & 0 & 0 & -4 & x & -4 & 0 & 0 & 0 & 0 & \cdots & 0 & 0 \\ 0 & \cdots & 0 & 0 & 0 & -4 & x & -6 & 0 & 0 & 0 & \cdots & 0 & 0 \\ 0 & \cdots & 0 & 0 & 0 & 0 & -6 & x & -6 & 0 & 0 & \cdots & 0 & -6 \\ 0 & \cdots & 0 & 0 & 0 & 0 & 0 & -6 & x & -4 & 0 & \cdots & 0 & 0 \\ 0 & \cdots & 0 & 0 & 0 & 0 & 0 & 0 & -4 & x & -4 & \cdots & 0 & 0 \\ \cdot & \cdots & \cdot & \cdot & \cdot & \cdot & \cdot & \cdot & \cdot & \cdot & \cdot & \cdot & \cdot & \cdot \\ 0 & \cdots & 0 & 0 & 0 & 0 & 0 & 0 & 0 & 0 & 0 & -4 & x & -4 \\ 0 & \cdots & 0 & 0 & 0 & 0 & 0 & -6 & 0 & 0 & 0 & 0 & -4 & x \end{vmatrix}.$$

Calculating this determinant, we get the desired result. ■

The following result states the recurrence formula for the edge-Zagreb characteristic polynomial of the complete bipartite graphs:

Theorem 5 *The recurrence formula for the edge-Zagreb characteristic polynomial of the complete bipartite graph $K_{r,s}$ obtained by means of the edge-Zagreb matrix is*

$$P_{K_{r,s}}(x) = x P_{K_{r,s-1}}(x) - \left[(rs)^3 - (r(s-1))^3\right] x^{r+s-2}$$

where $s \geq 2$.

Proof It can be observed that $P_{K_{r,s}}(x) = x^{r+s} - (rs)^2(rs)x^{r+s-2}$ after using elementary row and column operations. Therefore, since $P_{K_{r,s-1}}(x) = x^{r+s-1} - (r(s-1))^2(r(s-1))x^{r+s-3}$, we have $P_{K_{r,s}}(x)=x P_{K_{r,s-1}}(x)-\left[(rs)^3 - (r(s-1))^3\right] x^{r+s-2}$ which completes the proof. ■

3 Sum-Edge Adjacency Matrix and Recurrence Relations of Sum-Edge Characteristic Polynomials of Some Graph Types

Analogous to Sect. 2, in this section, we define the sum-edge adjacency matrix of a graph $G = (V, E)$ with n vertices. After that, we give some recurrence relations of the sum-edge characteristic polynomials of P_n, C_n, S_n, $T_{r,s}$ and $K_{r,s}$ without proofs as proofs can be done in similar fashion with the proofs in Sect. 2.

First of all, let us define the sum-edge adjacency matrix $S(G)$ of a graph G. It is a symmetric matrix and if $S(G) = [s_{ij}]_{n\times n}$, then it is defined as follows, [5]:

$$s_{ij} = \begin{cases} d(v_i) + d(v_j), & \text{if the vertices } v_i \text{ and } v_j \text{ are adjacent vertices} \\ 0, & \text{otherwise.} \end{cases}$$

We begin with the study of path graph P_n and then continue with the other frequently used graph classes:

Corollary 6 *The recurrence formula for the sum-edge characteristic polynomial of the path graph P_n, cycle graph C_n, star graph S_n, tadpole graph $T_{r,s}$ and complete bipartite graph $K_{r,s}$ obtained by means of the sum-edge matrix are*

$$P_{P_n}(x) = x P_{P_{n-1}}(x) - 16 P_{P_{n-2}}(x),$$

where $n > 4$;

$$P_{C_n}(x) = x P_{C_{n-1}}(x) - 16 P_{C_{n-2}}(x) + 2^{2n-1}x - 2^{2n+2},$$

where $n \geq 5$;

$$P_{S_n}(x) = x P_{S_{n-1}}(x) - \left(3n^2 - 5n + 2\right) x^{n-2}$$

where $n \geq 3$;

$$P_{T_{r,s}}(x) = xP_{T_{r,s-1}}(x) - 16P_{T_{r,s-2}}(x)$$

where $s > 3$*; and*

$$P_{K_{r,s}}(x) = xP_{K_{r,s-1}}(x) - \left(4r^2s + 3rs^2 + r^3 - 2r^2 - 3rs + r\right)x^{r+s-2}$$

where $s \geq 2$*, respectively.*

References

1. R.B. Bapat, *Graphs and Matrices* (Springer, Berlin, 2014)
2. I. Gutman, N. Trinajstic, Graph theory and molecular orbitals III. Total π-electron energy of alternant hydrocarbons. Chem. Phys. Lett. **17**, 535–538 (1972)
3. F. Celik, I.N. Cangul, Formulae and recurrence relations on spectral polynomials of some graphs. Adv. Stud. Contemp. Math. **27**(3), 325–332 (2017)
4. C. Yamac, M.S. Oz, I.N. Cangul, Edge-Zagreb indices of graphs. TWMS J. App. Eng. Math. **10**(1), 1–10 (2020)
5. D. Janežič, A. Miličević, S. Nikolić, N. Trinajstić, *Graph Theoretical Matrices in Chemistry* (CRC Press, Taylor and Francis Group, 2015)

Effects of Radiation on MHD Flow with Induced Magnetic Field

Lipika Panigrahi and J. P. Panda

Abstract We have analyzed numerically the impact of radiation and viscous dissipation on steady MHD flow over a stretching sheet with an induced magnetic field. The nonlinear equations are transferred to ODE with the help of a similar transformation. Further, transferred equations are solved using the fourth-order Runge–Kutta method with the help of Matlab BVP4C programming. Comparisons with the earlier results have been made and good agreements were found. Surface transport phenomena such as skin friction, Nusselt number are discussed besides the three boundary layers. Some important results are found: The magnetic parameter β decelerates velocity, and accelerates temperature profiles; The radiation parameter(R) enhances the thermal boundary layer thickness for both cases $a/c < 1$ and $a/c < 1$.

Keywords Induced magnetic field · Radiation · Stretching sheet · Viscous dissipation · MHD flow

1 Introduction

Thermal radiation is an important utilization in some mechanical applications, for example, glass creation and furthermore in space innovation, for example, comical flight aerodynamics, etc. The presentation of thermal radiation impacts on heat transfer opens another area of research with tremendous useful applications. Jalilpour et al. [1] have analyzed the effect of thermal radiation on the non-orthogonal stagnation flow of a nanofluid. One author studied the impact of Casson fluid over a stretching surface on thermal radiation and binary chemical reaction by Abbas et al. [2]. Misra and Sinha [3] have studied the radiation effect on MHD flow of blood in stretching motion.

L. Panigrahi · J. P. Panda (✉)
Department of Mathematics, Veer Surendra Sai University of Technology, Burla 768018, Odisha, India
e-mail: jpanda123@gmail.com

L. Panigrahi
e-mail: lipikapanigrahi4@gmail.com

S. K. Paikray et al. (eds.), *New Trends in Applied Analysis and Computational Mathematics*, Advances in Intelligent Systems and Computing 1356,
https://doi.org/10.1007/978-981-16-1402-6_10

The examination of stretching surface is applied widely in industrial manufacturing process such as extrusion of polymers, cooling of metallic plates, paper production, etc. The biomagnetic and hydromagnetic fluid flow through a stretching surface is a significant application in bioengineering, clinical applications, and metalworking procedures, respectively. Mahapatra and Gupta [4] and Nazar et al. [5] investigated the stagnation-point flow with different pertinent parameters through stretching surface. Further, studied the boundary layer flow over a moving surface for both the cases of assisting and opposing flows by Ishak et al. [6].

The induced magnetic field has gotten impressive enthusiasm attributable to its utilization in numerous logical and innovative marvels, for instance, in MHD vitality generator frameworks and magnetohydrodynamic boundary layer control technologies. In literature review, some researchers analyzed the induced magnetic toward stretching sheet with helping of strong numerical methods. Effect of induced magnetic field and heat transfer over a stretching sheet have studied by Ali et al. [7] and this paper extended using Casson fluid by Aziz et al. [8].

The main aim of this work is to explore the radiation and viscous dissipation effects on stagnation-point flow with induced magnetic field. This work is solved by numerical method with help of Matlab. The present results are compared with some authors which are studied in literatuire review and it is shown that the result is in good agreement with other work.

2 Basic Equations

We have analyzed the steady two-dimensional incompressible fluid on stagnation-point flow towards a moving sheet. The effect of the radiation and viscous dissipation is considered Ali et al. [9]. The applied magnetic field is parallel to the surface. We take thermal radiation terms from Mishra and Sinha [3] article. The modeled governing equations are Cowling [10]:

$$\nabla \cdot \vec{\mathrm{V}} = 0, \nabla \cdot \vec{\mathrm{H}} = 0 \tag{1}$$

$$(\mathrm{V} \cdot \nabla)\vec{\mathrm{V}} - \frac{\mu}{4\pi\rho}\left(\vec{\mathrm{H}} \cdot \nabla\right)\vec{\mathrm{H}} = -\frac{1}{\rho}\nabla \mathrm{P} + \nu\nabla^2\vec{\mathrm{V}} \tag{2}$$

$$(\nabla \times \mathrm{V}) \times \vec{\mathrm{H}} + \mu_e \nabla^2 \vec{\mathrm{H}} = 0 \tag{3}$$

$$(\mathrm{V} \cdot \nabla)\vec{T} = \alpha\nabla^2\vec{T} + \frac{1}{\rho C_p}\nabla\vec{q}_r \tag{4}$$

where $\mathrm{P} = \mathrm{p} + \frac{\mu|\mathrm{H}|^2}{8\pi}$ = MHD pressure (Fig. 1).

T (fluid temperature), p (fluid pressure), μ (magnetic permeability of the fluid), ν (kinematic viscosity), σ (electric conductivity of the fluid flow), ρ (fluid density), α (thermal diffusivity fluid), and $\mu_e = \frac{1}{4\pi\sigma}$ (magnetic diffusivity of fluid).

Using boundary layer approximations, Eqs. (1)–(4) become

$$\left.\begin{aligned} \frac{\partial U}{\partial X} + \frac{\partial U}{\partial Y} &= 0 \\ \frac{\partial H_X}{\partial X} + \frac{\partial H_Y}{\partial Y} &= 0 \end{aligned}\right\} \quad (5)$$

$$U\frac{\partial U}{\partial X} + V\frac{\partial U}{\partial Y} = \nu\frac{\partial^2 U}{\partial Y^2} + \frac{\mu_e}{\rho}\left(H_X\frac{\partial H_X}{\partial X} + H_Y\frac{\partial H_Y}{\partial Y}\right) + \left(U_e\frac{\partial U_e}{\partial X} - \frac{\mu_e}{\rho}H_e\frac{\partial H_e}{\partial Y}\right) \quad (6)$$

$$U\frac{\partial H_X}{\partial X} + V\frac{\partial H_Y}{\partial Y} - H_X\frac{\partial U}{\partial X} - H_Y\frac{\partial U}{\partial Y} = \eta_0\frac{\partial^2 H_X}{\partial Y^2} \quad (7)$$

$$U\frac{\partial T}{\partial X} + V\frac{\partial T}{\partial Y} = \alpha_0\frac{\partial^2 T}{\partial Y^2} + \frac{\nu}{C_p}\left(\frac{\partial U}{\partial Y}\right)^2 - \frac{1}{\rho C_p}\frac{\partial q_r}{\partial Y} \quad (8)$$

H_X and H_Y are the components of induced magnetic in x and y directions and $U_e(X)$ and $H_e(X)$ are the velocity and magnetic field, respectively. The boundary conditions are

$$\left.\begin{aligned} U = U_w = cx,\ V = 0,\ \frac{\partial H_X}{\partial Y} = H_Y = 0,\ T = T_w, \quad & \text{At } Y = 0 \\ U = U_e = ax,\ H = H_e = H_0 x,\ T = T_\infty, \quad & \text{At } Y \to \infty \end{aligned}\right\} \quad (9)$$

where a, c are the constants, and H_0 = uniform magnetic field, infinity upstream.

$$\left.\begin{aligned} \eta = Y\sqrt{\frac{c}{\nu}},\ U = cXf'(\eta),\ V = -(c\nu)^{\frac{1}{2}}f(\eta),\ H_X = H_0 X g'(\eta), \\ H_Y = -(c\nu)^{\frac{1}{2}}g'(\eta),\ \theta(\eta) = \frac{T - T_\infty}{T_w - T_\infty}, \end{aligned}\right\} \quad (10)$$

With help of the similar transformations to Eqs. (5)–(8), we get the following equations:

$$f''' + f f'' - 2f'^2 + \frac{a^2}{c^2} + \beta\left(g'^2 - g g'' - 1\right) = 0 \quad (11)$$

$$\lambda g''' + f g'' - f'' g = 0 \quad (12)$$

$$\frac{1}{\Pr}(1+R)\theta'' + f\theta' + Ec\, f''^2 = 0 \tag{13}$$

Then, the boundary conditions (9) reduce to

$$\left.\begin{array}{l} f'(0) = 1,\ f(0) = 0,\ f'(\infty) = 0,\ g''(0) = 0, \\ g(0) = 0,\ g'(\infty) = \dfrac{a}{c},\ \theta(0) = 1,\ \theta(\infty) = 0, \end{array}\right\} \tag{14}$$

where

$$\left.\begin{array}{l} \Pr = \dfrac{\alpha_0}{\nu},\ \eta_0 = \dfrac{1}{\mu\,\sigma},\ \lambda = \dfrac{\eta_0}{\nu},\ \beta = \dfrac{\mu_0\, H_0^2}{\rho\, c^2}, \\ Ec = \dfrac{c^2}{Cp\ T_0}, \alpha = \dfrac{a}{c},\ R = \dfrac{16\sigma^* T_\infty^3}{3kk^*} \end{array}\right\} \tag{15}$$

The magnetic parameter (β), is related to the Hartmann [11] number and magnetic Reynolds numbers Re and $\mathrm{Re_m}$:

$$\left.\begin{array}{l} \beta = \dfrac{Ha^2}{\mathrm{Re\,Re}_m},\ Ha = \mu H_0 l\sqrt{\dfrac{\sigma}{\mu}},\ \mathrm{Re} = \dfrac{(cl)l}{\nu}, \\ \mathrm{Re}_m = 4\pi U_\infty l\pi\sigma = \dfrac{(cl)l}{\mu_e} \end{array}\right\} \tag{16}$$

where l = distance of the stretching sheet. It may be noted that for $R = 0$ and $Ec = 0$ (without a Radiation and Eckert number), the work reduces to that of Ali et al. [9]. The skin friction coefficient and the local Nusselt number are

$$C_f = \frac{\tau_w}{\rho U_w^2},\ Nu = \frac{Xq_w}{k(T_w - T_\infty)}, \tag{17}$$

where,

$$\tau_w = \mu\left(\frac{\partial u}{\partial y}\right)_{Y=0},\ q_w = -k\left(\frac{\partial u}{\partial y}\right)_{Y=0} \tag{18}$$

where, $\boldsymbol{\tau}_w$ = wall heat flux, $\mathbf{q}_w$ = wall shear stress, k = thermal conductivity of the fluid.

3 Results and Discussion

We have solved the governing non-linear Eqs. (11)–(13) with B.C. (14) numerically using Runge–Kutta inbuilt BVP4C method. This method is important for all researchers because the nonlinear ODE/PDE equations are solved easily and also

Table 1 Skin friction coefficient ($f''(0)$) for $a/c = 3$ when $\beta = 0$

a/c	Mahapatra and Gupta	Nazar et al.	Ishak et al.	Ali et al.	Present
0.1	−0.969 4	−0.969 4	−0.969 4	−0.969 4	−0.969 4
0.2	−0.918 1	−0.918 1	−0.918 1	−0.918 1	−0.918 1
0.5	−0.667 3	−0.667 3	−0.667 3	−0.667 3	−0.667 3
2.0	2.017 5	2.017 6	2.017 5	2.017 5	2.017 5
3.0	4.729 3	4.729 6	4.729 4	4.729 3	4.729 3

gives good result. The boundary layer edge is chosen from 5 to 10. The skin friction coefficient is numerically obtained and analyzed with the previously distributed outcomes that appeared in Table 1 which is the best agreement.

Table 2 for fixed $\beta = 1$ and $a/c = 3$, $Pr = 0.72$ and $Ec = 0.1$ the local Nusselt $\mathrm{Re}_X^{\frac{-1}{2}} Nu$ reduces when the radiation effect increases.

As shown in Tables 3 and 4, when a/c >1, the skin friction coefficient ($\mathrm{Re}_X^{\frac{1}{2}} C_f$) falls and Nusselt number ($\mathrm{Re}_X^{\frac{-1}{2}} Nu$) rises when magnetic parameter (β) increases, while it shows the reverse phenomenon for the skin friction coefficient but same aspect observed in Nusselt number for a/c <1. On the other hand, the skin friction coefficient and Nusselt number reduces when radiation parameter (*R*) enhances for both cases.

The impact of the magnetic parameter β on velocity distributions, induced magnetic, and temperature profiles is depicted in Figs. 2, 3, and 4. The velocity distribution falls when β increases due to Lorentz force and g' profiles is reduced by β. The opposite effect is detected for the temperature profiles.

Table 2 Variation of Nusselt number

R	0.07	0.5	2	6.8	10
$\mathrm{Re}_X^{\frac{-1}{2}} Nu$	0.7383	0.6463	0.4857	0.3302	0.2942

Table 3 Variation of skin friction coeffcient and Nusselt number, ($a/c = 3$ and $Pr = 0.72$)

B	*R*	*Ec*	$\mathrm{Re}_X^{\frac{1}{2}} C_f$	$\mathrm{Re}_X^{\frac{-1}{2}} Nu$
0.5	0.5	0.1	4.72780796	0.59923460
8	0.5	0.1	4.70548316	0.60025411
0.5	**1**	0.2	4.727807955	0.38489916
0.5	**8**	0.2	4.727807954	0.28866996
0.1	1	**0.1**	4.728987476	0.55230177
0.1	1	**0.4**	4.728987473	0.04976661

Table 4 Variation of skin friction coeffcient and Nusselt number, ($a/c = 0.5$ and $Pr = 0.72$)

B	R	Ec	$\mathrm{Re}_X^{\frac{1}{2}} C_f$	$\mathrm{Re}_X^{\frac{-1}{2}} Nu$
0.5	0.5	0.1	– 0.666090895031662	0.458540931698125
2	0.5	0.1	– 1.068095342713774	0.153202111292846
0.5	**1**	0.2	– 0.666090895262369	0.391958490996971
0.5	**8**	0.2	– 0.666090895271880	0.242216935156161
0.1	1	**0.1**	– 0.667029889038610	0.397521329770668
0.1	1	**0.4**	– 0.667029889057074	0.380385863177244

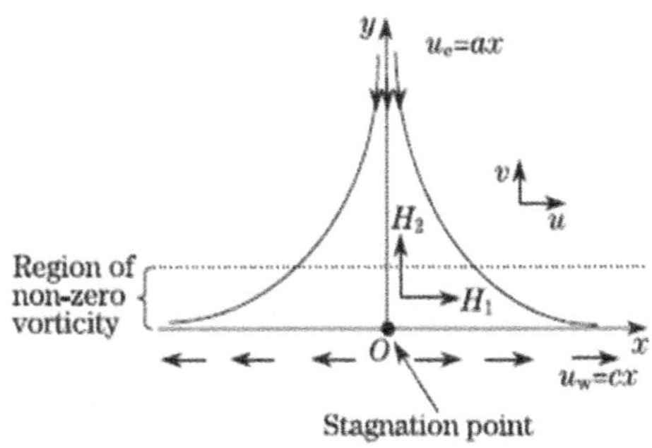

Fig. 1 Flow Geometry

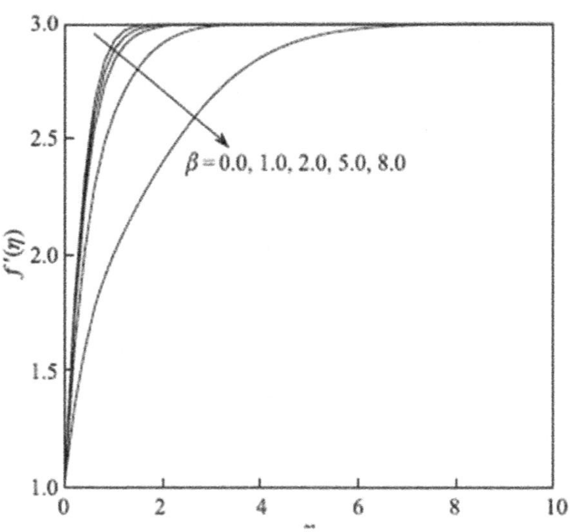

Fig. 2 Velocity profiles for magnetic field

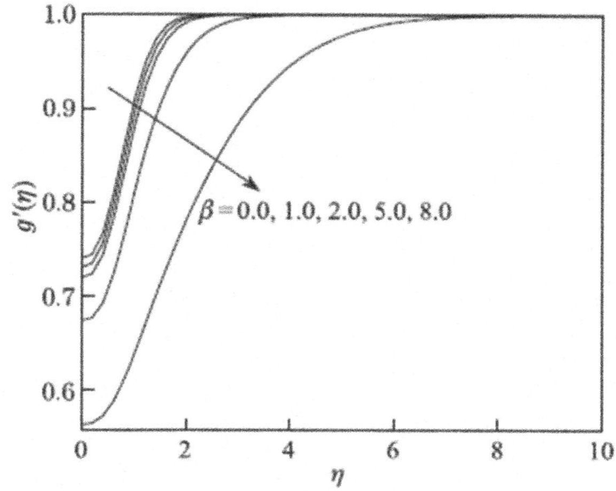

Fig. 3 g' profiles for magnetic field

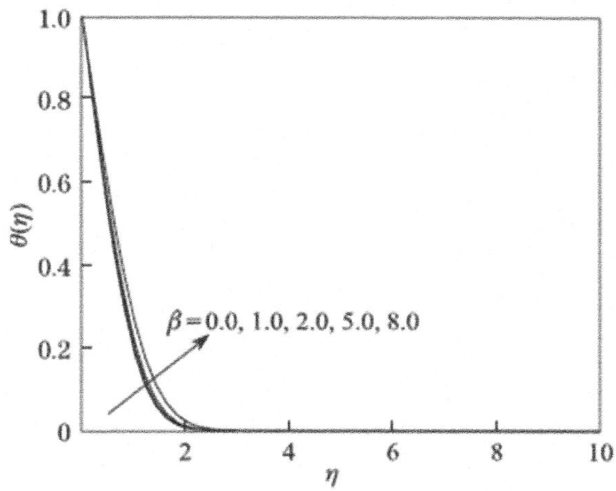

Fig. 4 Temperature profiles for magnetic field

For $a/c = 0.5$ and $Pr = 0.72$, the magnetic parameter β on the velocity, induced magnetic field, and temperature profiles are shown in Figs. 5, 6, and 7. The magnetic parameter accelerates velocity and g' and decelerates temperature profiles.

In Figs. 8, 9, and 10, we take $Pr = 0.72$ and $\beta = 0.1$ for various a/c. From Fig. 8, it is noticed that the stretching parameter (a/c) enhances the velocity distributions.

From Cowling [9], c means stretching of the sheet, when the values of an increase are compared with c implies that the straining motion is the rise and close to the stagnation region which accelerates the external stream. Hence, increase in a/c is to thin the boundary layer.

Fig. 5 Effect of magnetic field on velocity profiles

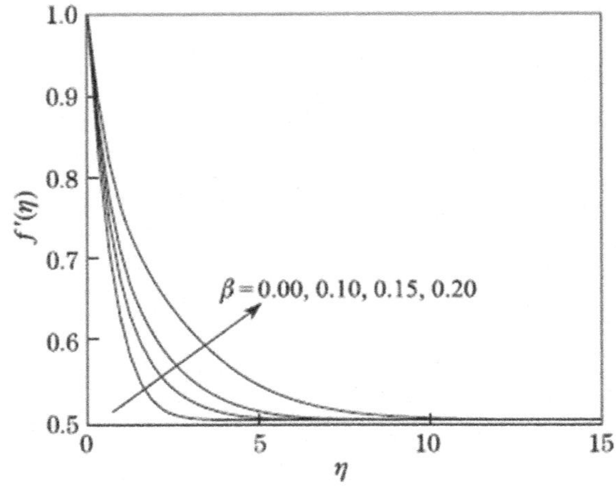

Fig. 6 Effect magnetic field on of induced magnetic(g') profiles

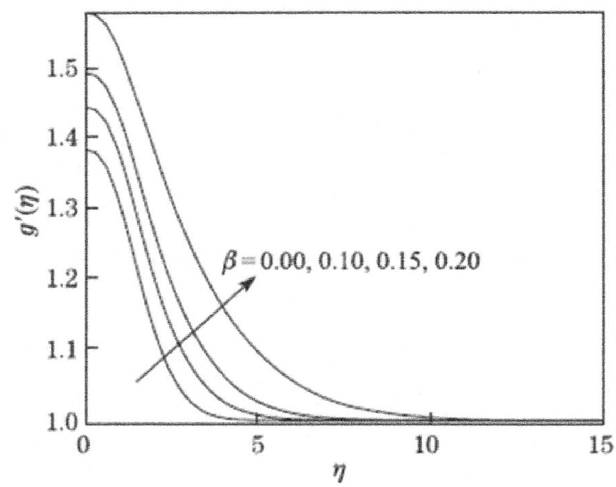

Figure 9 shows that the value of a/c reduces g' profiles but in case of *a/c* > 1, it enhances after a certain point. Further, from Fig. 10, it shows the effect of stretching parameter (a/c) on the temperature profile. Hence, we observed that a/c reduces thermal boundary layer thickness.

Figures 11 and 12 depicts tempterature profiles and the thermal boundary layer thickness enhance with an increase in R because heat to the fluid is increase with R for both $a/c = 3$ and $a/c = 0.5$.

Ratio of kinetic energy to enthalpy is called Eckert number. Eckert number rises Kinetic energy which causes the temperature is to accelerate as evident from Fig. 13.

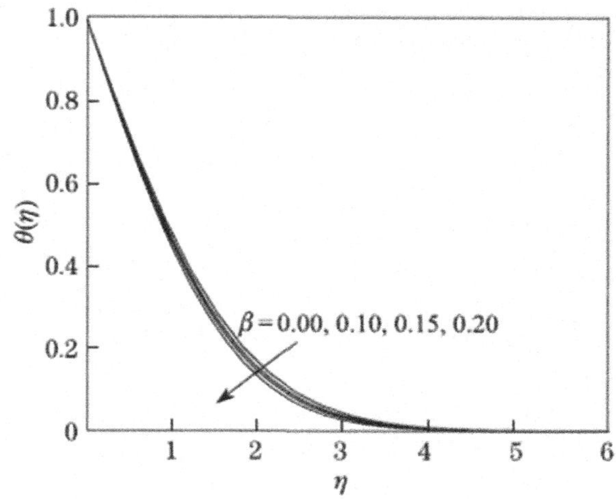

Fig. 7 Effect of magnetic field on temperature profiles

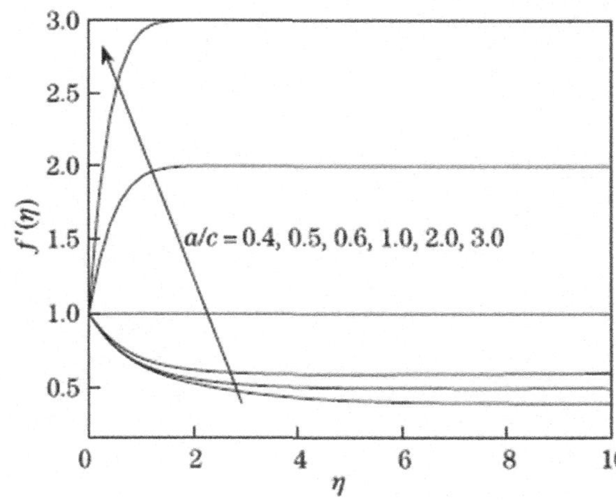

Fig. 8 Effect of *a/c* on velocity profiles

4 Conclusions

The above investigation brings out the following conclusions.

- The magnetic parameter β decelerates velocity, induced magnetic field, and accelerates temperature profiles.
- The magnetic parameter β reduces skin friction coefficients for a/c > 1 and the opposite sign is noticed for a/c < 1.
- The radiation parameter(R) enhances the thermal boundary layer thickness for both cases a/c < 1 and a/c < 1.
- The velocity distribution is enhanced when the stretching parameter increases.

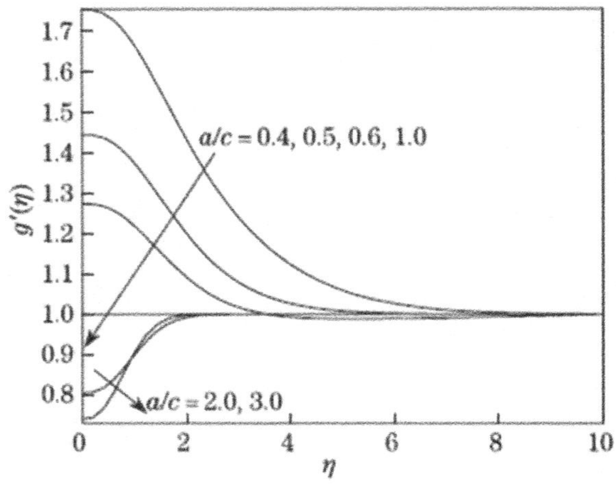

Fig. 9 Induced magnetic profiles for various values of *a/c*

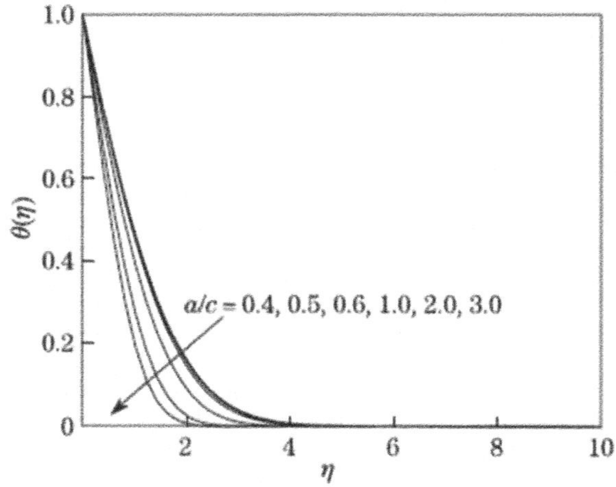

Fig. 10 Effect of *a/c* on temperature profiles

- The stretching parameter (a/c) reduces profiles but in the case of a/c > 1, it enhances after a certain point.
- In the case of the temperature profile, it is found that a/c reduces thermal boundary layer thickness.

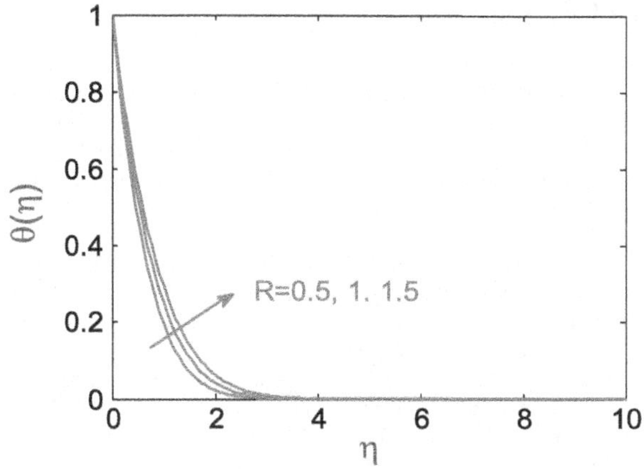

Fig. 11 Temperature profiles for several values of R when $a/c = 3.0$ and $\beta = 0.5$

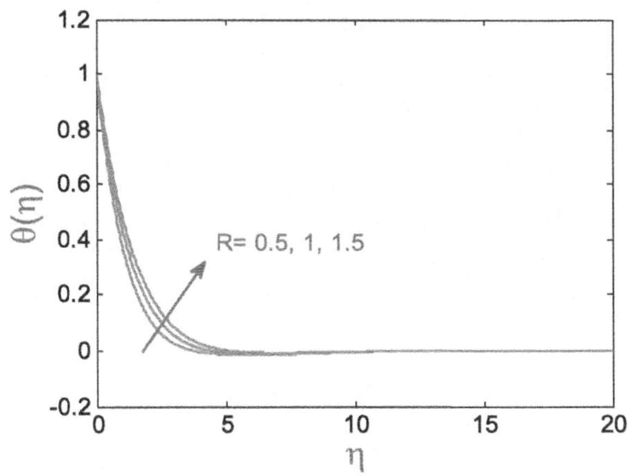

Fig. 12 Temperature profiles for several values of R when $a/c = 0.5$ and $\beta = 0.5$

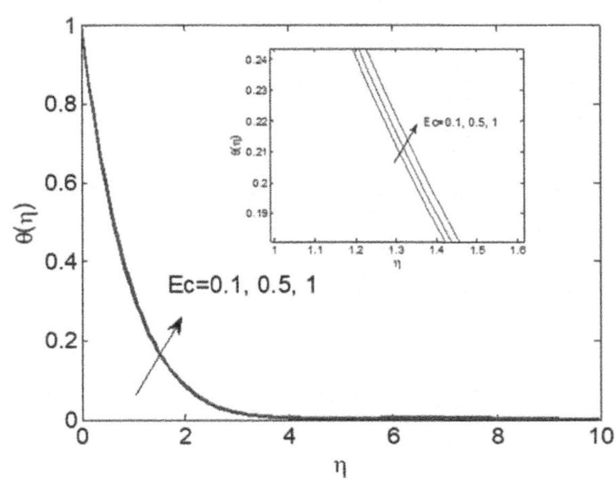

Fig. 13 Temperature profiles for several values of Ec

References

1. B. Jalilpour, S. Jafarmadar, M.M. Rashidi, D.D. Ganji, R. Rahime, A.B. Shotorban, MHD non-orthogonal stagnation point flow of a nanofluid towards astretching surface in the presence of thermal radiation. Ain Shams Engineering Journal **9**, 1671–1681 (2018)
2. Abbas, Z., Sheikh,M., Motsa, S.S. Numerical solution of binary chemical reaction on stagnation point flow of Casson fluid over a stretching/shrinking sheet with thermalradiation. Energy, 95 (2016) 12–20
3. J.C. Misra, A. Sinha, Effect of thermal radiation on MHD flow of blood and heat transfer in a permeable capillary in stretching motion. Heat Mass Transfer **49**, 617–628 (2013)
4. T.R. Mahapatra, A.S. Gupta, Heat tansfer in stagnation-point flow towards a stretching sheet. Heat Mass Transf. **38**, 517–521 (2002)
5. R. Nazar, N. Amin, D. Filip, I. Pop, Unsteady boundary layer flow in the region of the stagnation point on a stretching sheet. Int. J. Eng. Sci. **42**, 1241–1253 (2004)
6. A. Ishak, R. Nazar, I. Pop, Mixed convection boundary layers in the stagnation-point flow towards a stretching vertical sheet. Meccanica **41**, 509–518 (2006)
7. F.M. Ali, R. Nazar, N.M. Arifin, I. Pop, MHD boundary layer flow and heat transfer over a stretching sheet with induced magnetic field. Heat Mass Transfer **47**, 155–162 (2011)
8. El-Aziz, M. A., AfifyA. A. Influences of Slip Velocity and Induced Magnetic Field onMHD Stagnation-Point Flow and Heat Transfer of Casson Fluid over a Stretching Sheet. Mathematical Problems in Engineering, 9402836 (2018) 11, https://doi.org/10.1155/2018/9402836
9. F.M. Ali, R. Nazar, N.M. Arifin, I. Pop, MHD stagnation-point flow and heat transfer towards stretching sheet with induced magnetic field. Appl. Mathematics and Mechanics **32**(4), 409–418 (2011)
10. T.G. Cowling, *Magnetohydrodynamics* (Interscience Publication, New York, 1957)
11. J. Hartmann, Hg-dynamics I, theory of the laminar flow of an electrically conducting liquid in a homogenous magnetic field. Matematisk-Fysiske Meddelelser **15**(6), 1–28 (1937)

Various Approximate Multiplicative Inverse Lie $\star$-Derivations

B. V. Senthil Kumar, Khalifa Al-Shaqsi, and Hemen Dutta

Abstract In this investigation, various approximate multiplicative inverse Lie $\star$-derivations are determined in the framework of normed $\star$-algebras pertinent to Ulam–Hyers stability theory. These results are applied to achieve other classical stabilities by taking different upper bounds. The results obtained are compared with concluding remarks.

Keywords Reciprocal functional equation · Lie derivation · Lie $\star$-derivation · Generalized Ulam–Hyers stability

2010 Mathematics Subject Classification 39B52 · 39B62 · 39B82

1 Introduction

The foundation for the growth of research in solving stability problems of mathematical equations is laid down through an excellent question posed in [31]. It can be expressed as "Let a mapping 'a' satisfies the equation $a(u + v) = a(u) + a(v)$. Is it possible to exist an approximate solution near to the exact solution of this equation?" In other words, more precisely, this question is mathematically formulated as "Let X_1 be a group and X_2 be a metric group with a metric d equipped on X_2. Let $\epsilon > 0$ be given. Then does there exist a $\delta > 0$ such that if a mapping $a : X_1 \longrightarrow X_2$ satisfies $d(a(uv), a(u)a(v)) < \delta$ for all $u, v \in X_1$, then a homomorphism $A : X_1 \longrightarrow X_2$

B. V. Senthil Kumar (✉) · K. Al-Shaqsi
Department of Information Technology, University of Technology and Applied Sciences, Nizwa 611, Oman
e-mail: senthilkumar@nct.edu.om

K. Al-Shaqsi
e-mail: khalifa.alshaqsi@nct.edu.om

H. Dutta
Department of Mathematics, Gauhati University, Guwahati 781014, Assam, India
e-mail: hemen_dutta08@rediffmail.com

S. K. Paikray et al. (eds.), *New Trends in Applied Analysis and Computational Mathematics*, Advances in Intelligent Systems and Computing 1356,
https://doi.org/10.1007/978-981-16-1402-6_11

exists and satisfies the approximation $d(a(u), A(u)) < \epsilon$ for all $u \in X_1$?" If there is a substantial solution to this problem, then the stability of the multiplicative functional equation or homomorphism $a(uv) = a(u)a(v)$ holds. This question is answered by various mathematicians in different forms. For detailed literature about this research field, one can refer to [9, 10, 17, 18, 22]. Later on, stability problems of various functional equations are solved in numerous published papers. More information regarding stability problems are available in [1, 2, 8, 13, 15, 19].

The stability problems pertaining to quadratic $\star$-derivations in the setting of Banach $C^\star$-algebra are available in [11]. The stability properties using fixed point method are discussed in [16, 32]. The basic notions of cubic Lie derivations and their stability problems are obtained in [7]. The stabilities connected with various Lie derivations such as quadratic and quartic are studied respectively in [12, 20].

Quite recently, many interesting stability results of multiplicative inverse functional equations are dealt in [3, 6, 14, 21, 23–25, 27]. The applications and interpretations of these type of multiplicative inverse functional equations are brought out in [4, 5, 26, 28–30].

These recent novels and remarkable results stimulated us to investigate the stabilities of various multiplicative inverse Lie $\star$-derivations through direct method. In this entire investigation, unless or otherwise specified, let us assume that E to be a complex normed $\star$-algebra and F to be a Banach E-module with norms represented by $\|\cdot\|$. Also, we use the following set defined as

$$\Gamma = \{\eta \in \mathbb{C} : |\eta| = 1\}$$

in the main results of this investigation.

We consider the following equations to determine approximate (additive, quadratic, cubic, quartic) Lie $\star$-derivations:

$$h(2\alpha + \beta) + h(2\alpha - \beta) = \frac{4h(\alpha)h(\beta)^2}{4h(\beta)^2 - h(\alpha)^2}, \tag{1.1}$$

where $h : E \longrightarrow F$ is a mapping with $h(\alpha) \neq 0, 4h(\beta)^2 - h(\alpha)^2 \neq 0$, for all $\alpha, \beta \in E$,

$$h(2\alpha + \beta) + h(2\alpha - \beta) = \frac{2h(\alpha)h(\beta)[h(\alpha) + 4h(\beta)]}{[4h(\beta) - h(\alpha)]^2}, \tag{1.2}$$

where $h : E \longrightarrow F$ is a mapping with $h(\alpha) \neq 0, 4h(\beta) - h(\alpha) \neq 0$, for all $\alpha, \beta \in E$,

$$h(2\alpha + \beta) + h(2\alpha - \beta) = \frac{4h(\alpha)h(\beta)\left[4h(\beta) + 3h(\alpha)^{2/3}h(\beta)^{1/3}\right]}{\left[4h(\beta)^{2/3} - h(\alpha)^{2/3}\right]^3}, \tag{1.3}$$

where $h : E \longrightarrow F$ is a mapping with $h(\alpha) \neq 0$, $4h(\beta)^{2/3} - h(\alpha)^{2/3} \neq 0$, for all $\alpha, \beta \in E$ and

$$h(2\alpha+\beta)+h(2\alpha-\beta)=\frac{2h(\alpha)h(\beta)\left[h(\alpha)+16h(\beta)+24h(\alpha)^{1/2}h(\beta)^{1/2}\right]}{\left[4h(\beta)^{1/2}-h(\alpha)^{1/2}\right]^4}, \tag{1.4}$$

where $h : E \longrightarrow F$ is a mapping with $h(\alpha) \neq 0$, $4h(\beta)^{1/2} - h(\alpha)^{1/2} \neq 0$, for all $\alpha, \beta \in E$. The multiplicative inverse functions $h(\alpha) = \frac{1}{\alpha}$, $h(\alpha) = \frac{1}{\alpha^2}$, $h(\alpha) = \frac{1}{\alpha^3}$ and $h(\alpha) = \frac{1}{\alpha^4}$ are solution of Eqs. (1.1), (1.2), (1.3) and (1.4), respectively.

2 Stabilities of Multiplicative Inverse Lie ⋆-Derivations

In this section, we investigate the stabilities of multiplicative inverse additive Lie ⋆-derivations. Let us impose the ensuing definitions which are used for our investigation.

Definition 2.1 A mapping $h : E \longrightarrow F$ is called a multiplicative inverse homogenous mapping if $h(k\alpha) = \frac{1}{k}h(\alpha)$, for all $\alpha \in E$ and $k \in \mathbb{C}$.

Definition 2.2 A multiplicative inverse homogenous mapping $h : E \longrightarrow F$ is called a multiplicative inverse derivation if

$$h(\alpha\beta) = h(\alpha)\frac{1}{\beta} + \frac{1}{\alpha}h(\beta)$$

for all $\alpha, \beta \in E$.

Definition 2.3 A multiplicative inverse homogenous mapping h is called a multiplicative inverse Lie derivation if

$$h([\alpha,\beta]) = \left[h(\alpha), \frac{1}{\beta}\right] + \left[\frac{1}{\alpha}, h(\beta)\right]$$

for all $\alpha, \beta \in E$, where $[\alpha, \beta] = \alpha\beta - \beta\alpha$.

Definition 2.4 A multiplicative inverse Lie derivation h is called a multiplicative inverse Lie ⋆-derivation if $h(\alpha^\star) = h(\alpha)^\star$ for all $\alpha \in E$.

For the sake of obtaining the results in a simple form, let us consider for a mapping $h : E \longrightarrow F$, the following operators:

$$\Delta_k h(\alpha,\beta) = h(2k\alpha+k\beta) + h(2k\alpha-k\beta) - \frac{4h(k\alpha)h(k\beta)^2}{4h(k\beta)^2 - h(k\alpha)^2}$$

and

$$\Delta h(\alpha,\beta) = h([\alpha,\beta]) - \left[h(\alpha), \frac{1}{\beta}\right] - \left[\frac{1}{\alpha}, h(\beta)\right]$$

for all $\alpha, \beta \in E$, $k \in \mathbb{C}$.

Theorem 2.5 *Let $\rho = \pm 1$. Assume that there exists a function $\phi : E \longrightarrow [0, \infty)$ with a mapping $h : E \longrightarrow F$ such that*

$$\Phi(\alpha, \beta) = \sum_{j=0}^{\infty} 3^{j\left(\frac{\rho+1}{2}\right)} \phi\left(3^{j\left(\frac{\rho-1}{2}\right)}\alpha, 3^{j\left(\frac{\rho-1}{2}\right)}\beta\right) < \infty, \tag{2.1}$$

$$\|\Delta_k h(\alpha, \beta)\| \leq \phi(\alpha, \beta), \tag{2.2}$$

$$\|\Delta h(\alpha, \beta)\| \leq \phi(\alpha, \beta) \tag{2.3}$$

for all $\alpha, \beta \in E$ and $k \in \Gamma$. If the mapping $r \mapsto h(r\alpha)$ from $\mathbb{R}$ to F is continuous, then there exists a unique multiplicative inverse Lie derivation $H : E \longrightarrow F$ such that

$$\|h(\alpha) - H(\alpha)\| \leq \Phi(\alpha, \alpha) \tag{2.4}$$

for all $\alpha \in E$.

Proof First, we prove the case for $\rho = 1$. Considering $k = 1$ and $\beta = \alpha$ in the inequality (2.2), we obtain

$$\|3h(3\alpha) - h(\alpha)\| \leq 3\phi(\alpha, \alpha) \tag{2.5}$$

for all $\alpha \in E$. Proceeding with the induction for a positive integer m, we find that

$$\left\|3^t h\left(3^t\alpha\right) - 3^s h\left(3^s\alpha\right)\right\| \leq 3\sum_{j=s}^{t-1} 3^j \phi\left(3^j\alpha, 3^j\alpha\right) \tag{2.6}$$

for all $\alpha \in E$ and all $t > s \geq 0$. The inequality (1.2) implies that the sequence $\{3^m h\left(3^m\alpha\right)\}$ to be Cauchy and since Y is complete, we notice that it is convergent. This implies to define a mapping $H : E \longrightarrow F$ as

$$H(\alpha) = \lim_{m\to\infty} 3^m h\left(3^\alpha\right) \tag{2.7}$$

where $\alpha \in E$. By choosing $t = m > 0$, $us = 0$ in (2.6), we arrive at

$$\left\|3^m h\left(3^m\alpha\right) - h(\alpha)\right\| \leq 3\sum_{j=0}^{m-1} 3^j \phi\left(3^j\alpha, 3^j\alpha\right) \tag{2.8}$$

for all $\alpha \in E$. By taking $n \to \infty$, we find that (2.4) holds. Now,

$$\|\Delta_k h(\alpha, \beta)\| = \lim_{m\to\infty} \left\|3^m \Delta_k h(\alpha, \beta)\right\| \leq \lim_{m\to\infty} 3^m \phi\left(3^m\alpha, 3^m\alpha\right) = 0 \tag{2.9}$$

for all $\alpha, \beta \in E$ and $k \in \Gamma$. Taking $k = 1$ in (2.9), we obtain that H is a multiplicative inverse mapping.

From (2.9), it follows that $\Delta_k H(\alpha, \beta) = 0$. Hence, $H(3k\alpha) = \frac{1}{3k}H(\alpha)$, for all $\alpha \in E, k \in \Gamma$.

Let $\alpha_0 \in E$ be a fixed element. Now, we claim that $H(r\alpha_0) = \frac{1}{r}H(\alpha_0)$, for all $r(\neq 0) \in \mathbb{R}$. Let $\gamma \in F^\star$ be a linear functional, where $F^\star$ represents the dual space of F. From this, a function $\psi : \mathbb{R} - \{0\} \longrightarrow \mathbb{R}$ can be defined as

$$\psi(r) = \gamma(H(r\alpha_0)) \tag{2.10}$$

where $r(\neq 0) \in \mathbb{R}$. It can be easily verified that ψ is multiplicative inverse function. Now, define

$$\psi_m(r) = \gamma\left(3^m h\left(3^m r\alpha_0\right)\right) \tag{2.11}$$

for $m \geq 0$. It is easy to note that ψ is a pointwise limit of a sequence of continuous functions ψ_m, $m \geq 0$. Hence, ψ is a continuous multiplicative inverse function and

$$\psi(r) = \frac{1}{r}\psi(1) \tag{2.12}$$

where $r(\neq 0) \in \mathbb{R}$. On the other side,

$$\gamma(H(r\alpha_0)) = \psi(r) = \frac{1}{r}\psi(1) = \frac{1}{r}\gamma(H(\alpha_0)) = \gamma\left(\frac{1}{r}H(\alpha_0)\right) \tag{2.13}$$

where $r(\neq 0) \in \mathbb{R}$. Since the linear functional $\gamma \in F^\star$ is arbitrary, we conclude that

$$H(r\alpha_0) = \frac{1}{r}H(\alpha_0) \tag{2.14}$$

where $r(\neq 0) \in \mathbb{R}$. Now, suppose $u \in \mathbb{C}$. Then it is clear that $\frac{u}{|u|} \in \Gamma$, $\frac{|u|}{3} \in \mathbb{R}$, and hence, we have

$$\begin{aligned} H(u\alpha_0) = H\left(3\frac{u}{|u|}\left(\frac{|u|}{3}\alpha_0\right)\right) = \frac{1}{3}\left(\frac{|u|}{u}\right)H\left(\frac{|u|}{3}\alpha_0\right) &= \frac{1}{3}\frac{|u|}{u}\frac{3}{|u|}H(\alpha_0) \\ &= \frac{1}{u}H(\alpha_0). \end{aligned} \tag{2.15}$$

Since $\alpha_0 \in E$, this implies that H is multiplicative inverse homogenous. Next, let us prove that H is a multiplicative inverse Lie derivation. Using (2.3), we find that

$$\|\Delta H(\alpha, \beta)\| = \lim_{m\to\infty} 3^m \left\|\Delta h(3^m\alpha, 3^m\beta)\right\| \leq \lim_{m\to\infty} 3^m \phi(3^m\alpha, 3^m\alpha) = 0 \tag{2.16}$$

for all $\alpha, \beta \in E$. Using multiplicative inverse homogenity of H along with (2.16), we conclude that g is a multiplicative inverse Lie derivation. Finally, we assert the uniqueness of H. Assume that there exists another multiplicative inverse Lie derivation $H' : E \longrightarrow F$ which satisfies (2.4). Then

$$\begin{aligned}\left\|H(\alpha) - H'(\alpha)\right\| &= \lim_{m\to\infty} 3^m \left\|H(3^m\alpha) - H'(3^m\alpha)\right\| \\ &\le \lim_{m\to\infty} 2\cdot 3^m \Phi(3^m\alpha, 3^m\alpha) \\ &= \lim_{m\to\infty} 2\sum_{j=0}^{\infty} 3^{m+j}\phi\left(3^{m+j}\alpha, 3^{m+j}\alpha\right) \\ &= \lim_{m\to\infty} 2\sum_{p=m}^{\infty} 3^{p}\phi\left(3^{p}\alpha, 3^{p}\alpha\right) = 0, \end{aligned} \tag{2.17}$$

which implies that $H(\alpha) = H'(\alpha)$, for all $\alpha \in E$ and hence H is unique. The result for the case $\rho = -1$ can be obtained with the similar arguments as in the case $\rho = 1$ and this completes the proof. □

In the following corollaries, let $h : E \longrightarrow F$ be a mapping. Also, let us assume the mapping $r \mapsto h(r\alpha)$ from $\mathbb{R} \longrightarrow F$ is continuous for each fixed $\alpha \in E$. The following corollary is a direct application of Theorem 2.5 when $\rho = -1$ and the result presented is associated with the stability involving a positive constant $\frac{2}{3}\mu$ as an upper bound connected with Hyers' result.

Corollary 2.6 *Suppose h satisfies the inequalities (2.2) and (2.3) with $\phi(\alpha, \beta) = \frac{2}{3}\mu$, where $\mu \ge 0$, for all $\alpha, \beta \in E$. Then there exists a unique multiplicative inverse Lie derivation $H : E \longrightarrow F$ such that*

$$\|h(\alpha) - H(\alpha)\| \le \mu$$

for all $\alpha \in E$.

In the sequel, using the result of Theorem 2.5, we obtain the results pertaining to various stabilities of T. Rassias (upper bound with sum of powers of norms) and J. Rassias (upper bound with product of powers of norms) and J. Rassias (upper bound with mixed product-sum of powers of norms).

Corollary 2.7 *Suppose h satisfies the inequalities (2.2) and (2.3) with $\phi(\alpha, \beta) = \lambda_1\left(\|\alpha\|^q + \|\beta\|^q\right)$, where $\lambda_1 \ge 0$, $q \ne -1$, for all $\alpha, \beta \in E$. Then there exists a unique multiplicative inverse Lie derivation $H : E \longrightarrow Y$ such that*

$$\|h(\alpha) - H(\alpha)\| \le \begin{cases} \frac{6\lambda_1}{1-3^{q+1}}\|\alpha\|^q, & \text{for } q < -1 \\ \frac{6\lambda_1\cdot 3^q}{3^{q+1}-1}\|\alpha\|^q, & \text{for } q > -1 \end{cases}$$

for all $\alpha \in E$.

Corollary 2.8 *Suppose h satisfies the inequalities (2.2) and (2.3) with $\phi(\alpha, \beta) = \lambda_2\|\alpha\|^{q/2}\|\beta\|^{q/2}$, where $\lambda_2 \ge 0, q \ne -1$, for all $\alpha, \beta \in E$. Then there exists a unique multiplicative inverse Lie derivation $H : E \longrightarrow F$ such that*

$$\|h(\alpha) - H(\alpha)\| \leq \begin{cases} \frac{3\lambda_2}{1-3^{q+1}} \|\alpha\|^q, & for\ q < -1 \\ \frac{3\lambda_2 \cdot 3^q}{3^{q+1}-1} \|\alpha\|^q, & for\ q > -1 \end{cases}$$

for all $\alpha \in E$.

Corollary 2.9 *Suppose h satisfies the inequalities (2.2) and (2.3) with* $\phi(\alpha, \beta) = \lambda_3 \left(\|\alpha\|^{q/2} \|\beta\|^{q/2} + (\|\alpha\|^q + \|\alpha\|^q)\right)$, *where* $\lambda_3 \geq 0$, $q \neq -1$, *for all* $\alpha, \beta \in E$. *Then there exists a unique multiplicative inverse Lie derivation* $H : E \longrightarrow F$ *such that*

$$\|h(\alpha) - H(\alpha)\| \leq \begin{cases} \frac{9\lambda_3}{1-3^{q+1}} \|\alpha\|^q, & for\ q < -1 \\ \frac{9\lambda_3 \cdot 3^q}{3^{q+1}-1} \|\alpha\|^q, & for\ q > -1 \end{cases}$$

for all $\alpha \in E$.

3 Stabilities of Multiplicative Inverse Quadratic Lie ⋆-Derivations

In this section, we determine various stabilities of the multiplicative inverse quadratic Lie ⋆-derivations. We impose the following basic definitions connected with ⋆-normed algebras which will be useful to prove the main results.

Definition 3.1 A mapping $h : E \longrightarrow F$ is called a multiplicative inverse quadratic homogenous mapping if $h(k\alpha) = \frac{1}{k^2}h(\alpha)$, for all $\alpha \in E$ and $k \in \mathbb{C}$.

Definition 3.2 If $h : E \longrightarrow F$ is a multiplicative inverse quadratic homogenous mapping, then it is called a multiplicative inverse quadratic derivation if

$$h(\alpha\beta) = h(\alpha)\frac{1}{\beta^2} + \frac{1}{\alpha^2}h(\beta)$$

for all $\alpha, \beta \in E$.

Definition 3.3 A multiplicative inverse quadratic homogenous mapping h is called a multiplicative inverse quadratic Lie derivation if

$$h([\alpha, \beta]) = \left[h(\alpha), \frac{1}{\beta^2}\right] + \left[\frac{1}{\alpha^2}, h(\beta)\right]$$

for all $\alpha, \beta \in E$, where $[\alpha, \beta] = \alpha\beta - \beta\alpha$.

Definition 3.4 A multiplicative inverse quadratic Lie derivation h is called a multiplicative inverse quadratic Lie ⋆-derivation if $h(\alpha^\star) = h(\alpha)^\star$ for all $\alpha \in E$.

We define the following difference operators to arrive at the results in an easy manner. For a mapping $h : E \longrightarrow F$, let the difference operators be as follows:

$$\Delta_k h(\alpha,\beta) = h(2k\alpha + k\beta) + h(2k\alpha - k\beta) - \frac{2h(k\alpha)h(k\beta)[h(k\alpha)+4h(k\beta)]}{[4h(k\beta)-h(k\alpha)]^2}$$

and

$$\Delta h(\alpha,\beta) = h(\alpha,\beta]) - \left[h(\alpha), \frac{1}{\beta^2}\right] - \left[\frac{1}{\alpha^2}, h(\beta)\right]$$

for all $\alpha, \beta \in E, k \in \mathbb{C}$.

Theorem 3.5 *Let $\rho = \pm 1$. Assume that a function $\phi : E \longrightarrow [0,\infty)$ exists with a mapping $h : E \longrightarrow F$ such that*

$$\Phi(\alpha,\beta) = \sum_{j=0}^{\infty} 9^{i(\frac{\rho+1}{2})}\phi\left(3^{j(\frac{\rho-1}{2})}\alpha, 3^{j(\frac{\rho-1}{2})}\beta\right) < \infty, \tag{3.1}$$

$$\|\Delta_k h(\alpha,\beta)\| \le \phi(\alpha,\beta), \tag{3.2}$$

$$\|\Delta h(\alpha,\beta)\| \le \phi(\alpha,\beta) \tag{3.3}$$

for all $\alpha, \beta \in E$ and $k \in \Gamma$. If the mapping $r \mapsto h(r\alpha)$ from $\mathbb{R}$ to F is continuous, then a unique multiplicative inverse quadratic Lie derivation $H : E \longrightarrow F$ exists such that

$$\|h(\alpha) - H(\alpha)\| \le \Phi(\alpha,\alpha) \tag{3.4}$$

for all $\alpha \in E$.

Proof This theorem is also first proved for the case $\rho = 1$. Considering $k = 1$ and $\beta = \alpha$ in the inequality (3.2), we find that

$$\|9h(3\alpha) - h(\alpha)\| \le 9\phi(\alpha,\alpha) \tag{3.5}$$

for all $\alpha \in E$. By the technique of induction, we obtain the following inequality:

$$\left\|9^t h\left(3^t\alpha\right) - 9^s h\left(3^s\alpha\right)\right\| \le 9\sum_{j=s}^{t-1} 9^j \phi\left(3^j\alpha, 3^j\alpha\right) \tag{3.6}$$

for all $\alpha \in E$ and all $t > s \ge 0$. By the application of the condition (3.1), we find that the sequence $\{9^m h\left(3^m\alpha\right)\}$ is Cauchy and since F is complete, this sequence is convergent. Therefore, we can define a mapping $H : E \longrightarrow F$ by

$$H(\alpha) = \lim_{m\to\infty} 9^m h\left(3^m\alpha\right) \tag{3.7}$$

where $\alpha \in E$. Assuming $t = m > 0$, $s = 0$ in (3.6), we get

$$\left\|9^m h\left(3^m\alpha\right) - h(\alpha)\right\| \le 9\sum_{j=0}^{m-1} 9^j \phi\left(3^i\alpha, 3^j\alpha\right) \tag{3.8}$$

for all $\alpha \in E$. Taking the $m \to \infty$, we find that the inequality (3.4) holds. The rest of the proof completes with the akin arguments as in Theorem 2.5. □

In the following corollaries, let $h : E \longrightarrow F$ be a mapping. Also, the mapping $r \mapsto h(r\alpha)$ from $\mathbb{R}$ to F is continuous, for each fixed $\alpha \in E$. Using Theorem 3.5 when $\rho = -1$, we obtain the following result.

Corollary 3.6 *Suppose h satisfies the inequalities (3.2) and (3.3) with $\phi(\alpha, \beta) = \frac{8}{9}\mu$, where $\mu \geq 0$, for all $\alpha, \beta \in E$. Then there exists a unique multiplicative inverse quadratic Lie derivation $H : E \longrightarrow F$ such that*

$$\|h(\alpha) - H(\alpha)\| \leq \mu$$

for all $\alpha \in E$.

In the following corollaries, by applying the result of Theorem 3.5, we present the stabiliies with different upper bounds in the inequalities (3.2) and (3.3) connected with other classical stabilities.

Corollary 3.7 *Suppose h satisfies the inequalities (3.2) and (3.3) with $\phi(\alpha, \beta) = \lambda_1 \left(\|\alpha\|^q + \|\beta\|^q\right)$, where $\lambda_1 \geq 0$, $q \neq -2$, for all $\alpha, \beta \in E$. Then there exists a unique multiplicative inverse quadratic Lie derivation $H : E \longrightarrow F$ such that*

$$\|h(\alpha) - H(\alpha)\| \leq \begin{cases} \frac{18\lambda_1}{1-3^{q+2}} \|\alpha\|^q, & \text{for } q < -2 \\ \frac{18\lambda_1 \cdot 3^q}{3^{q+2}-1} \|\alpha\|^q, & \text{for } q > -2 \end{cases}$$

for all $\alpha \in E$.

Corollary 3.8 *Suppose h satisfies the inequalities (3.2) and (3.3) with $\phi(\alpha, \beta) = \lambda_2 \|\alpha\|^{q/2} \|\beta\|^{q/2}$, where $\lambda_2 \geq 0$, $q \neq -2$, for all $\alpha, \beta \in E$. Then there exists a unique multiplicative inverse quadratic Lie derivation $H : E \longrightarrow F$ such that*

$$\|h(\alpha) - H(\alpha)\| \leq \begin{cases} \frac{9\lambda_2}{1-3^{q+2}} \|\alpha\|^q, & \text{for } q < -2 \\ \frac{9\lambda_2 \cdot 3^q}{3^{q+2}-1} \|\alpha\|^q, & \text{for } q > -2 \end{cases}$$

for all $\alpha \in E$.

Corollary 3.9 *Suppose h satisfies the inequalities (3.2) and (3.3) with $\phi(\alpha, \beta) = \lambda_3 \left(\|\alpha\|^{q/2} \|\beta\|^{q/2} + \left(\|\alpha\|^q + \|\beta\|^q\right)\right)$, where $\lambda_3 \geq 0$, $q \neq -2$, for all $\alpha, \beta \in E$. Then there exists a unique multiplicative inverse quadratic Lie derivation $H : E \longrightarrow F$ such that*

$$\|h(\alpha) - H(\alpha)\| \leq \begin{cases} \frac{27\lambda_3}{1-3^{q+2}} \|\alpha\|^q, & \text{for } q < -2 \\ \frac{27\lambda_3 \cdot 3^q}{3^{q+2}-1} \|\alpha\|^q, & \text{for } q > -2 \end{cases}$$

for all $\alpha \in E$.

4 Stabilities of Multiplicative Inverse Cubic Lie ⋆-Derivations

In this section, we prove the classical stability results of the multiplicative inverse cubic Lie ⋆-derivations. We use the following definitions related to ⋆-normed algebras which will be useful to prove the main results.

Definition 4.1 A mapping $h : E \longrightarrow F$ is called a multiplicative inverse cubic homogenous mapping if $h(k\alpha) = \frac{1}{k^3}h(\alpha)$, for all $\alpha \in E$ and $k \in \mathbb{C}$.

Definition 4.2 If $h : E \longrightarrow F$ is a multiplicative inverse cubic homogenous mapping, then it is called a multiplicative inverse cubic derivation if

$$h(\alpha\beta) = h(\alpha)\frac{1}{\beta^3} + \frac{1}{\alpha^3}h(\beta)$$

for all $\alpha, \beta \in E$.

Definition 4.3 A multiplicative inverse cubic homogenous mapping h is called a multiplicative inverse cubic Lie derivation if

$$h\left([\alpha, \beta]\right) = \left[h(\alpha), \frac{1}{\beta^3}\right] + \left[\frac{1}{\alpha^3}, h(\beta)\right]$$

for all $\alpha, \beta \in E$, where $[\alpha, \beta] = \alpha\beta - \beta\alpha$.

Definition 4.4 A multiplicative inverse cubic Lie derivation h is called a multiplicative inverse cubic Lie ⋆-derivation if $h(\alpha^\star) = h(\alpha)^\star$ for all $\alpha \in E$.

The following difference operators are defined to prove the results in a simpler way. For a mapping $h : E \longrightarrow F$, let the difference operators be defined as follows:

$$\Delta_k h(\alpha, \beta) = h(2k\alpha + k\beta) + h(2k\alpha - k\beta) - \frac{4h(k\alpha)h(k\beta)\left[4h(k\beta) + 3h(k\alpha)^{2/3}h(k\beta)^{2/3}\right]}{\left[4h(k\beta)^{2/3} - h(k\alpha)^{2/3}\right]^3}$$

and

$$\Delta h(\alpha, \beta) = h(\alpha, \beta]) - \left[h(\alpha), \frac{1}{\beta^3}\right] - \left[\frac{1}{\alpha^3}, h(\beta)\right]$$

for all $\alpha, \beta \in E, k \in \mathbb{C}$.

Theorem 4.5 *Let $\rho = \pm 1$. Assume that a function $\phi : E \longrightarrow [0, \infty)$ exists with a mapping $h : E \longrightarrow F$ such that*

$$\Phi(\alpha, \beta) = \sum_{j=0}^{\infty} 27^{i\left(\frac{\rho+1}{2}\right)}\phi\left(3^{j\left(\frac{\rho-1}{2}\right)}\alpha, 3^{j\left(\frac{\rho-1}{2}\right)}\beta\right) < \infty, \tag{4.1}$$

$$\|\Delta_k h(\alpha, \beta)\| \leq \phi(\alpha, \beta), \tag{4.2}$$

$$\|\Delta h(\alpha, \beta)\| \leq \phi(\alpha, \beta) \tag{4.3}$$

for all $\alpha, \beta \in E$ and $k \in \Gamma$. If the mapping $r \mapsto h(r\alpha)$ from $\mathbb{R}$ to F is continuous, then a unique multiplicative inverse cubic Lie derivation $H : E \longrightarrow F$ exists such that

$$\|h(\alpha) - H(\alpha)\| \leq \Phi(\alpha, \alpha) \tag{4.4}$$

for all $\alpha \in E$.

Proof This theorem is also first proved for the case $\rho = 1$. Considering $k = 1$ and $\beta = \alpha$ in the inequality (4.2), we find that

$$\|27h(3\alpha) - h(\alpha)\| \leq 27\phi(\alpha, \alpha) \tag{4.5}$$

for all $\alpha \in E$. Using induction method, we find that

$$\left\|27^t h\left(3^t\alpha\right) - 27^s h\left(3^s\alpha\right)\right\| \leq 27\sum_{j=s}^{t-1} 27^j \phi\left(3^j\alpha, 3^j\alpha\right) \tag{4.6}$$

for all $\alpha \in E$ and all $t > s \geq 0$. By condition (4.1), we find that the sequence $\{27^m h\left(3^m\alpha\right)\}$ is Cauchy and since F is complete, this sequence is convergent. Therefore, we can define a mapping $H : E \longrightarrow F$ by

$$H(\alpha) = \lim_{m\to\infty} 27^m h\left(3^m\alpha\right) \tag{4.7}$$

where $\alpha \in E$. Assuming $t = m > 0$, $s = 0$ in (4.6), we get

$$\left\|27^m h\left(3^m\alpha\right) - h(\alpha)\right\| \leq 27\sum_{j=0}^{m-1} 27^j \phi\left(3^i\alpha, 3^j\alpha\right) \tag{4.8}$$

for all $\alpha \in E$. Taking the $m \to \infty$, we find that the inequality (4.4) holds. The rest of the proof completes with the akin arguments as in Theorem 2.5. □

In the following corollaries, let $h : E \longrightarrow F$ be a mapping. Also, the mapping $r \mapsto h(r\alpha)$ from $\mathbb{R}$ to F is continuous, for each fixed $\alpha \in E$. Using Theorem 4.5 when $\rho = -1$, we obtain the following result.

Corollary 4.6 *Suppose h satisfies the inequalities (4.2) and (4.3) with $\phi(\alpha, \beta) = \frac{26}{27}\mu$, where $\mu \geq 0$, for all $\alpha, \beta \in E$. Then there exists a unique multiplicative inverse cubic Lie derivation $H : E \longrightarrow F$ such that*

$$\|h(\alpha) - H(\alpha)\| \leq \mu$$

for all $\alpha \in E$.

In the following corollaries, by applying the result of Theorem 4.5, we present the stabiliies with different upper bounds in the inequalities (4.2) and (4.3) connected with other classical stabilities.

Corollary 4.7 *Suppose h satisfies the inequalities (4.2) and (4.3) with* $\phi(\alpha, \beta) = \lambda_1 \left(\|\alpha\|^q + \|\beta\|^q\right)$, *where* $\lambda_1 \geq 0$, $q \neq -3$, *for all* $\alpha, \beta \in E$. *Then there exists a unique multiplicative inverse cubic Lie derivation* $H : E \longrightarrow F$ *such that*

$$\|h(\alpha) - H(\alpha)\| \leq \begin{cases} \frac{54\lambda_1}{1-3^{q+3}} \|\alpha\|^q, & for\, q < -3 \\ \frac{54\lambda_1 \cdot 3^q}{3^{q+3}-1} \|\alpha\|^q, & for\, q > -3 \end{cases}$$

for all $\alpha \in E$.

Corollary 4.8 *Suppose h satisfies the inequalities (4.2) and (4.3) with* $\phi(\alpha, \beta) = \lambda_2 \|\alpha\|^{q/2} \|\beta\|^{q/2}$, *where* $\lambda_2 \geq 0, q \neq -3,$ *for all* $\alpha, \beta \in E$. *Then there exists a unique multiplicative inverse cubic Lie derivation* $H : E \longrightarrow F$ *such that*

$$\|h(\alpha) - H(\alpha)\| \leq \begin{cases} \frac{27\lambda_2}{1-3^{q+3}} \|\alpha\|^q, & for\, q < -3 \\ \frac{27\lambda_2 \cdot 3^q}{3^{q+3}-1} \|\alpha\|^q, & for\, q > -3 \end{cases}$$

for all $\alpha \in E$.

Corollary 4.9 *Suppose h satisfies the inequalities (4.2) and (4.3) with* $\phi(\alpha, \beta) = \lambda_3 \left(\|\alpha\|^{q/2} \|\beta\|^{q/2} + \left(\|\alpha\|^q + \|\beta\|^q\right)\right)$, *where* $\lambda_3 \geq 0$, $q \neq -3$, *for all* $\alpha, \beta \in E$. *Then there exists a unique multiplicative inverse cubic Lie derivation* $H : E \longrightarrow F$ *such that*

$$\|h(\alpha) - H(\alpha)\| \leq \begin{cases} \frac{81\lambda_3}{1-3^{q+3}} \|\alpha\|^q, & for\, q < -3 \\ \frac{81\lambda_3 \cdot 3^q}{3^{q+3}-1} \|\alpha\|^q, & for\, q > -3 \end{cases}$$

for all $\alpha \in E$.

5 Stabilities of Multiplicative Inverse Quartic Lie $\star$-Derivations

In this section, we prove the classical stability results of the multiplicative inverse quartic Lie $\star$-derivations. We use the following definitions related to $\star$-normed algebras which will be useful to prove the main results.

Definition 5.1 A mapping $h : E \longrightarrow F$ is called a multiplicative inverse quartic homogenous mapping if $h(k\alpha) = \frac{1}{k^4} h(\alpha)$, for all $\alpha \in E$ and $k \in \mathbb{C}$.

Definition 5.2 If $h : E \longrightarrow F$ is a multiplicative inverse quartic homogenous mapping, then it is called a multiplicative inverse quartic derivation if

$$h(\alpha\beta) = h(\alpha)\frac{1}{\beta^4} + \frac{1}{\alpha^4}h(\beta)$$

for all $\alpha, \beta \in E$.

Definition 5.3 A multiplicative inverse quartic homogenous mapping h is called a multiplicative inverse quartic Lie derivation if

$$h([\alpha, \beta]) = \left[h(\alpha), \frac{1}{\beta^4}\right] + \left[\frac{1}{\alpha^4}, h(\beta)\right]$$

for all $\alpha, \beta \in E$, where $[\alpha, \beta] = \alpha\beta - \beta\alpha$.

Definition 5.4 A multiplicative inverse quartic Lie derivation h is called a multiplicative inverse quartic Lie ⋆-derivation if $h(\alpha^\star) = h(\alpha)^\star$ for all $\alpha \in E$.

The following difference operators are defined to prove the results in a simpler way. For a mapping $h : E \longrightarrow F$, let the difference operators be defined as follows:

$$\Delta_k h(\alpha, \beta) = h(2k\alpha + k\beta) + h(2k\alpha - k\beta) - \frac{2h(k\alpha)h(k\beta)\left[h(k\alpha) + 16h(k\beta) + 24h(k\alpha)^{1/2}h(k\beta)^{1/2}\right]}{\left[4h(k\beta)^{1/2} - h(k\alpha)^{1/2}\right]^4}$$

and

$$\Delta h(\alpha, \beta) = h(\alpha, \beta]) - \left[h(\alpha), \frac{1}{\beta^4}\right] - \left[\frac{1}{\alpha^4}, h(\beta)\right]$$

for all $\alpha, \beta \in E, k \in \mathbb{C}$.

Theorem 5.5 *Let $\rho = \pm 1$. Assume that a function $\phi : E \longrightarrow [0, \infty)$ exists with a mapping $h : E \longrightarrow F$ such that*

$$\Phi(\alpha, \beta) = \sum_{j=0}^{\infty} 81^{i\left(\frac{\rho+1}{2}\right)} \phi\left(3^{j\left(\frac{\rho-1}{2}\right)}\alpha, 3^{j\left(\frac{\rho-1}{2}\right)}\beta\right) < \infty, \tag{5.1}$$

$$\|\Delta_k h(\alpha, \beta)\| \leq \phi(\alpha, \beta), \tag{5.2}$$

$$\|\Delta h(\alpha, \beta)\| \leq \phi(\alpha, \beta) \tag{5.3}$$

for all $\alpha, \beta \in E$ and $k \in \Gamma$. If the mapping $r \mapsto h(r\alpha)$ from $\mathbb{R}$ to F is continuous, then a unique multiplicative inverse quartic Lie derivation $H : E \longrightarrow F$ exists such that

$$\|h(\alpha) - H(\alpha)\| \leq \Phi(\alpha, \alpha) \tag{5.4}$$

for all $\alpha \in E$.

Proof This theorem is also first proved for the case $\rho = 1$. Considering $k = 1$ and $\beta = \alpha$ in the inequality (5.2), we find that

$$\|81h(3\alpha) - h(\alpha)\| \leq 81\phi(\alpha, \alpha) \tag{5.5}$$

for all $\alpha \in E$. By induction method, we lead to

$$\left\|81^t h\left(3^t\alpha\right) - 81^s h\left(3^s\alpha\right)\right\| \leq 81 \sum_{j=s}^{t-1} 81^j \phi\left(3^j\alpha, 3^j\alpha\right) \tag{5.6}$$

for all $\alpha \in E$ and all $t > s \geq 0$. Using the condition (5.1), we observe that the sequence $\{81^m h\left(3^m\alpha\right)\}$ is Cauchy and since F is complete, this sequence is convergent. Therefore, we can define a mapping $H : E \longrightarrow F$ by

$$H(\alpha) = \lim_{m\to\infty} 81^m h\left(3^m\alpha\right) \tag{5.7}$$

where $\alpha \in E$. Assuming $t = m > 0$, $s = 0$ in (5.6), we get

$$\left\|81^m h\left(3^m\alpha\right) - h(\alpha)\right\| \leq 81 \sum_{j=0}^{m-1} 81^j \phi\left(3^i\alpha, 3^j\alpha\right) \tag{5.8}$$

for all $\alpha \in E$. Taking the $m \to \infty$, we find that the inequality (5.4) holds. The rest of the proof completes with the akin arguments as in Theorem 2.5. □

In the following corollaries, let $h : E \longrightarrow F$ be a mapping. Also, the mapping $r \mapsto h(r\alpha)$ from $\mathbb{R}$ to F is continuous, for each fixed $\alpha \in E$. Using Theorem 4.5 when $\rho = -1$, we obtain the following result.

Corollary 5.6 *Suppose h satisfies the inequalities (5.2) and (5.3) with* $\phi(\alpha, \beta) = \frac{80}{81}\mu$, *where* $\mu \geq 0$, *for all* $\alpha, \beta \in E$. *Then there exists a unique multiplicative inverse quartic Lie derivation* $H : E \longrightarrow F$ *such that*

$$\|h(\alpha) - H(\alpha)\| \leq \mu$$

for all $\alpha \in E$.

In the following corollaries, by applying the result of Theorem 5.5, we present the stabiliies with different upper bounds in the inequalities (5.2) and (5.3) connected with other classical stabilities.

Corollary 5.7 *Suppose h satisfies the inequalities (5.2) and (5.3) with* $\phi(\alpha, \beta) = \lambda_1\left(\|\alpha\|^q + \|\beta\|^q\right)$, *where* $\lambda_1 \geq 0$, $q \neq -4$, *for all* $\alpha, \beta \in E$. *Then there exists a unique multiplicative inverse quartic Lie derivation* $H : E \longrightarrow F$ *such that*

$$\|h(\alpha) - H(\alpha)\| \leq \begin{cases} \frac{162\lambda_1}{1-3^{q+4}} \|\alpha\|^q, & for\, q < -4 \\ \frac{162\lambda_1 \cdot 3^q}{3^{q+4}-1} \|\alpha\|^q, & for\, q > -4 \end{cases}$$

for all $\alpha \in E$.

Corollary 5.8 *Suppose h satisfies the inequalities (5.2) and (5.3) with* $\phi(\alpha, \beta) = \lambda_2 \|\alpha\|^{q/2} \|\beta\|^{q/2}$, *where* $\lambda_2 \geq 0, q \neq -4$, *for all* $\alpha, \beta \in E$. *Then there exists a unique multiplicative inverse quartic Lie derivation* $H : E \longrightarrow F$ *such that*

$$\|h(\alpha) - H(\alpha)\| \leq \begin{cases} \frac{81\lambda_2}{1-3^{q+4}} \|\alpha\|^q, & for\, q < -4 \\ \frac{81\lambda_2 \cdot 3^q}{3^{q+4}-1} \|\alpha\|^q, & for\, q > -4 \end{cases}$$

for all $\alpha \in E$.

Corollary 5.9 *Suppose h satisfies the inequalities (5.2) and (5.3) with* $\phi(\alpha, \beta) = \lambda_3 \left(\|\alpha\|^{q/2} \|\beta\|^{q/2} + \left(\|\alpha\|^q + \|\beta\|^q\right)\right)$, *where* $\lambda_3 \geq 0$, $q \neq -4$, *for all* $\alpha, \beta \in E$. *Then there exists a unique multiplicative inverse quartic Lie derivation* $H : E \longrightarrow F$ *such that*

$$\|h(\alpha) - H(\alpha)\| \leq \begin{cases} \frac{243\lambda_3}{1-3^{q+4}} \|\alpha\|^q, & for\, q < -4 \\ \frac{243\lambda_3 \cdot 3^q}{3^{q+4}-1} \|\alpha\|^q, & for\, q > -4 \end{cases}$$

for all $\alpha \in E$.

6 Concluding Remarks

We close this investigation with the following concluding remarks. So far, the stability results are obtained by many mathematicians by considering quadratic, cubic, and quartic Lie ⋆-derivations. This is our first attempt to determine the stabilities of multiplicative inverse, multiplicative inverse quadratic, multiplicative inverse cubic, and multiplicative inverse quartic Lie ⋆-derivations. From the results obtained in this investigation, we conclude that the stabilities hold good for various multiplicative inverse functional equations via Lie ⋆-derivations. Also, the stability results obtained by considering the product of powers of norms in Corollaries 2.7, 3.7, 4.7, and 5.7 give better approximations of Lie ⋆-derivations when compared with the other stabilities in each section.

Acknowledgements The first two authors are supported by The Research Council, Oman (Under Project proposal ID: BFP/RGP/CBS/18/099). The third author is supported by the SERB-MATRICS Scheme, India (F. No.: MTR/2020/000534).

References

1. S. Alshybani, S.M. Vaezpour, R. Saadati, Stability of the sextic functional equation in various spaces. J. Inequal. Spec. Funct. **9**(4), 8–27 (2018)
2. T. Aoki, On the stability of the linear transformation in Banach spaces. J. Math. Soc. Jpn. **2**, 64–66 (1950)
3. A. Bodaghi, B.V. Senthil Kumar, Estimation of inexact reciprocal-quintic and reciprocal-sextic functional equations. Mathematica **49**(82)(1–2) 3–14 (2017)
4. H. Dutta, B.V. Senthil Kumar, Geometrical elucidations and approximation of some functional equations in numerous variables. Proc. Indian Natl. Sci. Acad. **85**(3), 603–611 (2019)
5. H. Dutta, B.V. Senthil Kumar, Classical stabilities of an inverse fourth power functional equation. J. Interdiscip. Math. **22**(7), 1061–1070 (2019)
6. A. Ebadian, S. Zolfaghari, S. Ostadbashi, C. Park, Approximation on the reciprocal functional equation in several variables in matrix non-Archimedean random normed spaces. Adv. Differ. Equ. **314**, 1–13 (2015)
7. A. Fošner, M. Fošner, Approximate cubic Lie derivations. Abstr. Appl. Anal. 1–15 (2013). Art. ID 425784
8. Z. Gajda, On the stability of additive mappings. Int. J. Math. Math. Sci. **14**, 431–434 (1991)
9. P. Găvruta, A generalization of the Hyers-Ulam-Rassias stability of approximately additive mapppings. J. Math. Anal. Appl. **184**, 431–436 (1994)
10. D.H. Hyers, On the stability of the linear functional equation. Proc. Natl. Acad. Sci. USA **27**, 222–224 (1941)
11. S. Jang, C. Park, Approximate $\star$-derivations and approximate quadratic $\star$-derivations on $C^\star$-algebra. J. Inequal. Appl. 1–13 (2011). Art. ID 55
12. D. Kang, H. Koh, A fixed point approach to the stability of quartic Lie $\star$ -derivations. Korean J. Math. **24**(4), 587–600 (2016)
13. G.H. Kim, H.Y. Shin, Hyers-Ulam stability of quadratic functional equations on divisible square-symmetric groupoid. Int. J. Pure Appl. Math. **112**(1), 189–201 (2017)
14. S.O. Kim, B.V. Senthil Kumar, A. Bodaghi, Stability and non-stability of the reciprocal-cubic and reciprocal-quartic functional equations in non-Archimedean fields. Adv. Differ. Equ. **77**, 1–12 (2017)
15. A. Najati, J.R. Lee, C. Park, T.M. Rassias, On the stability of a Cauchy type functional equation. Demonstr. Math. **51**, 323–331 (2018)
16. C. Park, A. Bodaghi, On the stability of $\star$ -derivations on Banach $\star$-algebras. Adv. Differ. Equ. **138**, 1–11 (2012)
17. T.M. Rassias, On the stability of the linear mapping in Banach spaces. Proc. Am. Math. Soc. **72**, 297–300 (1978)
18. J.M. Rassias, On approximately of approximately linear mappings by linear mappings. J. Funct. Anal. USA **46**, 126–130 (1982)
19. J.M. Rassias, Solution of the Ulam stability problem for cubic mappings. Glasnik Matematicki Ser. III **36**(56), 63–72 (2001)
20. J.M. Rassias, M. Arunkumar, S. Karthikeyan, Lagrange's quadratic functional equation connected with homomorphisms and derivations on Lie $C^\star$-algebras: Direct and fixed point methods. Malaya J. Mat. S(1), 228–241 (2015)
21. K. Ravi, B.V. Senthil Kumar, Ulam-Gavruta-Rassias stability of Rassias reciprocal functional equation. Glob. J. Appl. Math. Sci. **3**(1–2), 57–79 (2010)
22. K. Ravi, M. Arunkumar, J.M. Rassias, On the Ulam stability for the orthogonally general Euler-Lagrange type functional equation. Int. J. Math. Sci. **3**(8), 36–47 (2008)
23. K. Ravi, J.M. Rassias, B.V. Senthil Kumar, Ulam stability of reciprocal difference and adjoint functional equations. Aust. J. Math. Anal. Appl. **8**(1), 1–18 (2011). Art. 13
24. K. Ravi, E. Thandapani, B.V. Senthil Kumar, Stability of reciprocal type functional equations. Pan Am. Math. J. **21**(1), 59–70 (2011)

25. B.V. Senthil Kumar, H. Dutta, Non-Archimedean stability of a generalized reciprocal-quadratic functional equation in several variables by direct and fixed point methods. Filomat **32**(9), 3199–3209 (2018)
26. B.V. Senthil Kumar, H. Dutta, Fuzzy stability of a rational functional equation and its relevance to system design. Int. J. Gen. Syst. **48**(2), 157–169 (2019)
27. B.V. Senthil Kumar, H. Dutta, Approximation of multiplicative inverse undecic and duodecic functional equations. Math. Meth. Appl. Sci. **42**, 1073–1081 (2019)
28. B.V. Senthil Kumar, H. Dutta, S. Sabarinathan, Approximation of a system of rational functional equations of three variables. Int. J. Appl. Comput. Math. **5**(3), 1–16 (2019)
29. B.V. Senthil Kumar, H. Dutta, S. Sabarinathan, Fuzzy approximations of a multiplicative inverse cubic functional equation. Soft Comput. **24**, 13285–13292 (2020)
30. B.V. Senthil Kumar, K. Al-Shaqsi, H. Dutta, Classical stabilities of multiplicative inverse difference and adjoint functional equations. Adv. Differ. Equ. **215**, 1–9 (2020)
31. S.M. Ulam, *Problems in Modern Mathematics (Chap. VI)* (Wiley-Interscience, New York, 1964)
32. S.Y. Yang, A. Bodaghi, K.A.M. Atan, Approximate cubic ⋆ -derivations on Banach ⋆ -algebra. Abst. Appl. Anal. 1–12 (2012). Art ID 684179

Analysis and Computation of Reactive Second-Grade Fluid Flow with Variable Viscosity Within Porous Couette Device

Sukanya Padhi and Itishree Nayak

Abstract The main objective of this paper is to examine the influence on velocity and temperature profiles of an unsteady, reactive, incompressible, second-grade fluid with low electrical conductivity and variable viscosity within porous channel with asymmetric cooling at the walls. The highly coupled non-linear partial differential equations are generated and reduced to a system of algebraic equations using fully implicit finite difference scheme. The resulting system is then solved by damped-Newton method. The adopted numerical method observes second-order convergence and is highly stable along with ensuring error reduction after every iteration depending on the choice of damping applied. Lastly, the effect of different physical parameters on velocity and temperature profile is discussed and demonstrated graphically with the aid of MATLAB.

Keywords Reactive second-grade fluid · Variable viscosity · Convective cooling · Finite difference method · Damped-newton method

AMS Subject Classification 76A05

Nomenclature:

(1) u—Fluid axial velocity.
(2) V—Injective velocity.
(3) T—Temperature.
(4) T_a—Ambient temperature.
(5) T_0—Fluid initial temperature.
(6) t'—Time.
(7) ρ—Density.
(8) σ—Fluid electrical conductivity.

S. Padhi (✉) · I. Nayak
Veer Surendra Sai University of Technology, Burla 768018, Odisha, India
e-mail: sukanya.padhi23@gmail.com

I. Nayak
e-mail: itii.nayak@gmail.com

S. K. Paikray et al. (eds.), *New Trends in Applied Analysis and Computational Mathematics*, Advances in Intelligent Systems and Computing 1356,
https://doi.org/10.1007/978-981-16-1402-6_12

(9) B_0—Electromagnetic Induction.
(10) k—Thermal conductivity coefficient.
(11) C_p—Specific heat at constant pressure.
(12) h_i $(i = 1, 2)$—Heat transfer coefficients.
(13) Q—Heat of reaction.
(14) A—Rate constant.
(15) E—Activation energy.
(16) R—Universal gas constant.
(17) C_0—Initial concentration of the reacting species.
(18) a—Channel half width.
(19) l—Planck's number.
(20) K—Boltzmann constant.
(21) ν—Vibration frequency.
(22) α_1—Material coefficient.
(23) m—Numerical constant.
(24) μ'—Temperature-dependent viscosity.
(25) b—Viscosity variation parameter.
(26) μ_0—Initial fluid dynamic viscosity at temperature T_0.
(27) λ—Frank–Kamenetskii parameter.
(28) Pr—Prandtl number.
(29) $Bi_i(i = 1, 2)$—Biot numbers.
(30) ϵ—Activation energy parameter.
(31) α—Second-grade elastic parameter.
(32) δ—Variable viscosity parameter.
(33) Ω—Viscous heating parameter.
(34) Θ_a—Ambient temperature.
(35) Ha—Hartmann number.
(36) Re—Reynolds number.

1 Introduction

The study of flow and heat transfer of non-Newtonian fluids has been of great practical importance due to its excessive use in engineering processes, industries and agriculture and therefore a great deal of research work has been devoted to the study of non-Newtonian fluids [2, 5, 23]. The governing equations of non-Newtonian fluids of higher order have more complexity and highly non-linear thus making the process of obtaining accurate solution more tedious. A special subclass of non-Newtonian fluids are the second-grade fluids and many authors working in the area of fluid dynamics have studied different models for steady and unsteady flows of second-grade fluid [4, 7, 9, 10, 25], and have obtained accurate numerical solution to the governing equations with very efficient numerical methods having higher order convergence [15, 17].

The fluid flow phenomena is determined by many physical properties of a fluid, one such being viscosity. Viscosity is the measure of internal fluid friction that offers resistance to fluid flow. Viscosity of a liquid decreases with increasing temperature while it increases for gases. Earlier many problems in fluid dynamics were solved by assuming the viscosity to be constant, but in the past few years many interesting properties of fluid flow with variable viscosity have come to light. Massoudi and Phuoc [14] discussed the flow of a second-grade non-Newtonian fluid with variable viscosity. Okoya [18] examined the steady, reactive, 1-D plane Couette flow of third-grade fluid flow between parallel plates, the lower plate apparently being at rest, while the upper plate was subjected to uniform motion. The critical and transitional values of flow parameters were determined by numerical integration and the solution was obtained for Arrhenius, Sensitised and Bimolecular reaction types. He further extended the study in [19] where he adopted the HPM to solve the flow problem. Later, Akinbobola and Okaya [1] studied the steady flow of non-Newtonian second-grade fluid under the influence of temperature-dependent viscosity and thermal conductivity.

The combined effects of heat and mass transfer with chemical reaction are of great importance due to its implementations in chemical and hydro-metallurgical industries. The study is useful for designing equipments used in chemical processing, fog dispersion, moisture distribution over agricultural land, prevention of damage to crops due to freezing and food processing units. Salawu et al. [22] investigated the exothermic chemical reaction of time-dependent Poiseuille flow with viscous dissipation through porous channels under different chemical kinetics and concluded that temperature rises with increasing values of Frank–Kamenetskii parameter. Mahmoud [12] performed a theoretical study on Walters' B liquid past a stretching sheet in porous medium with variable viscosity and chemical reaction and numerically solved the resulting equation using shooting method. Veerakrishna and Reddy [24] investigated the time-dependent MHD reactive flow of second-grade fluid through porous medium in a rotating parallel-plate channel with Arrhenius reaction rate and derived an analytical solution to the flow problem through Laplace transformation. Ramesh and Ojjela [20] analysed the entropy generation of a viscoelastic second-grade nanofluid through porous channels with periodic suction and injection while maintaining distinct temperatures at the boundary of the plates. Hayat et al. [8] discussed the heat and mass transfer of second-grade fluid flow in presence of chemical reaction.

The principal objective of this paper is to study the unsteady MHD flow of an incompressible reactive second-grade fluid with variable viscosity in the presence of uniform transverse magnetic field. The mathematical formulation of the problem has been done in Sect. 2 followed by Sect. 3, which focuses on finding a solution for velocity field and temperature field. Section 4 illustrates the influence on velocity and temperature with variations in physical parameters. Section 5 culminates the paper.

2 Description of the Problem

The constitutive equation for a second-grade fluid is represented as

$$T = -pI + \mu A_1 + \alpha_1 A_2 + \alpha_2 A_1^2. \quad (1)$$

T—stress tensor at a point.
I—identity tensor.
p—pressure.
μ—usual viscosity.
α_1, α_2—material constants.

The kinematic tensors A_1 and A_2 are given by

$$A_1 = L + L^T, \ A_2 = \frac{dA_1}{dt} + A_1 L + L^T A_1, \ L = \nabla V,$$

where ∇ is the gradient operator and V is the velocity.

With the stated assumptions in [6] taken into consideration, the following relation holds:

$$\mu \geq 0, \ \alpha_1 \geq 0, \ |\alpha_1 + \alpha_2| = 0. \quad (2)$$

The flow problem is depicted in Fig. 1. Using the above-mentioned stress Eq. (1), the equation of motion and energy with variable viscosity can be written as

$$\rho\left(\frac{\partial u'}{\partial t'} + V\frac{\partial u'}{\partial y'}\right) = \frac{\partial}{\partial y'}\left(\mu'(T)\frac{\partial u'}{\partial y'}\right) + \alpha_1\left(\frac{\partial^3 u'}{\partial y'^2 \partial t'} + V\frac{\partial^3 u'}{\partial y'^3}\right) - \sigma B_0^2 u'. \quad (3)$$

$$\rho C_p\left(\frac{\partial T}{\partial t'} + V\frac{\partial T}{\partial y'}\right) = k\frac{\partial^2 T}{\partial y'^2} + \mu'(T)\left(\frac{\partial u'}{\partial y'}\right)^2 + QC_0 A\left(\frac{KT}{\nu l}\right)^m exp\left(-\frac{E}{RT}\right) + \sigma B_0^2 u'^2. \quad (4)$$

The desired boundary conditions to which (3) and (4) is subjected to are as follows

$$u'(y', 0) = 0, \quad T(y', 0) = T_0. \quad (5)$$

$$u'(0, t') = U, \quad -k\frac{\partial t}{\partial y'}(0, t') = h_1(T(0, t') - T_a), \ \text{for } t' > 0. \quad (6)$$

$$u'(a, t') = 0, \quad -k\frac{\partial t}{\partial y'}(a, t') = h_2(T(a, t') - T_a), \ \text{for } t' > 0. \quad (7)$$

The temperature-dependent viscosity can be expressed as

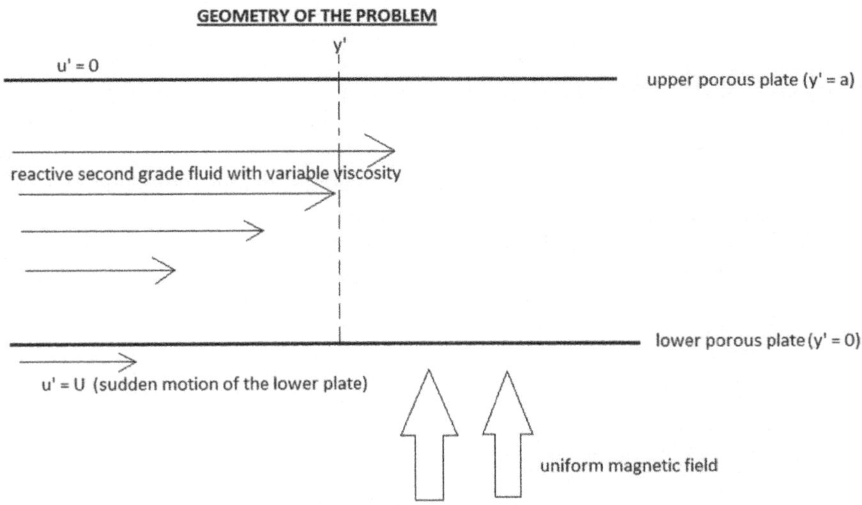

Fig. 1 Flow diagram

$$\mu'(T) = \mu_0\, exp(-b(T - T_0)). \tag{8}$$

We introduce the non-dimensional variables and parameters as follows:
$y = \frac{y'}{a}$, $\delta = \frac{bRT_0{}^2}{E}$, $u = \frac{u'}{U}$, $\theta = \frac{E(T-T_0)}{RT_0^2}$, $\theta_a = \frac{E(T_a - T_0)}{RT_0^2}$, $\alpha = \frac{\alpha_1}{\rho a^2}$, $Bi_1 = \frac{ah_1}{k}$, $Bi_2 = \frac{ah_2}{k}$, $Pr = \frac{aU\rho C_p}{k}$, $Ha^2 = \frac{\sigma B_0{}^2 a}{\rho U}$, $\mu = \frac{\mu'}{\mu_0}$, $\epsilon = \frac{RT_0}{E}$, $t = \frac{t'U}{a}$, $\lambda = \left(\frac{KT_0}{\nu l}\right)^m \frac{QEAa^2C_0 exp(-\frac{E}{RT})}{T_0^2 Rk}$, $\Omega = \left(\frac{\nu l}{KT_0}\right)^m \frac{\mu_0 U^2 exp(\frac{E}{RT})}{QAa^2C_0}$, $Re = \frac{V}{U}$.
and obtain the following dimensionless governing equation:

$$\frac{\partial u}{\partial t} + Re\frac{\partial u}{\partial y} = exp(-\delta\theta)\frac{\partial^2 u}{\partial y^2} - \delta\, exp(-\delta\theta)\frac{\partial \theta}{\partial y}\frac{\partial u}{\partial y} + \alpha\frac{\partial^3 u}{\partial y^2 \partial t} + Re\,\alpha\frac{\partial^3 u}{\partial y^3} - Ha^2 u. \tag{9}$$

$$Pr\frac{\partial \theta}{\partial t} + Re\, Pr\frac{\partial \theta}{\partial y} = \frac{\partial^2 \theta}{\partial y^2} + \lambda(1+\epsilon\theta)^m\, exp\left(\frac{\theta}{1+\epsilon\theta}\right) + \Omega\lambda\left[Ha^2u^2 + exp(-\delta\theta)\left(\frac{\partial u}{\partial y}\right)^2\right] \tag{10}$$

subjected to following initial and boundary conditions:

$$u(y,0) = 0,\ \theta(y,0) = 0. \tag{11}$$

$$u(0,t) = 1,\ \frac{\partial \theta}{\partial y}(0,t) = -Bi_1(\theta(0,t) - \theta_a),\ \text{for } t > 0. \tag{12}$$

$$u(1,t) = 0,\ \frac{\partial \theta}{\partial y'}(1,t) = -Bi_2(\theta(1,t) - \theta_a)\ \text{for } t > 0. \tag{13}$$

3 Numerical Solution

We choose to take up the following solution strategy to solve (9). We implement the implicit finite difference scheme of Crank–Nicolson type for discretisation in space as well as in time with a uniform mesh of space step h and time step k. The fully implicit finite difference scheme used here is unconditionally stable and provides second-order convergence in time as well as in space.

The derivatives at the nodes $(ih, j\Delta t)$, $i = 0, 1, \ldots, N+1$ *and* $j = 0, 1, \ldots, M-1$ are reckoned as

$$
\begin{aligned}
d_{1i}^{j} &= u_{i+1}^{j} - u_{i-1}^{j}. \\
e_{1i}^{j} &= \theta_{i+1}^{j} - \theta_{i-1}^{j}. \\
d_{2i}^{j} &= u_{i+1}^{j} - 2u_{i}^{j} + u_{i-1}^{j}. \\
e_{2i}^{j} &= \theta_{i+1}^{j} - 2\theta_{i}^{j} + \theta_{i-1}^{j}. \\
d_{3i}^{j} &= -u_{i-2}^{j} + 2u_{i-1}^{j} - 2u_{i+1}^{j} + u_{i-1}^{j}. \\
d_{3i}^{'j} &= -3u_{i-1}^{j} + 10u_{i}^{j} - 12u_{i+1}^{j} + 6u_{i+2}^{j} - u_{i+3}^{j}. \\
d_{3i}^{''j} &= u_{i-3}^{j} - 6u_{i-2}^{j} + 12u_{i-1}^{j} - 10u_{i}^{j} + 3u_{i+1}^{j}.
\end{aligned}
$$

We use the following notations to represent the difference approximations at nodes $(ih, j\Delta t), i = 1, 2, \ldots, N+1$ and $j = 0, 1, \ldots, M-1$ are taken as

$$
\begin{aligned}
\frac{\partial u}{\partial t} &\approx \frac{u_i^{j+1} - u_i^j}{\Delta t}. \\
\frac{\partial u}{\partial y} &\approx \frac{1}{4h}(d_{1i}^{j+1} + d_{1i}^{j}). \\
\frac{\partial \theta}{\partial y} &\approx \frac{1}{4h}(e_{1i}^{j+i} + e_{1i}^{j}). \\
\frac{\partial^2 \theta}{\partial y^2} &\approx \frac{1}{2h^2}(e_{2i}^{j+1} + e_{2i}^{j}). \\
\frac{\partial^2 u}{\partial y^2} &\approx \frac{1}{2h^2}(d_{2i}^{j+1} + d_{2i}^{j}). \\
\frac{\partial^3 u}{\partial y^3} &\approx \frac{1}{2h^3}(d_{3i}^{j+1} + d_{3i}^{j}), \quad i \neq 1, N. \\
\frac{\partial^3 u}{\partial y^2 \partial t} &\approx \frac{1}{h^2 \Delta t}(d_{2i}^{j+1} - d_{2i}^{j}).
\end{aligned}
\tag{14}
$$

At the interior points $(1, j\Delta t)$ and $(N, j\Delta t)$, the third derivative $\frac{\partial^3 u}{\partial y^3}$ is replaced, respectively, by $\frac{1}{2h^3}\left(d_{3i}^{'j+1} + d_{3i}^{'j}\right)$ and $\frac{1}{2h^3}\left(d_{3i}^{''j+1} + d_{3i}^{''j}\right)$. Using the differences obtained in (14), the governing equation of velocity is discretised as

$$\frac{u_i^{j+1}-u_i^j}{\Delta t}+\frac{Re}{4h}(d_{1i}^{j+1}+d_{1i}^j)-\frac{1}{2h^2}\exp\left(-\delta\frac{\theta_i^{j+1}+\theta_i^j}{2}\right)(d_{2i}^{j+1}+d_{2i}^j)$$
$$+\frac{\delta}{16h^2}\exp\left(-\delta\frac{\theta_i^{j+1}+\theta_i^j}{2}\right)(d_{1i}^{j+1}+d_{1i}^j)(e_{1i}^{j+1}+e_{1i}^j)-\frac{\alpha}{h^2\Delta t}(d_{2i}^{j+1}-d_{2i}^j)-\frac{\alpha Re}{2h^3}(d_{3i}^{j+1}+d_{3i}^j)$$
$$+Ha^2\frac{u_i^{j+1}+u_i^j}{2}=0 \tag{15}$$

for $i=2,3,\ldots,N-1,\quad j=1,2,\ldots,M$. The discretised form of the initial and boundary conditions for velocity is represented as

$$u_{N+1}^j=0,\ j=1,\ldots,M. \tag{16}$$

$$u_i^0=0,\ i=0,1,\ldots,N+1. \tag{17}$$

$$u_0^j=1. \tag{18}$$

Using the above system of equations, the residue matrix is represented as

$$R=\begin{bmatrix}R_1 & R_2 & \ldots & R_i & \ldots & R_N.\end{bmatrix}$$

The elements of the Jacobian matrix for $i=1,2,\ldots N$ and $j=1,2,\ldots,M$ are represented as follows:

$$J=\begin{bmatrix}\frac{\partial R_1}{\partial u_1^{j+1}} & \frac{\partial R_1}{\partial u_2^{j+1}} & \cdots & \frac{\partial R_1}{\partial u_N^{j+1}}\\ \frac{\partial R_2}{\partial u_1^{j+1}} & \frac{\partial R_2}{\partial u_2^{j+1}} & \cdots & \frac{\partial R_2}{\partial u_N^{j+1}}\\ \vdots & \ddots & \cdots & \vdots\\ \frac{\partial R_i}{\partial u_1^{j+1}} & \frac{\partial R_i}{\partial u_2^{j+1}} & \cdots & \frac{\partial R_i}{\partial u_N^{j+1}}\\ \vdots & \ddots & \cdots & \vdots\\ \frac{\partial R_N}{\partial u_1^{j+1}} & \frac{\partial R_N}{\partial u_2^{j+1}} & \cdots & \frac{\partial R_N}{\partial u_N^{j+1}}.\end{bmatrix}$$

The system of Eq. (15) along with the boundary conditions (16)–(18) is solved by using damped-Newton method which converges quadratically.

An approximate value of M is chosen in accordance with the algorithm stated in [11]. The iterations are repeated till the absolute difference of the successive solutions obtained at nodes $((N+1)h,\ j\Delta t)$ and $((N+2)h,\ j\Delta t)$ becomes less than ε.

A good initial guess is very important in order to achieve the desired level of accuracy. For a preferable assumption of initial velocity, Eq. (15) is organised in a tridiagonal form by equating α and m to zero and is solved by Gaussian elimination.

The damped-Newton method is carried through by computing the residuals and jacobians $(R_i, i = 1, \ldots, N)$, $\left(\left(\frac{\partial R_i}{\partial u_j}\right) \neq 0, i = 1, \ldots, N \text{ and } j = 1, \ldots, M\right)$, respectively. We consider the next approximation to be $x^{k+1} = \left(x^k + \frac{h}{2^i}\right)$ for that i, where

$$i = min\left(j : 0 \leq j \leq j_{max} \mid \| residue(x^k + \frac{h}{2^j}) \|_2 < \| residue(x^k) \|_2\right),$$

thus validating the error reduction in every iteration and convergence of the method as given in [3] precisely to five decimal places.

The dimensionless energy equation (10) in discretised form is represented as

$$Pr\left(\frac{\theta_i^{j+1} - \theta_i^j}{\Delta t}\right) + \frac{ReP_r}{4h}\left[\left(e_{1i}^{j+1} + e_{1i}^j\right)\right] - \frac{1}{2h^2}\left(e_{2i}^{j+1} + e_{2i}^j\right) - \lambda(1 + \epsilon\theta)^m \; exp\left(\frac{\theta_i^j}{1 + \epsilon\theta_i^j}\right)$$
$$- \Omega\lambda\left[Ha^2(u_i^j)^2 + exp\,(-\delta\theta)\left(\frac{1}{4h}(d_{1i}^{j+1} + d_{1i}^j)\right)^2\right] = 0. \quad (19)$$

The discretised form of the initial and boundary conditions for energy equation is represented as

$$\theta_i^0 = 0, \; i = 0, \ldots, N + 1. \quad (20)$$

$$\theta_0^j = \frac{2hB_{i1}\theta_a - 4\theta_1^j + \theta_2^j}{2B_{i1}h - 3} \; and \quad (21)$$

$$\theta_N^j = \frac{2hB_{i2}\theta_a + 4\theta_{N-1}^j - \theta_{N-2}^j}{2B_{i2}h + 3}. \quad (22)$$

Equation (19) is arranged in tridiagonal form and further solution is obtained by using exponential fitted scheme as given in [16]

$$\left(\frac{RePr}{4h\Delta t} - \frac{1}{2h^2}\right)\theta_{i+1}^{j+1} + \left(\frac{Pr}{\Delta t} + \frac{1}{h^2}\right)\theta_i^{j+1} + \left(-\frac{RePr}{4h\Delta t} - \frac{1}{2h^2}\right)\theta_{i-1}^{j+1}$$
$$= \left(-\frac{RePr}{4h\Delta t} + \frac{1}{2h^2}\right)\theta_{i+1}^j + \left(\frac{Pr}{\Delta t} - \frac{1}{h^2}\right)\theta_i^j + \left(\frac{RePr}{4h\Delta t} + \frac{1}{2h^2}\right)\theta_{i-1}^j$$
$$+ \lambda(1 + \epsilon\theta_i^j)^m exp\left(\frac{\theta_i^j}{1 + \epsilon\theta_i^j}\right) \quad (23)$$
$$+ \frac{exp(-\delta\theta_i^j)}{16h^2}\Omega\lambda\left(d_{1i}^{j+1} + d_{1i}^j\right)^2 + Ha^2(u_i^j)^2\Omega\lambda.$$

4 Results and Discussion

Unless otherwise stated, the parametric values will be considered as follows:
$Pr = 7, \theta_a = 0.1, Bi_1 = 0.1, Bi_2 = 1, m = 0.5, \epsilon = 0.1, \Omega = 0.1, \alpha = 0.1,$ $\delta = 0.001$ and $\lambda = 0.1$.

Figures 2 and 3 represent the effect of second-grade elastic parameter on fluid flow and temperature profile. The decreasing effect in velocity is due to the more prominent effect of elastic behaviour of the fluid which leads to enhancement of tensile stresses with increasing values of α as a result of which the fluid flow decreases. Elasticity

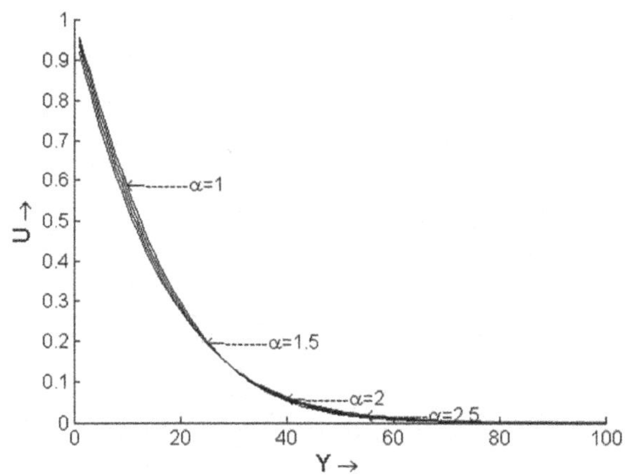

Fig. 2 Influence on velocity with variations in α

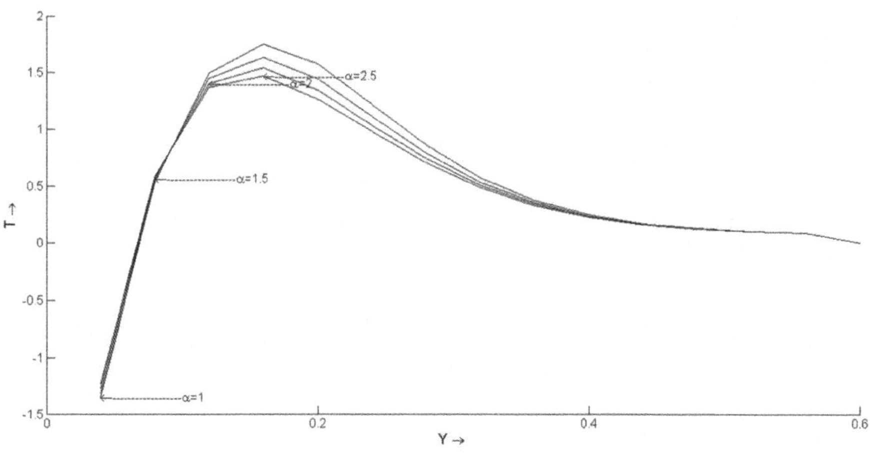

Fig. 3 Influence on temperature with variations in α

Fig. 4 Influence on velocity with variations in ϵ

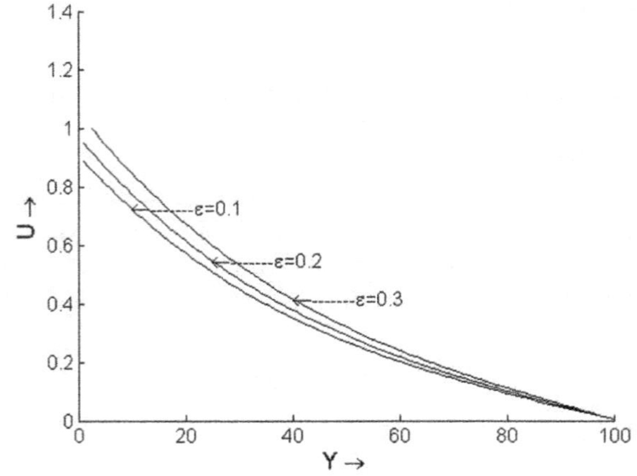

Fig. 5 Influence on temperature with variations in ϵ

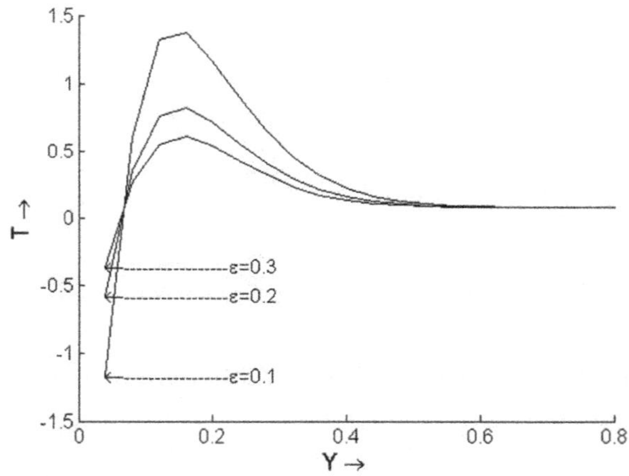

Fig. 6 Influence on velocity with variations in Ha

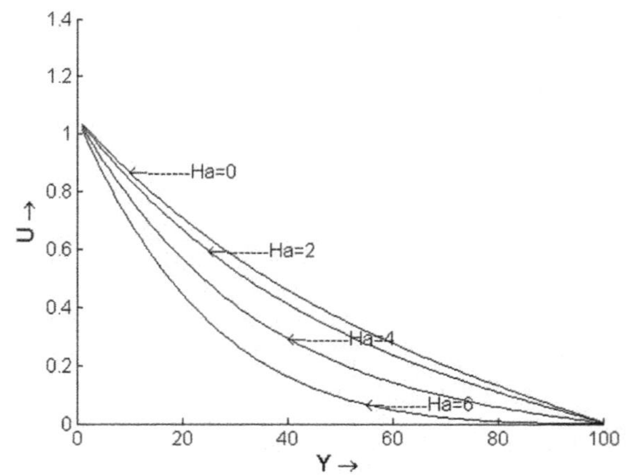

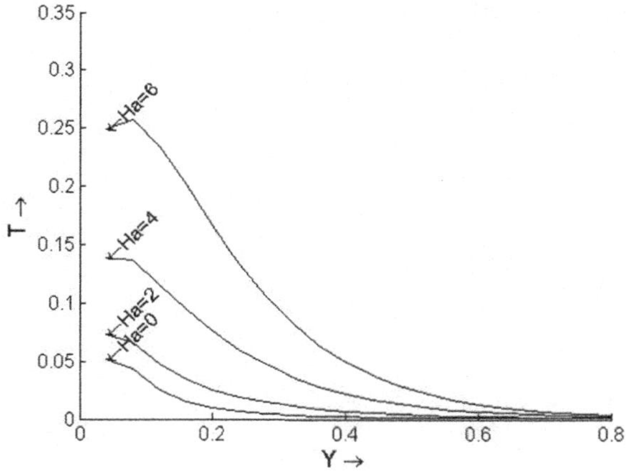

Fig. 7 Influence on temperature with variations in Ha

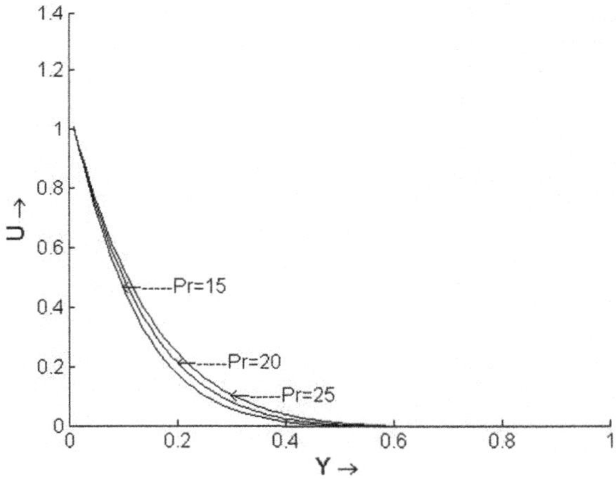

Fig. 8 Influence on velocity with variations in Pr

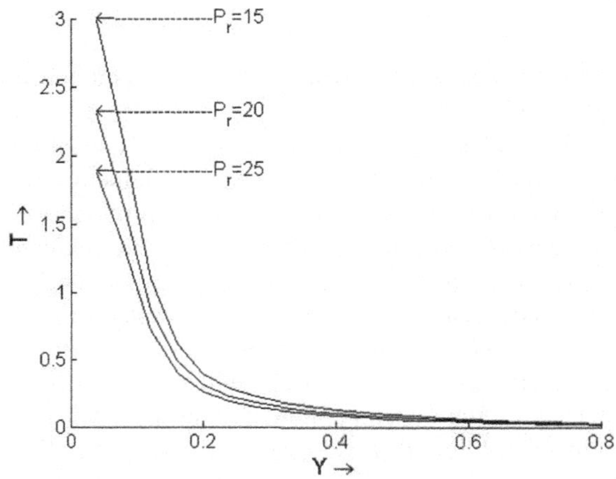

Fig. 9 Influence on temperature with variations in Pr

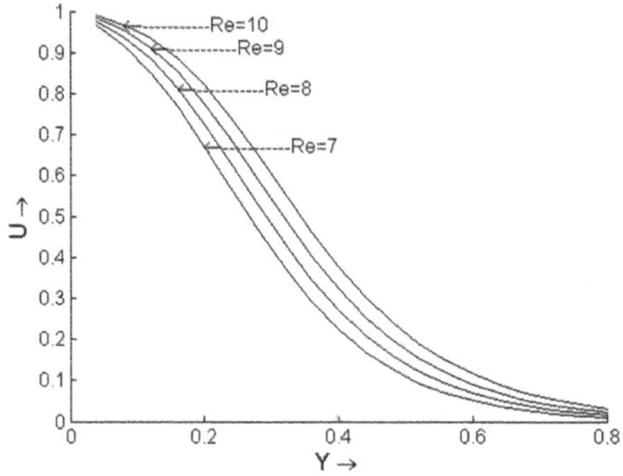

Fig. 10 Influence on velocity with variations in *Re*

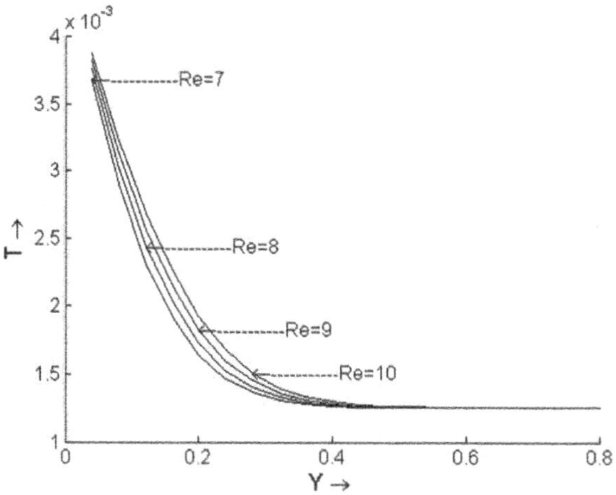

Fig. 11 Influence on temperature with variations in *Re*

of the fluid reduces the frictional heating and hence the overall temperature of the system decreases.

Figures 4 and 5 represent the influence of activation energy parameter ϵ on momentum and energy profiles. Increase in ϵ results in a rise in velocity profile and causes fall in temperature which is rather more pronounced.

Figures 6 and 7 depict that increase in Hartmann number Ha enhances the damping magnetic properties of the second-grade fluid and the flow is inhibited due to increased resistance. Consequently, velocity profile decreases. Energy of the system is enhanced due to ohmic heating.

Increase in the values of Pr enhances thermal diffusivity through porous plates and leads to fall in temperature as a result of which fluid viscosity decreases leading to increased fluid flow velocity as shown in Figs. 8 and 9.

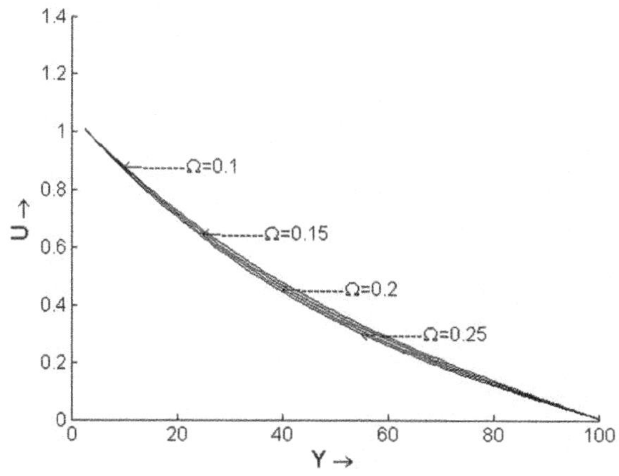

Fig. 12 Influence on velocity with variations in Ω

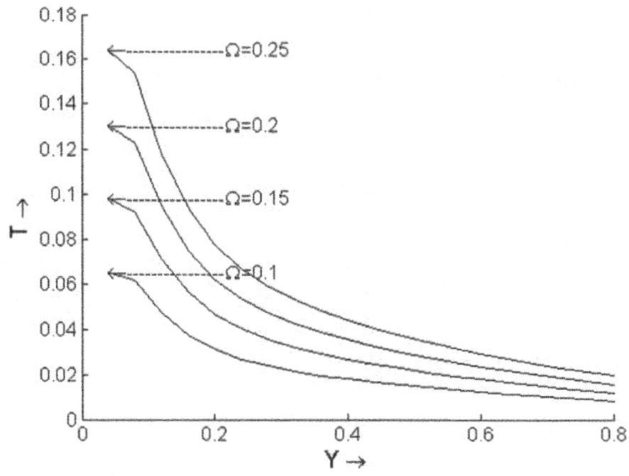

Fig. 13 Influence on temperature with variations in Ω

Reynolds number augments the fluid flow and temperature profiles. Increase in *Re* values decreases the fluid flow viscosity. The increased velocity due to reduced viscosity increases the viscous heating source terms in the energy equation, thus increasing fluid temperature as portrayed in Figs. 10 and 11.

The impact of viscous heating parameter Ω is depicted in Figs. 12 and 13. The kinetic energy of the fluid in motion is converted to internal energy causing rise in temperature. It is recognisable from the graph that velocity is a decreasing function of Ω.

With increase in the values of variable viscosity parameter δ, the phenomena of crossing over is noticed at two different points where the velocity of fluid initially increases, then appears to decrease after the first crossover point and finally increases after second crossover point. Due to the increase in viscous heating source terms and chemical reaction source terms, temperature profile increases as in shown Figs. 14 and 15.

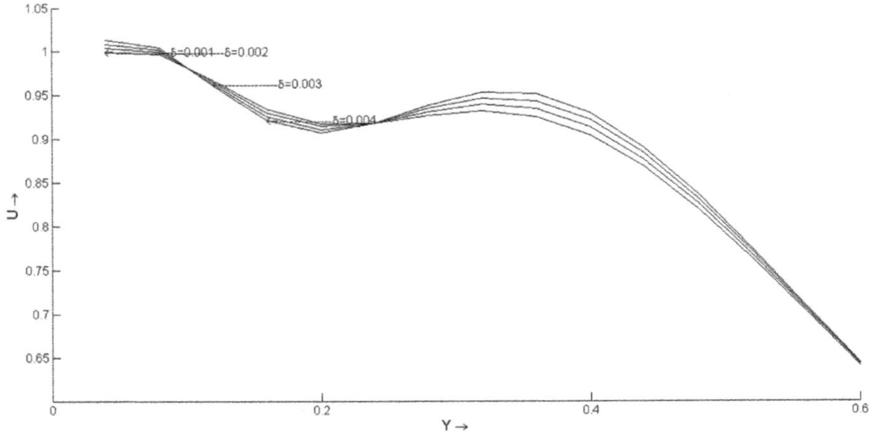

Fig. 14 Influence on velocity with variations in δ

Fig. 15 Influence on temperature with variations in δ

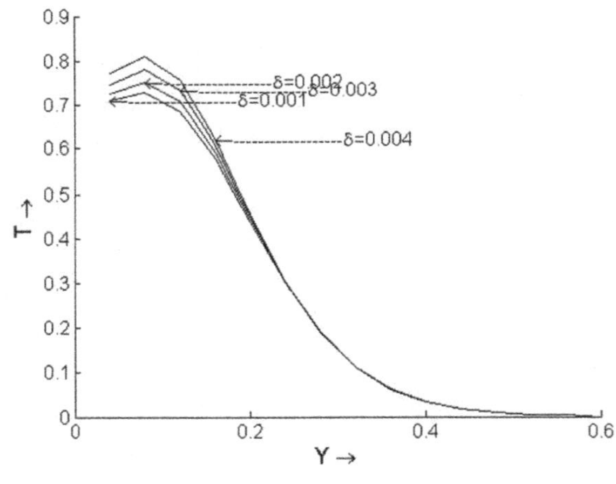

Figures 16 and 17 represent the impact of Frank–Kamenetskii parameter λ on velocity and temperature profile. λ enters the velocity equation implicitly through temperature/viscosity coupling and therefore its effect is not as prominent as in the case of fluid temperature and this appears marginal. Since λ is associated to exponentially increasing reaction source terms, heat generation increases, thus enhancing the overall fluid temperature. The values of λ are carefully controlled in order to avoid blow up solution as in [13, 21]. Figure 18 shows a transient decrease in the velocity profile of the second-grade fluid.

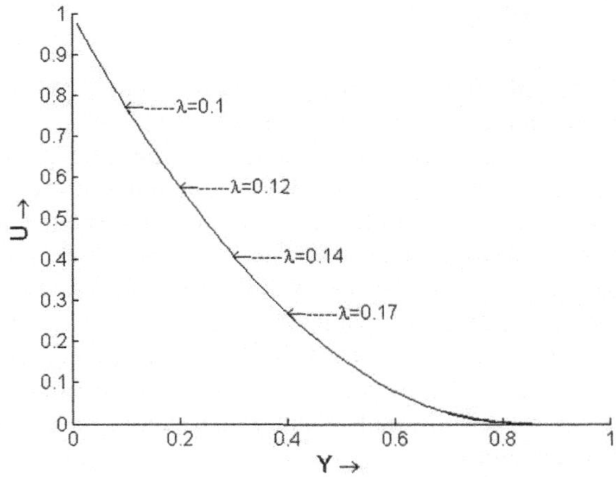

Fig. 16 Influence on velocity with variations in λ

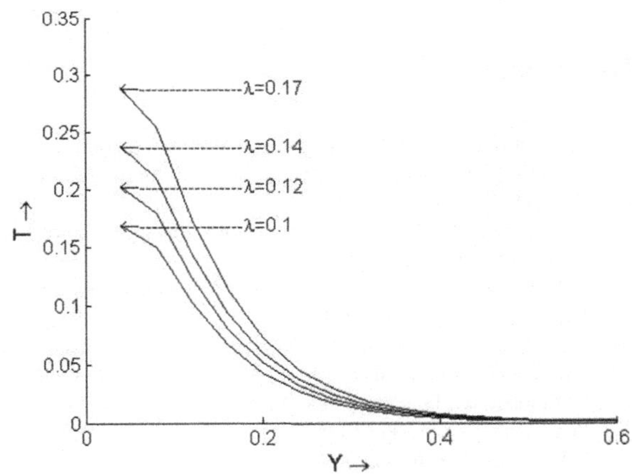

Fig. 17 Influence on temperature with variations in λ

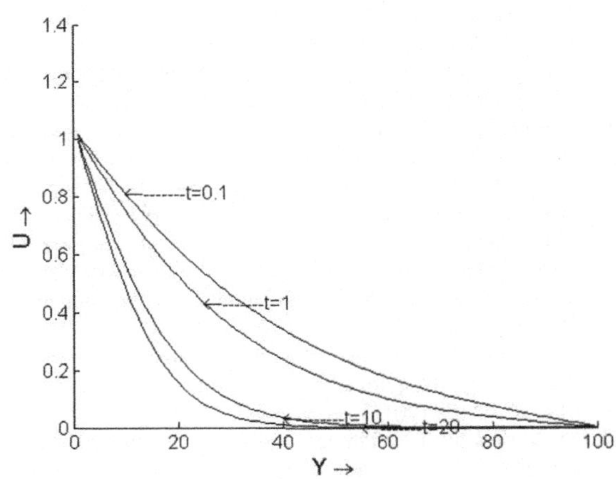

Fig. 18 Transient velocity profile

5 Conclusion

The significance and main advantage of used scheme is that it is valid for both small and large parametric values and repetition of calculation is not required for every change in boundary conditions. The residual error also decreases after every iteration according to the choice of damping unlike Newton's method. Due to the implementation of fully implicit finite difference scheme used in the paper, the solution is unconditionally stable and suffices the second-order convergence in time as well as in space. The observation of various graphical representations reveals that second-grade elastic parameter α and Reynolds number Re has completely contrasting effects on fluid velocity and temperature, whereas the impact of magnetic parameter Ha mirrors to the viscous heating parameter Ω. The variable viscosity parameter δ and Frank–Kamenetskii parameter λ contribute to enhance the fluid temperature.

References

1. T.E. Akinbobola, S.S. Okaya, The flow of second grade fluid over a stretching sheet with variable thermal conductivity and viscosity in the presence of heat source/sink. J. Niger. Math. Soc. **34**, 331–342 (2015)
2. P.D. Ariel, Flow of viscoelastic fluids through a porous channel-I. Int. J. Numer. Methods Fluids **17**, 605–633 (1993)
3. S. D. Conte, C. De Boor, *Elementary Numerical Analysis an Algorithmic Approach* (McGraw-Hill, inc., New-York, 1980)
4. M.E. Erdogan, C.E. Imrak, On unsteady unidirectional flows of a second grade fluid. Int. J. Non-Linear Mech. **40**, 1238–1251 (2005)
5. C. Fetecau, C. Fetecau, On some axial Couette flows of non-Newtonian fluids. J. Appl. Math. Phys. **56**, 1098–1106 (2005)
6. R.L. Fosdick, K.R. Rajagopal, Thermodynamics and stability of fluids of third grade. Proc. R. Soc. Lond.-A **339**, 351–377 (1980)
7. T. Hayat, N. Ahmed, M. Sajid, S. Asghar, On the MHD flow of a second grade fluid in a porous channel. Comput. Math. Appl. **54**, 407–414 (2007)
8. T. Hayat, M.I. Khan, A. Alsaedi, M. Waqas, Mechanism of chemical aspect in ferromagnetic flow of second grade liquid. Results Phys. **7**, 4162–4167 (2017)
9. T. Hayat, S.A. Khan, A. Alsaedi, Simulation and modeling of entropy optimized MHD flow of second grade fluid with dissipation effect. J. Mat. Res. Tech. **9**, 11993–12006 (2020)
10. I.E. Ireka, S.S. Okoya, Analysis of unsteady flow of second grade fluid with power law spatially distributed viscosity. Afr. Mat. **31**, 1175–1191 (2020)
11. M.K. Jain, *Numerical Solution of Differential equations* (Wiley Eastern, New Delhi, 1984)
12. M.A.A. Mahmoud, Chemical reaction and variable viscosity effects on flow and mass transfer of a non-Newtonian visco-elastic fluid past a stretching surface embedded in a porous medium. Meccanica **45**, 835–846 (2010)
13. O.D. Makinde, T. Chinyoka, Numerical study of unsteady hydromagnetic generalised Couette flow of a reactive third grade fluid with asymmetric convective cooling. Comput. Math. Appl. **61**, 1167–1179 (2011)
14. M. Massuodi, T.X. Phuoc, Flow of a generalized second grade non-Newtonian fluid with variable viscosity. Contin. Mech. Thermodyn. **16**, 529–538 (2004)

15. P. Mohapatra, S. Padhy, S. Tripathy, Finite difference analysis of the flow of a second order liquid past an infinite vertical plate in the presence of a magnetic field. Int. J. Appl. Mech. Eng. **6**, 71–90 (2001)
16. K.W. Morton, *Numerical Solution of Convection-Diffusion Problems* (Chapman and Hall, London, 1996)
17. J.C. Misra, G.C. Shit, S. Chandra, P.K. Kundu, Hydromagnetic flow and heat transfer of a second-grade viscoelastic fluid in a channel with oscillatory stretching walls: application to the dynamics of blood flow. J. Eng. Math. **69**, 91–100 (2011)
18. S.S. Okaya, Criticality and transition for a steady reactive plane couette flow of a viscous fluid. Mech. Res. Commun. **34**, 130–135 (2007)
19. S.S. Okaya, On the transition for a generalized Couette flow of a reactive third-grade fluid with viscous dissipation. Int. Commun. Heat Mass Transf. **35**, 188–196 (2008)
20. K. Ramesh, O. Ojjela, Entropy generation analysis of natural convective radiative second grade nanofluid flow between parallel plates in a porous medium. Appl. Math. Mech. **40**, 481–498 (2019)
21. L. Rundora, O.D. Makinde, Unsteady MHD flow of non-Newtonian fluid in a channel filled with a saturated porous medium with asymmetric Navier slip and convective heating. Appl. Math. Inf. Sci. Int. J. **12**, 483–493 (2018)
22. S.O. Salawu, N.K. Oladejo, M.S. Dada, Analysis of unsteady viscous dissipative poiseuille fluid flow of two-step exothermic chemical reaction through a porous channel with convective cooling. Ain Shams Eng. J. **10**, 565–572 (2019)
23. K. Vajravelu, J.R. Cannon, D. Rollins, Analytical and numerical solutions of nonlinear differential equations arising in non-Newtonian fluid flows. J. Math. Anal. Appl. **250**, 204–221 (2000)
24. M. Veerakrishna, G.S. Reddy, Unsteady MHD reactive flow of second grade fluid through porous medium in a rotating parallel plate channel. J. Anal. **27**, 103–120 (2019)
25. H. Xu, S. Wang, M. Zhao, Oscillatory flow of second grade fluid in a straight rectangular duct. J. Non-newtonian Fluid Mech. **279**, 104245 (2020)

Hopf Bifurcation and Stability Analysis of Delayed Lotka–Volterra Predator–Prey Model Having Disease for Both Existing Species

A. Ghasemabadi and M. H. Rahmani Doust

Abstract The study of biological problems using mathematical models not only has had significant advances but also it has attracted the attention of many scientists. Mathematical modeling of diseases enables one to predict when the disease occurs, and therefore it leads to the successful control of the diseases before it gets epidemics. This paper constructs a biological model in the mathematical aspect. In this paper, a delay predator–prey model is proposed with logistic growth in the prey population. It is assumed that this model includes an SIS infection in both prey and predator species. After dealing with disease in the prey population, we analyze the disease predator population. Moreover, we prove the existence of Hopf bifurcation for this system by analyzing characteristic equations. Then important threshold quantities are identified. Our theoretical study indicates that threshold quantities R_1 and R_2 are important when a transmissible disease runs among the prey population and the predators are disease, respectively.

Keywords Hopf bifurcation · Asymptotically stability · Prey–predator · Lotka–Volterra · Delay

AMS Subject Classification 34C23 · 34L20 · 92D40

A. Ghasemabadi
Esfarayen University of Technology, Esfarayen, North Khorasan, Iran
e-mail: ghasemabadi.math@gmail.com

M. H. Rahmani Doust (✉)
Department of Mathematics, University of Neyshabur, Neyshabur, Iran
e-mail: mh.rahmanidoust@neyshabur.ac.ir

S. K. Paikray et al. (eds.), *New Trends in Applied Analysis and Computational Mathematics*, Advances in Intelligent Systems and Computing 1356,
https://doi.org/10.1007/978-981-16-1402-6_13

1 Introduction

The Lotka–Volterra model is famous and well known which may help in so many areas of science especially in biology, ecology, etc. Some of the main applications may be seen as: marriage [8], war [1], genetics [28], biochemical reactors [2], computer games [9], and urban sewage purification systems. In fact, they can been simulated based on Lotka–Volterra equations. There is a great interest in differential equations with time delay, especially in ecological and biological models because in these models the reproduction is not instantaneous. The time delay has a destabilizing effect on the models of population dynamics, and often it has the ability to alter the dynamical behavior of a model system significantly.

An application of control and optimal treatment for predator–prey model is studied in [27]. Some of the Lotka–Volterra models having intraspecific or interspecific predator–prey, competition, and coexistence interactions are studied in [15, 16, 18, 21, 24, 25]. Some of the Lotka–Volterra models having harvested factor are analyzed in [17, 19, 20, 22]. The stability of solutions for the Gauss model concluding harvested factor is investigated in [29]. The behavior of solutions for the Gauss model having two preys and one predator is studied in [26]. The stable orbits for a prey–predator system are studied in [23]. Discussion of differential equations and dynamical system; mathematical modeling; and their applications in biology, ecology, epidemiology, marriage dynamic, etc. may be found in some text books such as [3, 5, 12, 14, 30]. The analysis and application of mathematical modeling for war equations may be found in [1]. One is able to see the stability and behavior of solution for complex model ecosystems in [13]. Readers can find the effects of nutrient recycling and food-chain length on resilience in [4]. The stability and bifurcation in a generalized delay prey–predator model, and Lotka–Volterra model with disease are analyzed in [6, 7]. A nonstandard finite difference scheme for a SEI epidemic model is analyzed in [11]. Indeed, the qualitative properties of the system such as positivity, stability of the equilibria, and Neimark–Sacker bifurcation are studied. The functional response of predators to prey density and its role in mimicry and population regulation [10].

The aim of the present study is the investigation of efficiency disease in both prey and predator species. When prey or predator gets disease, species is split out into two groups, the susceptible and the infected populations. Now, consider the following delay Lotka–Volterra equations:

$$\begin{cases} \dot{x} = \left(r \left(1 - \dfrac{x(t)}{K_1} \right) - a\, y(t) \right) x(t), \\ \dot{y} = (k\, a\, x(t-\tau) - d_2)\; y(t), \end{cases} \tag{1.1}$$

where the variables x and y are prey and predator densities, respectively. In the absence of predator, the growth rate of the prey is logistic, the predation rate is $a\,x\,y$. Constants k and K_1 are the efficiency predation into new predator and the prey-carrying capacity, respectively. The delay or lag can represent incubation periods, transport delays, gestation times, or simply lump complicated biological processes together, accounting only for the time required for these processes to occur. Transmissible disease in a species is rapidly becoming a major field of study in its own right. In most papers, the authors considered the case when disease is spreading among the predator or prey population only.

In Sect. 2, two models with delay time are analyzed. First, disease in the prey population is supposed. Some conditions have been obtained for local and global stability of the nonnegative equilibrium point. Moreover, we observe that the basic reproduction number R_1 is much more important in determining the behavior of the predator–prey system when the prey gets disease. In the second case, we consider the case that disease spreads among the predator population. The model considers the case here also studies its stability where an epidemic runs among the predator population but differs from previous models.

2 Main Results

In this section, we analyze a disease modeling the transmission of certain infectious disease among the population of animals. We assume that there are diseases in both existing species. The population is classified into the following classes:

- The number of susceptible animals $S(t)$.
- The number of infected who are infectious and can transmit the infection $I(t)$.

For system (1.1), SIS model in the prey population is supposed. The total size of the prey population is $x = S + I$, where S is the number of susceptible prey and I is the number of infectious prey. Similarly, the total size of the predator population is $y = S_2 + I_2$, where S_2 is the number of susceptible predators and I_2 is the number of infectious predators. We use the next-generation matrix approach for calculating the threshold. Therefore, the two epidemiological threshold quantities are

$$
\begin{aligned}
R_1 &= \frac{\beta_1}{\gamma_1 + d_1 + (1 - a_1)\, r\, d_2/k\, a\, K_1 + a\, y^*}, \\
R_2 &= \frac{\beta_2}{d_2 + \gamma_2}.
\end{aligned} \tag{2.1}
$$

Important threshold quantities R_1 and R_2 are identified as above. Our theoretical study indicates that threshold quantity R_1 is important when a transmissible disease runs among the preys' population. And also, threshold quantity R_2 is important when the predators' population is disease.

2.1 Model of Disease Prey

Now, consider a predator–prey SIS model with disease in the prey population as follows:

$$\begin{cases} \dot{S}(t) = \left[b_1 - \dfrac{a_1\, r\, x(t)}{K_1}\right] x(t) - \left[d_1 + (1-a_1)\dfrac{r\, x(t)}{K_1}\right] S(t) \\ \quad - a\, y(t)\, S(t) - \beta_1 \dfrac{S(t) I(t)}{x(t)} + \gamma_1 I(t) \\ \dot{I}(t) = \beta_1 \dfrac{S(t) I(t)}{x} - \gamma_1 I(t) - \left[d_1 + (1-a_1)\dfrac{r\, x(t)}{K_1}\right] I(t) \\ \quad - a\, y(t)\, I(t) \\ \dot{x} = \left[r(1 - \dfrac{x(t)}{K_1}) - a\, y(t)\right] x(t), \\ \dot{y} = [k\, a\, x(t-\tau) - d_2]\, y(t). \end{cases} \tag{2.2}$$

In the absence of disease, coefficient $b_1 - \frac{a_1 r x}{K_1}$ is the birth rate and $d_1 + (1-a_1)\frac{r x}{K_1}$ is death rate of the prey population, where

$$r = b_1 - d_1 \text{ and } 0 \leq a_1 < 1.$$

Since $x = S + I$, the last system can be reduced to the following three-dimensional system:

$$\begin{cases} \dot{I} = \left[\beta_1 \dfrac{x(t) - I(t)}{x(t)} - \gamma_1 - d_1 - (1-a_1)\dfrac{r\, x(t)}{K_1} - a\, y(t)\right] I(t) \\ \dot{x} = \left[r(1 - \dfrac{x(t)}{K_1}) - a\, y(t)\right] x(t), \\ \dot{y} = [k\, a\, x(t-\tau) - d_2]\, y(t). \end{cases} \tag{2.3}$$

The following equilibria exist for the above system:

$$E_{31} = (0, 0, 0), \quad E_{32} = (0, K_1, 0)$$
$$E_{33} = (0, x^*, y^*) \quad E_{34} = (I^*, x^*, y^*),$$

where x^*, y^*, I^* are defined as follows:

$$x^* = \frac{d_2}{k\, a},$$
$$y^* = \frac{r_1}{a}\left[1 - \frac{d_2}{k\, a\, K_1}\right],$$
$$I^* = x^* \left(1 - \frac{1}{R_1}\right).$$

A sufficient condition is obtained for local and global stability at the positive equilibrium point E_{34}. The related characteristic equation is determined as

$$(a_{11} - \lambda)[\lambda^2 - a_{22}\lambda - a_{23}\, a_{32}\, e^{-\lambda\tau}] = 0, \tag{2.4}$$

where

$$a_{11} = \frac{-\beta_1 I^*}{x^*}, \qquad a_{22} = \frac{-r\, x^*}{K_1},$$
$$a_{23} = -ax^*, \qquad a_{32} = k\, a\, y^*.$$

Theorem 2.1 *Suppose that $R_1 > 1$ and $\tau = 0$, then the equilibrium point $E_{34} = (I^*, x^*, y^*)$ for system (2.3) is locally asymptotically stable.*

Proof A solution of characteristic equation (2.4) is

$$\lambda_1 = -\beta_1 \left(1 - \frac{1}{R_1}\right) < 0.$$

Now consider $\tau = 0$, the following equation is obtained:

$$\lambda^2 - a_{22}\lambda - a_{23}\, a_{32} = 0.$$

By using the Routh–Hurwitz, this equation has the negative eigenvalues provided that

$$a_{22} < 0, \qquad a_{23}a_{32} < 0,$$

where

$$a_{22} = \frac{-rx^*}{K_1} < 0, \quad a_{23}a_{32} = -d_2 r_1 \left(1 - \frac{d_2}{k\, a\, K_1}\right).$$

Since y^* is positive, the inequality $a_{23}a_{32} < 0$ holds. Hence, the equilibrium point E_{34} is locally asymptotically stable. □

By substituting $\lambda = i\varphi$ in equation $\lambda^2 - a_{22}\lambda - a_{23}\, a_{32}\, e^{-\lambda\tau} = 0$, we have

$$\varphi^2 + a_{22}\varphi\, i + a_{23}a_{32}e^{-\varphi\, i\, \tau} = 0.$$

Separating the real and imaginary parts implies

$$\begin{cases} a_{23}\, a_{32}\cos\varphi\tau = -\varphi^2, \\ a_{23}\, a_{32}\sin\varphi\tau = a_{22}\,\varphi. \end{cases}$$

Adding up the squares of the corresponding sides of the above equations, we obtain the following equation:

$$\varphi^4 + a_{22}^2\varphi^2 - a_{23}^2\, a_{32}^2 = 0. \tag{2.5}$$

By using $a_{22}^4 + 4a_{23}^2\, a_{32}^2 > 0$, we see thus Eq. (2.5) has at least one positive root. Take $\varphi_0 = \sqrt{\mu_0}$, where

$$\mu_0 = \frac{-r^2\, x^{*^2}}{2K_1^2} + \frac{x^* \sqrt{r^4\, x^{*^2} + 4a^4\, k^2\, y^{*^2}}}{2K_1^2}. \tag{2.6}$$

The delay τ is used as bifurcation parameter. Let $\lambda(\tau) = w(\tau) + i\varphi(\tau)$ be the eigenvalue of (2.4) such that for some initial value of the bifurcation parameter τ_0, we have $w(\tau_0) = 0$ and $\varphi(\tau_0) = \varphi_0$. Therefore

$$\tau_j = \frac{1}{\varphi_0} \arccos\left(\frac{\varphi_0^2}{a_{23}\, a_{32}} + 2j\pi\right) \qquad for \qquad j = 0, 1, 2, \ldots$$

Suppose that

$$h(\mu) = \mu^3 + ((a_{11} + a_{22})^2 - 2\, a_{22}\, a_{11})\mu^2 + (a_{11}^2\, a_{22}^2 - a_{23}^2\, a_{32}^2)\, \mu + a_{11}^2\, a_{23}^2\, a_{32}^2.$$

We therefore may establish the following theorem:

Theorem 2.2 *Model (2.3) undergoes a Hopf bifurcation provided*

$$h'(\varphi_0^2) \neq 0.$$

Proof To establish the Hopf bifurcation at $\tau = \tau_0$, we show that $\frac{dRe\lambda(\tau)}{d\tau} \neq 0$. From (2.4) derivation with respect to τ, the following relation is obtained:

$$\begin{aligned}[3\lambda^2 - 2(a_{11} + a_{22})\lambda + a_{11}\, a_{22}]\frac{d\lambda}{d\tau} =& [\tau(a_{11}\, a_{23}\, a_{32} - a_{23}\, a_{32}\lambda) + a_{23}\, a_{32}]e^{-\lambda\tau}\frac{d\lambda}{d\tau} \\ &+ (a_{11}\, a_{23}\, a_{32} - a_{23}\, a_{32}\lambda)\lambda e^{-\lambda\tau},\end{aligned}$$

which implies

$$\begin{aligned}\left(\frac{d\lambda}{d\tau}\right)^{-1} =& \frac{3\lambda^2 - 2(a_{11} + a_{22})\lambda + a_{11}\, a_{22} - \tau(a_{11}\, a_{23}\, a_{32} - a_{23}\, a_{32}\lambda)e^{-\lambda\tau} - a_{23}\, a_{32}e^{-\lambda\tau}}{(a_{11}\, a_{23}\, a_{32} - a_{23}\, a_{32}\lambda)\lambda e^{-\lambda\tau}} \\ =& \frac{3\lambda^2 - 2(a_{11} + a_{22})\lambda + a_{11}\, a_{22}}{\lambda(a_{11}\, a_{23}\, a_{32} - a_{23}\, a_{32}\lambda)e^{-\lambda\tau}} + \frac{-a_{23}\, a_{32}}{(a_{11}\, a_{23}\, a_{32} - a_{23}\, a_{32}\lambda)\lambda} - \frac{\tau}{\lambda}.\end{aligned}$$

Thus,

$$
\begin{aligned}
&\left\{\left(Re\frac{d\lambda}{d\tau}\right)^{-1}\right\}_{\lambda=i\varphi_0}\\
&=\left\{Re\left[\frac{3\lambda^2-2(a_{11}+a_{22})\,\lambda+a_{11}\,a_{22}}{\lambda(a_{11}\,a_{23}\,a_{32}\;-a_{23}\,a_{32}\,\lambda)e^{-\lambda\tau}}+\frac{-a_{23}\,a_{32}}{(a_{11}\,a_{23}\,a_{32}\;-a_{23}\,a_{32}\,\lambda)\lambda}-\frac{\tau}{\lambda}\right]\right\}_{\lambda=i\varphi_0}\\
&=\left\{Re\left[\frac{a_{22}\,a_{11}-3\varphi_0^2-2(a_{11}+a_{22})\varphi_0 i}{(a_{22}\,a_{11}\varphi_0^2-\varphi_0^4)-i(a_{11}+a_{22})\varphi_0^3}\right.\right.\\
&\quad\left.\left.+\frac{-a_{23}\,a_{32}}{a_{23}\,a_{32}\varphi_0^2+a_{11}\,a_{23}\,a_{32}\varphi_0 i}\right]\right\}\\
&=\frac{3\varphi_0^4+(2(a_{11}+a_{22})^2-4a_{22}\,a_{11})\varphi_0^2+a_{22}^2\,a_{11}^2}{(a_{22}\,a_{11}\,\varphi_0-\varphi_0^3)^2+(a_{11}+a_{22})^2\varphi_0^4}\\
&\quad-\frac{(a_{23}\,a_{32})^2}{(a_{23}\,a_{32})^2\varphi_0^2+(a_{11}\,a_{23}\,a_{32})^2}\\
&=\frac{h'(\mu_0)}{(a_{23}\,a_{32})^2\varphi_0^2+(a_{11}\,a_{23}\,a_{32})^2},
\end{aligned}
$$

where $\mu_0=\varphi_0^2$ and

$$
h(\mu)=\mu^3+((a_{11}+a_{22})^2-2\,a_{22}\,a_{11})\mu^2+(a_{11}^2\,a_{22}^2-a_{23}^2\,a_{32}^2)\,\mu+a_{11}^2\,a_{23}^2\,a_{32}^2.
$$

Thus,

$$
\left\{\left(Re\frac{d\lambda}{d\tau}\right)^{-1}\right\}_{\lambda=i\varphi_0}=\frac{h'(\mu_0)}{(a_{23}\,a_{32})^2\varphi_0^2+(a_{11}\,a_{23}\,a_{32})^2}.
$$

Therefore , if condition $h'(\mu_0)\neq 0$, then $\{(Re\frac{d\lambda}{d\tau})^{-1}\}_{\lambda=i\varphi_0}\neq 0$ and the transversality condition holds, and hence Hopf bifurcation occurs at $\tau=\tau_0$. □

2.2 Model of Disease Predator

We now consider the following system:

$$
\begin{cases}
\dot{x}=\left[r(1-\dfrac{x(t)}{K_1})-a\,y(t)\right]x(t),\\
\dot{I}_2=\left[\beta_2\dfrac{y(t)-I_2(t)}{y(t)}-d_2-\gamma_2\right]I_2(t)\\
\dot{y}=[k\,a\,x(t-\tau)-d_2]\,y(t).
\end{cases}
\tag{2.7}
$$

The above system has interior equilibrium point $E_{43}=(x^*,I_2^*,y^*)$ where

$$x^* = \frac{d_2}{k\,a}, \quad y^* = \frac{r_1}{a}\left[1 - \frac{d_2}{k\,a\,K_1}\right], \quad I_2^* = y^*(1 - \frac{1}{R_2}).$$

Equilibrium point E_{43} is positive provided $R_2 > 1$ and $y^* > 0$. The following characteristic equation at point E_{43} is obtained:

$$(b_{22} - \lambda)\,(\lambda^2 - b_{11}\,\lambda - b_{13}\,b_{31}\,e^{-\lambda\tau}) = 0, \tag{2.8}$$

where

$$b_{11} = \frac{-r}{K_1}x^*, \quad b_{22} = -\beta_2\left(1 - \frac{1}{R_2}\right),$$
$$b_{13} = -a\,x^*, \quad b_{31} = k\,a\,y^*.$$

The following theorem which is proved in [7] shows that the equilibrium point $E^* = (x^*, y^*)$ is the globally asymptotically stable.

Theorem 2.3 *If* $\frac{d_2}{k\,a\,K_1} < 1$ *and* $\tau = 0$ *then the equilibrium point* $E^* = (x^*, y^*)$ *is locally asymptotically stable. Furthermore,* E_{13} *is globally asymptotically stable.*

We are now in a position that to construct the following theorem which proves at that interior equilibrium point E_{43} is locally and globally asymptotically stable.

Theorem 2.4 *Consider system (2.7). Moreover, assume that* $R_2 > 1$, $\frac{d_2}{k\,a\,K_1} < 1$ *and* $\tau = 0$, *then the equilibrium point* E_{43} *is locally and globally asymptotically stable.*

Proof A root of the characteristic equation (2.8) is

$$\lambda_1 = -\beta_2\left(1 - \frac{1}{R_2}\right) < 0.$$

By using the Routh–Hurwitz, we see that equation

$$\lambda^2 - b_{11}\lambda - b_{13}\,b_{31} = 0$$

has the negative eigenvalues provided that

$$b_{11} < 0, \quad b_{31}\,b_{13} < 0,$$

where

$$b_{11} = \frac{-rx^*}{K_1} < 0, \quad b_{31}b_{13} = -k\,a^2\,x^*\,y^* < 0.$$

Hence, the equilibrium point E_{34} is locally asymptotically stable.

We now see that condition $\frac{d_2}{k\,a\,K_1} < 1$ and Theorem (2.3) imply that $x \to x^*$ and $y \to y^*$ as $t \to \infty$. It is clear that $I_2 \to I_2^*$ when $R_2 > 1$. Since the equilibrium point

E_{43} is locally asymptotically stable, then E_{43} is globally asymptotically stable which complete the proof. □

Now consider the equation $\lambda^2 - b_{11}\,\lambda - b_{13}\,b_{31}\,e^{-\lambda\tau} = 0$. Substituting $\lambda = i\,\varphi$ in the last equation, we have

$$\varphi^2 + b_{11}\varphi i + b_{13}\,b_{31}(\cos(\varphi\,\tau) - i\,\sin(\varphi\,\tau)) = 0.$$

Also separating the real and imaginary parts implies

$$\begin{cases} b_{13}\,b_{31}\,\cos\varphi\tau = -\varphi^2, \\ b_{13}\,b_{31}\,\sin\varphi\tau = b_{11}\,\varphi. \end{cases}$$

Adding up the squares of the corresponding sides of the above equations, we obtain the following equation:

$$\varphi^4 + b_{11}^2\varphi^2 - b_{13}^2\,b_{31}^2 = 0. \tag{2.9}$$

By using $b_{11}^4 + 4b_{13}^2\,b_{31}^2 > 0$, thus Eq. (2.9) has at least one positive root. Take $\varphi_0 = \sqrt{\mu_0}$, where

$$\mu_0 = \frac{-r^2\,x^{*^2}}{2K_1^2} + \frac{x^*\sqrt{r^4\,x^{*^2} + 4a^4\,k^2\,y^{*^2}}}{2K_1^2}. \tag{2.10}$$

The delay τ is used as bifurcation parameter.

Now let $\lambda(\tau) = w(\tau) + i\varphi(\tau)$ be the eigenvalue for characteristic equation (2.8) such that for some initial value of the bifurcation parameter τ_0, we have $w(\tau_0) = 0$, and $\varphi(\tau_0) = \varphi_0$. Therefore,

$$\tau_j = \frac{1}{\varphi_0}\arccos\left(\frac{\varphi_0^2}{b_{13}\,b_{31}} + 2j\pi\right) \qquad for \qquad j = 0, 1, 2, \ldots$$

Now suppose that

$$h(\mu) = \mu^3 + ((b_{11} + b_{22})^2 - 2\,b_{22}\,b_{11})\mu^2 + (b_{11}^2\,b_{22}^2 - b_{13}^2\,b_{31}^2)\,\mu - b_{22}^2\,b_{13}^2\,b_{31}^2.$$

We therefore may establish the following theorem:

Theorem 2.5 *System (2.7) undergoes a Hopf bifurcation provided*

$$h'(\varphi_0^2) \neq 0.$$

Proof To establish the Hopf bifurcation at $\tau = \tau_0$, we show that $\frac{dRe\lambda(\tau)}{d\tau} \neq 0$. From (2.8) derivation with respect to τ, the following relation is obtained:

$$[3\lambda^2 - 2(b_{11} + b_{22})\lambda + b_{11}\, b_{22}]\frac{d\lambda}{d\tau} = [\tau(b_{22}\, b_{13}\, b_{31} - b_{13}\, b_{31}\lambda) + b_{13}\, b_{31}]e^{-\lambda\tau}\frac{d\lambda}{d\tau}$$
$$+ (b_{22}\, b_{13}\, b_{31} - b_{13}\, b_{31}\lambda)\lambda e^{-\lambda\tau}.$$

This gives

$$\left(\frac{d\lambda}{d\tau}\right)^{-1} = \frac{3\lambda^2 - 2(b_{11} + b_{22})\lambda + b_{11}\, b_{22} - \tau(b_{22}\, b_{31}\, b_{13} - b_{13}\, b_{31}\lambda)e^{-\lambda\tau} - b_{13}\, b_{31}e^{-\lambda\tau}}{(b_{22}\, b_{13}\, b_{31} - b_{13}\, b_{31}\lambda)\lambda e^{-\lambda\tau}}$$
$$= \frac{3\lambda^2 - 2(b_{11} + b_{22})\lambda + b_{11}\, b_{22}}{\lambda(b_{22}\, b_{13}\, b_{31} - b_{13}\, b_{31}\lambda)e^{-\lambda\tau}} + \frac{-b_{13}\, b_{31}}{(b_{22}\, b_{13}\, b_{31} - b_{13}\, b_{31}\lambda)\lambda} - \frac{\tau}{\lambda}.$$

Thus,

$$\left\{\left(Re\frac{d\lambda}{d\tau}\right)^{-1}\right\}_{\lambda=i\varphi_0}$$
$$= \left\{Re\left[\frac{3\lambda^2 - 2(b_{11} + b_{22})\,\lambda + b_{11}\, b_{22}}{\lambda(b_{22}\, b_{13}\, b_{31}\; - b_{13}\, b_{31}\;\lambda)e^{-\lambda\tau}} + \frac{-b_{13}\, b_{31}}{(b_{22}\, b_{13}\, b_{31}\; - b_{13}\, b_{31}\;\lambda)\lambda} - \frac{\tau}{\lambda}\right]\right\}_{\lambda=i\varphi_0}$$
$$= \left\{Re\left[\frac{b_{11}\, b_{22} - 3\varphi_0^2 - 2(b_{11} + b_{22})\varphi_0 i}{(b_{22}\, b_{11}\varphi_0^2 - \varphi_0^4) - i(b_{11} + b_{22})\varphi_0^3}\right.\right.$$
$$\left.\left.+ \frac{-b_{13}\, b_{31}}{b_{13}\, b_{31}\varphi_0^2 + b_{22}\, b_{13}\, b_{31}\varphi_0 i}\right]\right\}$$
$$= \frac{3\varphi_0^4 + (2(b_{11} + b_{22})^2 - 4b_{22}\, b_{11})\varphi_0^2 + b_{22}^2\, b_{11}^2}{(b_{22}\, b_{11}\,\varphi_0 - \varphi_0^3)^2 + (b_{11} + b_{22})^2\varphi_0^4}$$
$$- \frac{(b_{13}\, b_{31})^2}{(b_{13}\, b_{31})^2\varphi_0^2 + (b_{22}\, b_{13}\, b_{31})^2}$$
$$= \frac{h'(\mu_0)}{(b_{13}\, b_{31})^2\varphi_0^2 + (b_{22}\, b_{13}\, b_{31})^2},$$

where $\mu_0 = \varphi_0^2$ and

$$h(\mu) = \mu^3 + ((b_{11} + b_{22})^2 - 2\, b_{22}\, b_{11})\mu^2 + (b_{11}^2\, b_{22}^2 - b_{13}^2\, b_{31}^2)\,\mu - b_{22}^2\, b_{13}^2\, b_{31}^2.$$

Thus,

$$\left\{\left(Re\frac{d\lambda}{d\tau}\right)^{-1}\right\}_{\lambda=i\varphi_0} = \frac{h'(\mu_0)}{(b_{13}\, b_{31})^2\varphi_0^2 + (b_{22}\, b_{31}\, b_{13})^2}.$$

Therefore, if condition $h'(\mu_0) \neq 0$, then $\{(Re\frac{d\lambda}{d\tau})^{-1}\}_{\lambda=i\varphi_0} \neq 0$ and the transversality condition holds, and hence Hopf bifurcation occurs at $\tau = \tau_0$. □

3 Concluding Remarks

We have proposed and analyzed a mathematical model that consists of two non-linear differential equations for six different populations, namely, susceptible prey S and infected prey I, predator y or prey x, susceptible predator S_2, infected predator I_2. In every case, a sufficient condition is obtained for local and global stability at the interior equilibrium point. We have investigated that there are two epidemiological threshold quantities (2.1) for the model. The point (x^*, y^*, I^*), where

$$x^* = \frac{d_2}{k\,a}, \quad y^* = r\left[1 - \frac{d_2}{k\,a\,K_1}\right]/a \qquad I^* = \frac{x^*}{\beta_1}(R_1 - 1) \tag{3.1}$$

are components of the equilibrium prey level in the presence of the predator when disease is absent in the predator.

When the predator population gets disease, the equilibrium point is given as follows:

$$x^* = \frac{d_2}{k\,a}, \quad y^* = r_1\left[1 - \frac{d_2}{k\,a\,K_1}\right]/a \qquad I_2^* = y^*\left(1 - \frac{1}{R_2}\right). \tag{3.2}$$

We also observed that the threshold R_1 is much more important in determining the behavior of the predator–prey system when the prey is disease. Equilibrium point E_{34} is locally and globally asymptotically stable for system (2.3) provided $R_1 > 1$. Also for said model, threshold R_2 is much more important when a transmissible disease runs among the predator species. Equilibrium point E_{43} is globally asymptotically stable provided $R_2 > 1$.

The existence of Hopf bifurcation is proved for models (2.3) and (2.7). Biologically, it implies that delay is crucial for a predator–prey system.

Acknowledgements The authors thank the International Conference on Advances in Mathematics and Computing (ICAMC—2020), Department of Mathematics, V.S.S. University of Technology and readers of Springer, for making this a success.

References

1. S. Beckerman, The equations of war. J. Curr. Anthropol. **32**(5), 636–640 (1991)
2. E. Bruce Nauman, *Chemical Reactor Design, Optimization and Scaleup* (Wiley, Hoboken, 2008)
3. S. Busenberg, M. Martelli, *Differential Equations Models in Biology, Epidemiology and Ecology*, Proceedings of a Conference held in Claremont California, 13–16 January 1990 (Springer, 1991)
4. D.L. DeAngelis, S.M. Bartell, A.L. Brenkert, Effects of nutrient recycling and food-chain length on resilience. Am. Nat. **134**(5), 778–805 (1989)
5. H.I. Freedman, *Deterministic Mathematical Models in Population Ecology*. Monograph Textbooks Pure Applied Mathematics, vol. 57 (Marcel Dekker, New York, 1980)

6. A. Ghasemabadi, Stability and bifurcation in a generalized delay prey-predator model. Nonlinear Dyn. **90**, 2239–2251 (2017)
7. A. Ghasemabadi, M.H. Rahmani Doust, Investigating the dynamics of Lotka-Volterra model with disease in the prey and predator species. Int. J. Nonlinear Anal. Appl. **10**(1) (2020). Article in Press
8. J.M. Gottman, J.D. Murray, C.C. Swanson, R. Tyson, R.K. Swanson, *The Mathematics of Marriage Dynamic Nonlinear Model* (The MIT Press, Cambridge, 2002)
9. C. Grabner, H. Hahn, U. Leopold-Wildburger, S. Pickl, Analyzing the sustainability of harvesting behavior and the relationship to personality traits in a simulated Lotka-Volterra biotope. Eur. J. Oper. Res. **193**, 761–767 (2009)
10. C.S. Holling, The functional response of predators to prey density and its role in mimicry and population regulation. Mem. Entomol. Soc. Can. **45**, 3–60 (1965)
11. R. Memmarbashi, F. Alipour, A. Ghasemabadi, A nonstandard finite difference scheme for a SEI epidemic model. J. Math. **49**(3), 133–147 (2017)
12. J.D. Murray, *Mathematical Biology Vol. I: An Introduction* (Springer, 2002)
13. L. Nunney, The stability of complex model ecosystems. Am. Nat. **115**(5), 639–649 (1980)
14. L. Perko, *Differential Equations and Dynamical System* (Springer, New York, 2000)
15. M.H. Rahmani Doust, Analysis of system of coexistence equation. Ultra Sci. **19**(1), 49–56 (2007)
16. M.H. Rahmani Doust, An analysis of system of competition equations, Ultra Sci. **19**(2), 393–400 (2007)
17. M.H. Rahmani Doust, The efficiency of harvested Lotka-Volterra predato-prey model. Caspian J. Math. Sci. **5**(1), 51–59 (2015)
18. M.H. Rahmani Doust, R. Rangarajan, A global analysis of Lotka-Volterra predator-prey model with interaspecies competition. J. Anal. Comput. (JAC) **4**(1), 43–50 (2008)
19. M.H. Rahmani Doust, F.Haghighifar, The Lotka-Volterra predator-prey system; having interspecific interactions or harvested factor. J. Intell. Syst. Res. (JISR) **5**(2), 105–111 (2011)
20. M.H. Rahmani Doust , F.Haghighifar, Two species Lotka-Volterra harvested model having competition interaspecific factor. J. Anal. Comput. (JAC) **7**(2), 105–112 (2011)
21. M.H. Rahmani Doust, S. Gholizade, An analysis of the modified Lotka-Volterra predator-prey equations. Gen. Math. Notes **25**(2), 1–5 (2014)
22. M.H. Rahmani Doust, F. Haghighifarm, The stability of some systems of harvested Lotka-Volterra predator-prey equations. Caspian J. Math. Sci. **3**(1), 131–139 (2014)
23. M.H. Rahmani Doust, S. Gholizade, Prey-Predator system; having stable orbit. Glob. Anal. Discret. Math. **1**(1), 21–27 (2016)
24. M.H. Rahmani Doust, R. Rangarajan, M.N. Modoodi, Analysis of predator-prey equations with intraspecies coexistence. Ultra Sci. **20**(3), 819–824 (2008)
25. M.H. Rahmani Doust, F. Haghighifar, M.N. Modoodi, The Lotka-Volterra competition model. Proc. Jangjeon Math. Soc. **15**(3), 259–265 (2012)
26. M.H. Rahmani Doust, F. Haghighifar, V. Lokesha, The stability of gauss model having two-preys and one-predator. Proc. Jangjeon Math. Soc. **17**(3), 347–354 (2014)
27. M.H. Rahmani Doust, M. Shirazian, M. Shamsabadi, Application of control and optimal treatment for predator-prey model. IJNAO **10**(1) (2020)
28. L. Randy, L. Haupt, S.E. Haupt, *Practical Genetic Algorithms* (Wiley, Hoboken, 2004)
29. M. Saraj, M.H. Rahmani Doust, F. Haghighifar, The stability of gauss model; having harvested factor. Selcuk J. Appl. Math. **13**(2), 3–10 (2012)
30. P. Turchin, *Complex Population Dynamics: A Theoretical/Empirical Syn-thesis* (Princeton University Press, Princeton, 2003)

An EOQ Model Without Shortages with Uncertain Cost Associated with Some Fuzzy Parameters and Interval Parameters

Anuradha Sahoo and Arati Nath

Abstract In this paper, economic order quantity (EOQ) models without shortages for single item and multi-items are presented. Here, the holding cost of the item is a continuous function of the order quantity. The costs involved in this model are imprecise in nature. The main contributions of this research are as follows: The proposed EOQ model is discussed in two cases by describing the model in an uncertain environment. In case-1, EOQ models with fuzzy parameters (like ordering cost, holding cost, and unit product cost) are considered. Here all the fuzzy parameters are represented by trapezoidal fuzzy numbers. The said EOQ model is carried out by using the signed-distance method. In case-2, EOQ models with interval parameters (like ordering cost, holding cost, unit product cost, and the total money investment for the quantities) are considered. This proposed model is solved by using interval linear programming problem (ILPP) technique based on the best and the worst optimum values of the objective function. Numerical examples are given to exemplify the proposed model and also the results of different models are compared.

Keywords EOQ · Trapezoidal fuzzy number · Interval number · Signed-distance method · Interval linear programming problem (ILPP)

1 Introduction

An EOQ model in inventory management is one of the most useful models. The formulation of this EOQ model depends on various costs value and this cost cannot be fixed in reality. In real life, the costs of the items depend on many factors like quality of the item, stock level of the item, selling price, duration of storage, etc. Uncertainty arises in partially apparent or stochastic environments, along with illiteracy, idleness, or both in any number of grounds, together with insurance, philosophy,

A. Sahoo (✉) · A. Nath
Department of Mathematics, ITER, Siksha 'O' Anusandhan (Deemed To Be University), Bhubaneswar 751030, Odisha, India

S. K. Paikray et al. (eds.), *New Trends in Applied Analysis and Computational Mathematics*, Advances in Intelligent Systems and Computing 1356,
https://doi.org/10.1007/978-981-16-1402-6_14

physics, engineering, ecology, information science, metrology, finance, and meteorology. Nowadays, almost every real-world problem involves the cost as uncertain. Several researchers have studied the inventory models in uncertain environments in different situations. By using geometric programming (GP) approach, Kotb and Fergany [1] developed a multi-item EOQ model with varying holding cost and obtained the optimal solutions. But in our paper, we used their proposed EOQ model for both single item and multi-item and considered their model in uncertain environment by taking different parameters as fuzzy numbers and interval numbers. Also make a comparison of the optimal solution for both models.

Shaocheng [2] developed a fuzzy number and interval number linear programming. The problems of fuzzy number linear programming (FLP) with fuzzy number coefficients are approached in two ways, such as "fuzzy decisive set approach" and "Interval number linear programming approach," for several membership levels. And the problems of interval number linear programming (ILP) with interval number coefficients are approached by taking "minimum value range" and "maximum value range" inequalities as constraint conditions, reduced it into two classical linear programming, and obtained an optimal interval solution to it. By using the GP approach, Aboul-El-Ata and Kotb [3] proposed a multi-item EOQ model by considering holding cost to be a continuous function of the order quantity under varying holding cost. In this paper, a classical EOQ model and an EOQ model by taking holding costs as constant are derived without any constraint. Chinneck and Ramadan [4] proposed a new method to find the best and the worst optimum values in which some or all coefficients of linear programming are stated as intervals, where the point settings of the intervals coefficients that yield the range of the optimized objective function, and the coefficient settings give some insight into the likelihood of these extremes. To maximize the total profit and to determine the optimal order size, Chang [5] first proposed a model with a fuzzy defective rate. Then, they presented that model with defective rate and annual demand in a fuzzy environment. To find the estimate of total profit per unit time in the fuzzy sense and then derive the corresponding optimal lot size, they use a signed-distance method for fuzzy numbers. An inventory model without shortage is considered by taking ordering cost and holding cost in the fuzzy environment by Sayed and Aziz [6]. Triangular fuzzy numbers are used to obtain the optimum order quantity. Also signed-distance method is used for the defuzzification of discussed fuzzy model. A multi-item EOQ model with varying holding cost is developed by Kotb and Fergany [1] and they obtained the optimal solutions by using geometric programming (GP) approach. Dutta and Kumar [7] proposed an inventory model in a fuzzy environment where shortages are not allowed and determine the optimal total cost and the optimal order quantity. They use trapezoidal fuzzy numbers. The computation of economic order quantity is carried out through the defuzzification process by using the signed-distance method. An interval linear programming problem proposed by Allahdadi and Nehi [8] for finding the best and worst optimum values of the objective function by illustrating the weakness of the Shaocheng method by taking both equality and inequality constraints, when there is at least one equality constraint is present in their proposed model. Sahoo and Dash [9] consider purchasing cost as a fuzzy number and demand as a random variable in a fuzzy environment to formulate a single-period inventory model for multi-item newsboy problem where there is the occurrence of randomness and fuzziness. Buckley's minimization concept is used to obtain the expected profit and optimum order quantity. Rajalakshmi and Rosario [10] investigated a fuzzy inventory model with allowable shortage which is completely backlogged. We fuzzify the carrying cost, backorder cost, and ordering cost using triangular, trapezoidal, pentagonal fuzzy numbers to obtain the fuzzy total cost. The signed-distance method is used for defuzzification to estimate the total cost. A new methodology is developed by Ren et al. [11] for solving the interval bi-level linear programming problem. So, they considered all the coefficients of objective functions and constraints as interval numbers. The concept of a preference δ-optimal solution is also given in this paper. Later, the constructed deterministic nonlinear bi-level problem is solved with the help of estimation of distribution algorithm. Numerical examples are also provided to determine the effectiveness of the proposed model. Based on the binding constraint indices of the optimal solution to the linear programming (LP) model, Ashayerinasab et al. [12] developed a feasible system for linear equations. A new algorithm is also introduced in which an arbitrary characteristic model of the interval LP (ILP) model is chosen and solved. The proposed algorithm applies to large-scale problems. The solutions to several problems obtained by the new algorithm and a Monte Carlo simulation are also compared. Jayalakshmi [13] developed the "break and bound" method to find an optimal interval solution for fully interval linear programming. So, they use midpoint in their proposed model, to decompose the fully interval linear programming problems (FILPP) into three crisp linear programming problems (CLPP) with bounded inequality. The three CLPP are solved separately and by using its optimal solutions, an optimal interval solution to the given FILPP is also obtained. There is no restriction on the elements of the coefficient matrix in the proposed method. Later, the "break and bound" method is illustrated by some numerical examples. In case of interval linear programming (ILP), the best–worst cases (BWC) method and two-step method (TSM) do not ensure feasibility condition for, while the modified ILP (MILP), robust TSM (RTSM), improved TSM (ITSM), and three-step method (ThSM) guarantee feasibility condition, whose solution spaces may not be completely optimal. Allahdadi [14] proposes an improved ThSM (IThSM) for ILP problems, which ensures both feasibility and optimality conditions, which introduce an extra step to optimality. Fuzzy programming and chance-constrained programming technique developed by Sahoo and Dash [15] by considering the purchasing cost, selling price, salvage cost as type-1 fuzzy numbers, and the demand as a fuzzy random variable whose mean and variance are type-1 fuzzy numbers, to deal with uncertainty and ambiguous situation. They offered a chance-constrained programming approach in which the space area is treated as a fuzzy random variable for a single-period inventory model. They also obtained a solution process to solve a single-period inventory probabilistic model in a fuzzy environment.

The EOQ model with shortages reduces the assumption that shortages cannot occur. Since shortages are filled when inventory is restocked, the maximum inventory level does not reach order quantity, but instead a level equal to order quantity. In this paper, we consider an EOQ model in an uncertain environment under the limitation of the total number of orders and total annual holding cost where total money investment for the quantities is fixed. We use two types of methods to obtain the optimal solution of our proposed models, such as the signed-distance method and interval linear programming problem (ILPP) approach. In the signed-distance method, first, we defuzzify the fuzzy inventory model to get the crisp inventory model and then obtain the optimal solution of that crisp model using LINGO software. Here we considered ordering cost, holding cost, and unit product cost as symmetric trapezoidal fuzzy numbers in objective function as well as in constraint to achieve our goal. And in case of an interval linear programming problem (ILPP) approach, we find the best and the worst optimal values for the objective function by taking ordering cost, holding cost, unit product cost, and the total money investment for the quantities as interval numbers of our proposed inventory model. The proposed (best and worst cases) technique is employed in case-2 and the worst case optimal value is rendered as the optimal solution of the proposed model.

2 Mathematical Model

In this paper, an economic order quantity (EOQ) model is developed with varying holding costs under the limitation of the total number of orders and holding costs. Here we have considered different assumptions and different notations to construct an EOQ model for single item and multi-item to minimize the average total annual cost under the limitation of the total number of orders and holding cost, where the money invested for the quantities is fixed.

(i) **Single-Item Economic Order Quantity (SEOQ) Model**:

Assumption:

- The demand rate is uniform over time for each product.
- The production rate is finite and constant for each product.
- Shortages are not allowed.
- The time horizon is finite.
- The holding cost of the item is a continuous function of the order quantity q. So, the holding cost of the item is $h(q) = hq^a$, where $h > 0$ and $0 \leq a < 1$.

Notation:

d = Demand per unit time.

q = Number of order quantity per unit time.

$h(q)$ = Holding cost per unit item per unit time.

o = The ordering cost per unit item per unit time.

p = Annual production rate.

$n =$ Number of Items per unit time.

$n_o =$ Number of orders per year.

$l_1 =$ Limitation of the total number of orders.

$l_2 =$ Limitation of annual holding cost.

$c =$ Total annual cost.

Our main objective is to minimize the average total annual cost. We know that the sum of total annual ordering cost and total annual holding cost (or carrying cost) is recognized as the annual cost.

That is,

$$\text{Total average annual cost} = \frac{do}{q} + \frac{\left(1 - \frac{d}{p}\right)qh(q)}{2} = \frac{do}{q} + \frac{\left(1 - \frac{d}{p}\right)hq^{a+1}}{2},$$

where

$$\text{Total ordering cost} = \frac{\text{Demand} * \text{Ordering cost per unit item per unit time}}{\text{Order quantity}} = \frac{do}{q}$$

$$\text{Total holding cost} = \frac{\left(1 - \frac{\text{Demand}}{\text{Annual production rate}}\right) * \text{Order quantity} * \text{Holding cost per unit item per unit time}}{2} = \frac{\left(1 - \frac{d}{p}\right)qh(q)}{2}$$

The limitation of the total number of orders and holding costs for the proposed inventory model are given as follows:

Number of orders constraint: $\frac{dn_0}{q} \leq l_1$.

Holding cost constraint: $\frac{(1-\frac{d}{p})qh}{2} \leq l_2$.

Hence, a single-item inventory model with varying holding costs corresponding to the number of orders and holding cost constraints is given as follows:

$$\text{SEOQ}: \ \text{Min } c = \frac{do}{q} + \frac{(1 - \frac{d}{p})hq^{a+1}}{2}$$

$$\text{s.t.} \frac{dn_0}{q} \leq l_1$$

$$\frac{(1 - \frac{d}{p})qh}{2} \leq l$$

$$q > 0$$

(ii) **Multi-item Economic Order Quantity (MEOQ) Model**:

Assumption:

- The demand rate is uniform over time for each product.
- The production rate is finite and constant for each product.
- Shortages are not allowed.
- The time horizon is finite.
- The holding cost of *i*th item is a continuous function of the order quantity q.

So, the holding cost of the *i*th item is $h_i(q_i) = h_i\ {q_i}^a$, where $h > 0$ and $0 \leq\ a < 1$.

Notation:

d_i = Demand per unit time.

q_i = Number of order quantity per unit time.

$h_i(q_i)$ = Holding cost per unit item per unit time.

o_i = The ordering cost per unit item per unit time.

p_i = Annual production rate.

n = Number of items per unit time.

n_o = Number of orders per year.

l_1 = Limitation of the total number of orders.

l_2 = Limitation of annual holding cost.

c = Total annual cost.

Our main purpose is to construct a model for multi-items in which the average total annual cost is minimum. We know that the sum of total annual ordering cost and total annual holding cost (or carrying cost) is recognized as the annual cost.

That is,

$$\text{total average annual cost} = \sum_{i=1}^{n} \frac{d_i o_i}{q_i} + \frac{\left(1 - \frac{d_i}{p_i}\right) q_i\ h_i(q_i)}{2} = \sum_{i=1}^{n} \frac{d_i o_i}{q_i} + \frac{\left(1 - \frac{d_i}{p_i}\right) h_i\ q_i^{a+1}}{2},$$

where

$$\text{total ordering cost} = \frac{\text{Demand} * \text{Ordering cost per unit item per unit time}}{\text{Order quantity}} = \sum_{i=1}^{n} \frac{d_i o_i}{q_i},$$

total holding cost

$$= \frac{\left(1 - \frac{\text{Demand}}{\text{Annual production rate}}\right) * \text{Order quantity} * \text{Holding cost per unit item per unit time}}{2}$$

$$= \sum_{i=1}^{n} \frac{\left(1 - \frac{d_i}{p_i}\right) q_i \; h_i(q_i)}{2}.$$

The limitation of the total number of orders and holding costs for the proposed inventory model are given as follows:

Number of orders constraint: $\sum_{i=1}^{n} \frac{d_i n_o}{q_i} \leq l_1$.

Holding cost constraint: $\sum_{i=1}^{n} \frac{(1-\frac{d_i}{p_i})h_i q_i}{2} \leq l_2$.

Hence, a multi-item inventory model with varying holding costs with some restrictions like the number of orders and holding cost is described as follows:

$$\text{MEOQ : Min } c = \sum_{i=1}^{n} \frac{d_i o_i}{q_i} + \frac{(1 - \frac{d_i}{p_i}) q_i h_i^{a+1}}{2}$$

$$\text{s.t.} \sum_{i=1}^{n} \frac{d_i n_o}{q_i} \leq l_1$$

$$\sum_{i=1}^{n} \frac{(1 - \frac{d_i}{p_i}) h_i q_i}{2} \leq l_2$$

$$q_i > 0; i = 1, 2, 3, \ldots, n.$$

In real-life situations, uncertainty arises. So, we considered parameters as the trapezoidal fuzzy number and interval number to describe an inventory EOQ model for single items and multi-items. In this paper, we described the above model in two cases as follows:

Case 1:

(i) **Fuzzy Single-Item Economic Order Quantity (FSEOQ) Model**:

An EOQ model is considered in fuzzy environment by taking different parameters as fuzzy. Here we considered different notations for our proposed fuzzy inventory model for single item.

Notations:

$\tilde{h}(q)$ = Fuzzy holding cost per unit item per unit time.

$\tilde{o}(q)$ = Fuzzy ordering cost.

$\tilde{c}$ = Fuzzy total annual cost.

When uncertainty arises in real-life situation, the previously discussed single-item EOQ model becomes fuzzy single-item EOQ (FSEOQ). So, the fuzzy single-item EOQ model in an uncertain environment can be described as follows:

$$\text{FSEOQ}: \ \text{Min}\tilde{c} = \frac{d\tilde{o}}{q} + \frac{(1-\frac{d}{p})\tilde{h}q^{a+1}}{2}$$

$$\text{s.t.}\frac{dn_0}{q} \leq l_1$$

$$\frac{(1-\frac{d}{p})\tilde{h}q}{2} \leq l_2$$

$$q > 0$$

(ii) **Fuzzy Multi-Item Economic Order Quantity (FMEOQ) Model**:

An EOQ model is considered in fuzzy environment by taking different parameters as fuzzy. Here we considered different notations for our proposed fuzzy inventory model for multi-item.

Notations:

$\tilde{h}_i\ (q)$ = Fuzzy holding cost per unit item per unit time.

$\tilde{o}_i$ = Fuzzy ordering cost.

$\tilde{c}$ = Fuzzy total annual cost.

When uncertainty arises in real-life situation, the previously discussed multi-item EOQ model becomes fuzzy multi-item EOQ (FMEOQ). So, the fuzzy multi-item EOQ model in an uncertain environment can be described as follows:

$$\text{FMEOQ}: \ \text{Min}\ \tilde{c} = \sum_{i=1}^{n} \frac{d\tilde{o}_i}{q_i} + \frac{(1-\frac{d_i}{p_i})\tilde{h}_i q_i^{a+1}}{2}$$

$$\text{s.t.}\sum_{i=1}^{n} \frac{d_i n_0}{q_i} \leq l_1$$

$$\sum_{i=1}^{n} \frac{(1-\frac{d_i}{p_i})\tilde{h}_i q_i}{2} \leq l_2$$

$$q_i > 0;\ i = 1, 2, 3, \ldots, n.$$

3 Methodology

To obtain the optimal solutions, we used signed-distance method for the defuzzification of our proposed fuzzy models. The fuzzy parameters described in the said models are represented by trapezoidal fuzzy numbers for both single item and multi-item.

3.1 Methodology for Interval EOQ Models

To solve interval EOQ models, we used interval linear programming problem (ILP) approach for both single item and multi-item by considering different parameters as interval numbers.

Trapezoidal Fuzzy Number

Trapezoidal fuzzy number represented by four points is given as follows:

$$\tilde{A} = (a_1, a_2, a_3, a_4).$$

Membership function of trapezoidal fuzzy number is defined by (Fig. 1)

$$\mu_{\tilde{A}}(x) = \begin{cases} 0 & ;\ x < a_1 \\ \frac{x-a_1}{a_2-a_1} & ;\ a_1 \le x \le a_2 \\ 1 & ;\ a_2 \le x \le a_3 \\ \frac{a_4-x}{a_4-a_3} & ;\ a_3 \le x \le a_4 \\ 0 & ;\ x > a_4 \end{cases}$$

Now, α-cut of this trapezoidal fuzzy number is given by

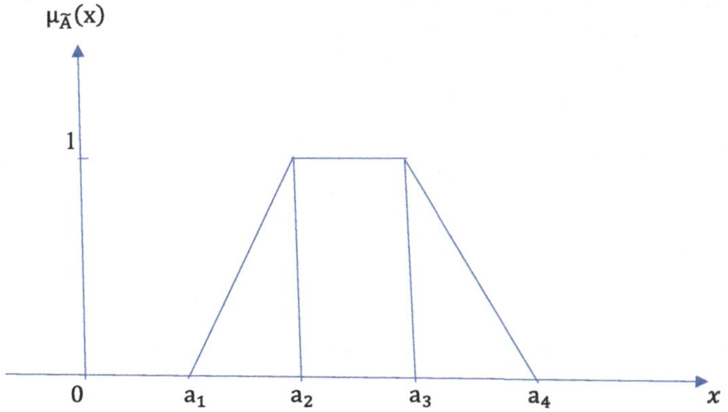

Fig. 1 Trapezoidal fuzzy number

$$\tilde{A}_\alpha = [(a_2 - a_1)\alpha + a_1, -(a_4 - a_3) + a_4].$$

When $a_2 = a_3$, the trapezoidal fuzzy number coincides with a triangular fuzzy number.

$$\tilde{A}_\alpha = \left[a_1^{(\alpha)}, a_3^{(\alpha)}\right]$$
$$= [(a_2 - a_1)\alpha + a_1, -(a_3 - a_2) + a_3]$$

Signed-Distance Method [8]:

Let, $\tilde{A} = (a_1, a_2, a_3, a_4)$ be a trapezoidal fuzzy number. Then the signed distance of $\tilde{A}$ is defined as follows:

$$d\left(\tilde{A}, 0\right) = \frac{1}{2}\int_0^1 [A_L(\alpha) + A_R(\alpha)]d\alpha,$$

where

$$A_L(\alpha) = a_1 + (a_2 - a_1)\alpha$$

$$A_R(\alpha) = a_4 - (a_4 - a_3)\alpha.$$

To defuzzify the above fuzzy models, here we have used the signed- distance method. Here, we considered ordering cost and holding cost as trapezoidal fuzzy numbers to minimize the total annual cost under the limitation of the number of orders and holding cost for single item and multi-item, where the budget is fixed. By using the signed-distance method, our fuzzy models become crisp. Then the crisp models can be solved by using an optimization technique. But here we have used LINGO software to obtain the optimal solution.

The model FSEOQ can also be written as

$$\text{FSEOQ}' : \text{Min}\tilde{c} = (m_1, m_2, m_3, m_4)(\text{say})$$

$$\text{s.t.} \frac{dn_0}{q} \leq l_1$$

$$(r_1, r_2, r_3, r_4) \leq l_2$$

$$q > 0.$$

The model FMEOQ can also be written as

$$\text{FMEOQ}' : \text{Min } \tilde{c} = (m_1, m_2, m_3, m_4)(\text{say})$$

$$\text{s.t.} \sum_{i=1}^{n} \frac{d_i n_o}{q_i} \le l_1$$

$$(r_1, r_2, r_3, r_4) \le l_2$$

$$q_i > 0; i = 1, 2, 3, \ldots, n.$$

Let, $\tilde{c} = (m_1, m_2, m_3, m_4)$ be a trapezoidal fuzzy number. Then the signed distance of $\tilde{c}$ is defined as follows:

$$d\left(\tilde{c}\left(\tilde{o}, \tilde{h}\right), 0\right) = \frac{1}{2} \int_0^1 [A_L(\alpha) + A_R(\alpha)] d\alpha$$

where

$$A_L(\alpha) = m_1 + (m_2 - m_1) + \alpha$$

$$A_R(\alpha) = m_4 - (m_4 - m_3)\alpha, \alpha \in [0, 1].$$

Let, $\tilde{r} = (r_1, r_2, r_3, r_4)$ be a trapezoidal fuzzy number. Then the signed distance of $\tilde{r}$ is defined as follows:

$$d\left(\tilde{r}\left(\tilde{h}\right), 0\right) = \frac{1}{2} \int_0^1 [A_L(\alpha) + A_R(\alpha)] d\alpha,$$

where

$$A_L(\alpha) = r_1 + (r_2 - r_1)(r_2 - r_1)\alpha$$

$$A_R(\alpha) = r_4 - (r_4 - r_3)\alpha, \alpha \in [0, 1].$$

Case 2:

Interval Number [8]:

Let Z be an interval number.

Then, Z can be represented as $[\underline{Z}, \overline{Z}]$, where $\underline{Z} \le \overline{Z}$.

If $\underline{Z} = \overline{Z}$, then Z will degenerate.

(i) **Interval Single-Item Economic Order Quantity (ISEOQ) Model**:

An EOQ model is also considered by taking different parameters as interval numbers. Here we considered different notations for our proposed inventory model for single item.

Notation:

$b =$ Unit product cost.

$B =$ Total money investment for the quantities.

$[\underline{c}, \overline{c}] =$ Interval total annual cost.

$[\underline{o}, \overline{o}] =$ Interval ordering cost.

$[\underline{h}, \overline{h}] =$ Interval holding cost per unit item per unit time.

$[\underline{b}, \overline{b}] =$ Interval unit product cost.

$[\underline{B}, \overline{B}] =$ Interval total money investment for the quantities.

Now considering budget as fixed number for the proposed SEOQ model, the budget for the proposed inventory model is given as follows:

Budget constraint: $bq = B$.

Hence, a single-item inventory model with varying holding costs corresponding to the number of orders and holding cost constraints where the total money investment for the quantities is fixed is given as follows:

$$\text{SEOQ}' : \text{ Min } c = \frac{do}{q} + \frac{\left(1 - \frac{d}{p}\right)hq^{a+1}}{2}$$

$$\text{s.t.} \frac{dn_0}{q} \leq l_1$$

$$\frac{(1 - \frac{d}{p})qh}{2} \leq l_2$$

$$bq = B$$

$$q > 0.$$

When uncertainty arises, the previously discussed SEOQ′ model also becomes the interval single-item EOQ. So, the interval single-item EOQ model in the uncertain environment by considering the ordering cost, holding cost, unit product cost, and the total money investment for the quantities as interval number can be described as follows:

$$\text{ISEOQ}: \text{ Min } c = \frac{d[\underline{o}, \bar{o}]}{q} + \frac{\left(1 - \frac{d}{p}\right)[\underline{h}, \bar{h}]q^{a+1}}{2}$$

$$\text{s.t.} \frac{dn_0}{q} \leq l_1$$

$$\frac{(1 - \frac{d}{p})[\underline{h}, \overline{h}]q}{2} \leq l_2$$

$$[\underline{b}, \bar{b}]q = [\underline{B}, \bar{B}]$$

$$q > 0$$

(ii) **Interval Multi-Item Economic Order Quantity (IMEOQ) Model**:

An EOQ model is also considered by taking different parameters as interval numbers. Here we considered different notations for our proposed inventory model for multi-item.

Notation:

$b_i =$ Unit product cost.

$B =$ Total money investment for the quantities.

$[\underline{c_I}, \overline{c_i}] =$ Interval total annual cost.

$[\underline{o_I}, \overline{o_i}] =$ Interval ordering cost.

$[\underline{h_I}, \overline{h_i}] =$ Interval holding cost per unit item per unit time.

$[\underline{b_I}, \overline{b_i}] =$ Interval unit product cost.

$[\underline{B}, \overline{B}] =$ Interval total money investment for the quantities.

Now considering budget as fixed number for the proposed MEOQ model, the budget for the proposed inventory model is given as follows:

Budget constraint: $\sum_{i=1}^{n} b_i q_i = B$.

Hence, a multi-item inventory model with varying holding costs corresponding to the number of orders and holding cost constraints where the total money investment for the quantities is fixed which is given as follows:

$$\text{MEOQ}': \text{ Min } c = \sum_{i=1}^{n} \frac{d_i o_I}{q_i} + \frac{(1 - \frac{d_i}{p_i})q_I h_i{}^{a+1}}{2}$$

$$\text{s.t.} \sum_{i=1}^{n} \frac{d_i n_o}{q_I} \leq l_1$$

$$\sum_{i=1}^{n} \frac{(1-\frac{d_i}{p_i})h_i q_i}{2} \le l_2$$

$$\sum_{i=1}^{n} b_i q_i = B$$

$$q_i > 0; i = 1, 2, 3, \ldots, n.$$

When uncertainty arises, the previously discussed MEOQ′ model also becomes the interval multi-item EOQ. So, the interval multi-item EOQ model in an uncertain environment by considering the ordering cost, holding cost, unit product cost, and the total money investment for the quantities as interval numbers can be described as follows:

$$\text{IMEOQ}: \text{ Min } c = \sum_{i=1}^{n} \frac{d_i[\underline{o_I}, \overline{o_i}]}{q_i} + \frac{\left(1-\frac{d_i}{p_i}\right)[\underline{h_I}, \overline{h_i}]q_i{}^{a+1}}{2}$$

$$\text{s.t.} \sum_{i=1}^{n} \frac{d_i n_o}{q_I} \le l_1$$

$$\sum_{i=1}^{n} \frac{\left(1-\frac{d_i}{p_I}\right)[\underline{h_I}, \overline{h_i}]q_i}{2} \le l_2$$

$$\sum_{i=1}^{n} [\underline{b_I}, \overline{b_i}]q_i[\underline{B}, \overline{B}]$$

$$q_i > 0; i = 1, 2, 3, \ldots, n.$$

Interval Linear Programming (ILP) Problem

The interval multi-item linear programming problem (IMLPP) can be written as follows:

$$\text{IMEOQ}: \text{ Min } c = \sum_{i=1}^{n} [\underline{c_I}, \overline{c_i}]x_i$$

$$\text{s.t.} \sum_{i=1}^{n} \left[\underline{a_{ij}}, \overline{a_{ij}}\right]x_i \ge \left[\underline{b_j}, \overline{b_j}\right]$$

$$\sum_{i=1}^{n} \left[\underline{a_{ij}}, \overline{a_{ij}}\right]x_i \le \left[\underline{b_j}, \overline{b_j}\right]$$

$$\sum_{i=1}^{n}\left[\underline{a_{ij}}, \overline{a_{ij}}\right]x_i = \left[\underline{b_j}, \overline{b_j}\right]$$

$$x_i \in \left(X^o,\ X^{sj}\right), j = 1, 2, 3, \ldots, m.$$

The best and worst values can be obtained by solving the following submodels.

The best value interval multi-item economic order quantity model (BIMEOQ) can be written as follows:

$$\text{BIMEOQ : Min } \underline{c} = \sum_{j=1}^{n} c_i{}' x_i$$

$$\text{s.t.} \sum_{i=1}^{n} a_{ij}{}' x_i \geq \underline{b_j}$$

$$\sum_{i=1}^{n} a_{ij}{}'' x_i \leq \overline{b_j}$$

$$\sum_{i=1}^{n} a_{ij}{}' x_i \geq \underline{b_j}$$

$$\sum_{i=1}^{n} a_{ij}{}'' x_i \leq \overline{b_j}$$

$$x_i \in X^{sj}, j = 1, 2, 3, \ldots, m.$$

The worst value interval multi-item economic order quantity models (WIMEOQ) can be written as follows:

$$\text{WIMEOQ1 : Min } \overline{c} = \sum_{i=1}^{n} c_i{}'' x_i$$

$$\text{s.t.} \sum_{i=1}^{n} a_{ij}{}'' x_i \geq \overline{b_j}$$

$$\sum_{i=1}^{n} a_{ij}{}' x_i \leq \underline{b_j}$$

$$\sum_{i=1}^{n} a_{ij}{}' x_i = \underline{b_j}$$

$$x_i \in X^{sj}, j = 1, 2, 3, \ldots, m.$$

and

$$\text{WIMEOQ2} : \text{Min } \overline{c} = \sum_{i=1}^{n} c_i{''} x_i$$

$$\text{s.t.} \sum_{i=1}^{n} a_{ij}{''} x_i \geq \overline{b_j}$$

$$\sum_{i=1}^{n} a_{ij}{'} x_i \leq \underline{b_j}$$

$$\sum_{j=1}^{n} a_{ij}{''} x_i = \overline{b_j}$$

$$x_i \in X^{sj}, j = 1, 2, 3, \ldots, m,$$

where $a_{ij}{'} = \begin{cases} \overline{a_{ij}}, x_i \geq 0 \\ \underline{a_{ij}}, x_i \leq 0 \end{cases}$, $a_{ij}{''} = \begin{cases} \underline{a_{ij}}, x_i \geq 0 \\ \overline{a_{ij}}, x_i \leq 0 \end{cases}$, $c_i{'} = \begin{cases} \overline{c_I}, x_i \geq 0 \\ \underline{c_I}, x_i \leq 0 \end{cases}$ and $c_i{''} = \begin{cases} \overline{c_I}, x_i \geq 0 \\ \underline{c_I}, x_i \leq 0 \end{cases}$.

4 Numerical Example

A company uses 1000 units of raw materials per unit time. Placing each order costs Rs.10 and holding cost Rs. 6 per item per year of the inventory, the unit product cost Rs. 4 and the total money investment for the quantities is Rs. 90,000. Limitation of annual holding cost Rs. 15,000 and limitation of annual ordering quantity is 5000. The number of orders per year is 100 and the annual production rate of the company is 2000. Find the optimum total annual cost quantity for a single-item and multi-item inventory models.

Solution:

Given, $d = 1000$ units, $o =$ Rs.10, $h =$ Rs. 6, $l_1 = 5000$ units, $l_2 =$ Rs.15,000, $p = 2000$, $b =$ Rs.4, and $B =$ Rs. 90,000.

General Economic Order Quantity Model:

Single-Item Economic Order Quantity (SEOQ) Model:

Consider, $a = 0.5$. By putting the given data, in SEOQ model, we have obtained the inventory model as follows:

$$\text{SEOQ} : \text{Min } c = \frac{10{,}000}{q} + 1.5q^{1.5}$$

$$\text{s.t.} \frac{100{,}000}{q} \leq 5000$$

$$1.5q \leq 15{,}000$$

$$q > 0.$$

By using LINGO software, the optimal solution of the model is as follows:

$$C = 579.0576, q = 28.7824.$$

Multi-Item Economic Order Quantity (MEOQ) Model:

Consider, $d_1 =$ 1000 units, $d_2 =$ 1030 units, $d_3 =$ 1060 units.

$o_1 =$ Rs. 10, $o =$ Rs.12, $o_3 =$ Rs. 14.

$h_1 =$ Rs. 6, $h_2 =$ Rs.8, $h_3 =$ Rs. 10.

$b_1 =$ Rs. 4, $b_2 =$ Rs.11, $b_3 =$ Rs. 21, and a = 0.5.

By putting the given data, in MEOQ model, we have obtained the inventory model as follows:

$$\text{MEOQ} : \text{ Min } c = \frac{10{,}000}{q_1} + 1.5{q_1}^{1.5} + \frac{12{,}360}{q_2} + 2.5{q_2}^{1.5} + \frac{14{,}540}{q_3} + 3.75{q_3}^{1.5}$$

$$\text{s.t.} \frac{100{,}000}{q_1} + \frac{103{,}000}{q_2} + \frac{106{,}000}{q_3} \leq 5000$$

$$1.5q_1 + 2.5q_2 + 3.75q_3 \leq 15{,}000$$

$$q_i > 0; i = 1, 2, 3, \ldots, n.$$

By using LINGO software, the optimal solution of the model is as follows:

$$C = 248.712, q_1 = 68.61456, q_2 = 60.24370, q_3 = 57.83308.$$

Case 1:

(i) **Fuzzy Single-Item Economic Order Quantity (FSEOQ) Model**:

Consider, $\tilde{o} =$ Rs. (10, 12, 14, 16), $\tilde{h} =$ Rs. (6, 8, 10, 14), $\tilde{b} =$ Rs. (4, 5, 7, 8), $a =$ 0.5, and $\alpha = 1$.

By putting the given data in FSEOQ model, we have

$$\text{FSEOQ}': \ \text{Min } c = \left(\frac{10{,}000}{q} + 1.5q^{1.5}, \frac{12{,}000}{q} + 2q^{1.5}, \frac{14{,}000}{q} + 2.5q^{1.5}, \frac{16{,}000}{q} + 3.5q^{1.5} \right)$$

$$\text{s.t.} \frac{100{,}000}{q} \le 5000$$

$$1.5q, 2q, 2.5q, 3.5q) \le 15{,}000$$

$$q > 0.$$

Now by using defuzzification technique the FSEOQ model is converted to CSEOQ model as follows:

$$\text{CSEOQ}: \ \text{Min } c = \frac{26{,}000}{q} + 4.5q^{1.5}$$

$$\text{s.t.} \frac{100{,}000}{q} \le 5000$$

$$4.5q \le 15{,}000$$

$$q > 0.$$

By using LINGO software, the optimal solution of the model is as follows:

$$c = 594.242, q = 27.18115$$

(ii) **Fuzzy Multi-Item Economic Order Quantity (FMEOQ) Model**:

Consider $\tilde{o}_1 =$ Rs. (10, 12, 14, 16), $\tilde{o}_2 =$ Rs. (12, 14, 16, 18), $\tilde{o}_3 =$ Rs. (14, 16, 18, 20),

$\tilde{h}_1 =$ Rs. (6, 8, 10, 14), $\tilde{h}_2 =$ Rs. (8, 12, 14, 16), $\tilde{h}_3 =$ Rs. (10, 14, 16, 20),

$\tilde{b}_1 =$ Rs. (4, 5, 7, 8), $\tilde{b}_2 =$ Rs. (11, 12, 14, 15), $\tilde{b}_3 =$ Rs. (21, 23, 25, 27), $a = 0.5$, and $\alpha = 1$.

By putting the given data in FMEOQ model, we have

$$\text{FMEOQ}': \ \text{Min } c = \left(\frac{10{,}000}{q_1} + 1.5{q_1}^{1.5} + \frac{12{,}120}{q_2} + 2{q_2}^{1.5} + \frac{14{,}280}{q_3} + 2.5{q_3}^{1.5}, \right.$$

$$\left.\frac{12{,}000}{q_1} + 2{q_1}^{1.5} + \frac{14{,}140}{q_2} + 3{q_2}^{1.5} + \frac{16{,}320}{q_3} + 3.5{q_3}^{1.5}, \frac{14{,}000}{q_1} + 2.5{q_1}^{1.5} + \frac{16{,}160}{q_2} + 3.5{q_2}^{1.5} + \frac{18{,}360}{q_3} + 4{q_3}^{1.5}, \frac{16{,}000}{q_1} + 3.5{q_1}^{1.5} + \frac{18{,}180}{q_2} + 4{q_2}^{1.5} + \frac{20{,}400}{q_3} + 5{q_3}^{1.5}\right)$$

$$\text{s.t.}\frac{100{,}000}{q_1} + \frac{103{,}000}{q_2} + \frac{106{,}000}{q_3} \leq 5000$$

$$(1.5q_1 + 2q_2 + 2.5q_3, 2q_1 + 3q_2 + 3.5q_3, 2.5q_1 + 3.5q_2 + 4q_3, 3.5q_1 + 4q_2 + 5q_3) \leq 15{,}000$$

$$q_i > 0; i = 1, 2, 3, \ldots, n.$$

Now by using defuzzification technique, the FMEOQ model is converted to CMEOQ model as follows:

$$\text{CMEOQ}: \text{ Min } c = \frac{26{,}000}{q_1} + 4.5{q_1}^{1.5} + \frac{30{,}300}{q_2} + 6.5{q_1}^{1.5} + \frac{34{,}680}{q_3} + 7.5{q_1}^{1.5}$$

$$\text{s.t.}\frac{100{,}000}{q_1} + \frac{103{,}000}{q_2} + \frac{106{,}000}{q_3} \leq 5000$$

$$4.5q_1 + 6.5q_2 + 7.5q_3 \leq 15{,}000$$

$$q_i > 0; i = 1, 2, 3, \ldots, n.$$

By using LINGO software, the optimal solution of the model is given as follows:

$$c = 10{,}377.09, q_1 = 68.61456, q_2 = 60.24370, q_3 = 57.83308.$$

Case 2:

(i) **Interval Single-Item Economic Order Quantity (ISEOQ) Model**:

Consider, $o =$ Rs. [10, 16], $h =$ Rs. [1, 14], $b =$ Rs. [5, 8], and $a = 0.5$.

Let an inventory model for single item with an interval inequality constraint is given by

$$\text{ISEOQ}: \text{ Min } c = \frac{1000[10, 16]}{q} + \frac{1}{4}[6, 14]q$$

$$\text{s.t.}\frac{1000[80,\,120]}{q} \leq 5000$$

$$\frac{1}{4}[1,14]q \leq 15{,}000$$

$$[5,8]q \geq [70{,}000,\,90{,}000]$$

$$q > 0.$$

Best Case ISEOQ (BISEOQ):

$$\text{BISEOQ}: \ \text{Min}\ \underline{c} = \frac{16{,}000}{q} + 3.5q$$

$$\text{s.t.}\frac{80{,}000}{q} \leq 5000$$

$$1.5q \leq 15{,}000$$

$$8q \leq 70{,}000$$

$$4q \leq 90{,}000$$

$$q > 0.$$

By using LINGO software, the optimal solution of the model is given as follows:

$$c = 30{,}626.83,\, q = 8750.$$

Worst Case ISEOQ (WISEOQ):

$$\text{WISEOQ1}: \text{Min}\ \overline{c} = \frac{16{,}000}{q} + 3.5q$$

$$\text{s.t.}\frac{120{,}000}{q} \leq 5000$$

$$3.5q \leq 15{,}000$$

$$8q \leq 70{,}000$$

$$q > 0.$$

By using LINGO software, the optimal solution of the model is infeasible.
And

$$\text{WISEOQ2 : Min}\overline{c} = \frac{16{,}000}{q} + 3.5q$$

$$\text{s.t.}\frac{120{,}000}{q} \leq 5000$$

$$3.5q \leq 15{,}000$$

$$4q \leq 90{,}000$$

$$q > 0.$$

By using LINGO software, the optimal solution of the model is given as follows:

$$c = 73.2864, q = 67.61234$$

(ii) **Interval Multi-Item Economic Order Quantity (IMEOQ) Model**:

Consider, $o_1 =$ Rs. [10, 16], $o_2 =$ Rs. [12, 18], $o_3 =$ Rs. [14, 20], $h_1 =$ Rs. [1, 14], $h_2 =$ Rs. [8, 16], $h_3 =$ Rs. [10, 20], $b_1 =$ Rs. [5, 8], $b_2 =$ Rs. [11, 15], $b_3 =$ Rs. [21, 27], and a = 0.

Let an inventory model for multi-item with an interval inequality constraint be given by

$$\text{IMEOQ}: \text{ Min } c = \frac{1000[10, 16]}{q_1} + \frac{1}{4}[6, 14]q_1 + \frac{1000[12, 18]}{q_2} + \frac{1}{4}[8, 16]q_2 + \frac{1000[14, 20]}{q_3} + \frac{1}{4}[10, 20]q_3$$

$$\text{s.t.}\frac{1000[80, 120]}{q_1} + \frac{1030[80, 120]}{q_2} + \frac{1060[80, 120]}{q_3} \leq 5000$$

$$\frac{1}{4}[1, 14]q_1 + \frac{1}{4}[8, 16]q_2 + \frac{1}{4}[10, 20]q_3 \leq 15{,}000$$

$$[5, 8]q_1 + [11, 15] + q_2 + [21, 27]q_3 \geq [70{,}000, 90{,}000]$$

$$q_i > 0; i = 1, 2, 3, \ldots, n.$$

Best Case IMEOQ (BIMEOQ):

$$\text{BIMEOQ}: \text{ Min } \underline{c} = \frac{16{,}000}{q_1} + 3.5q_1 + \frac{18{,}540}{q_2} + 4q_2 + \frac{21{,}200}{q_3} + 5q_3$$

$$\text{s.t.}\frac{80{,}000}{q_1} + \frac{82{,}400}{q_2} + \frac{84{,}800}{q_3} \le 5000$$

$$1.5q_1 + 2q_2 + 2.5q_3 \le 15{,}000$$

$$8q_1 + 15q_2 + 27q_3 \ge 70{,}000$$

$$4q_1 + 11q_2 + 21q_3 \le 90{,}000$$

$$q_i > 0; i = 1, 2, 3, \ldots, n.$$

By using LINGO software, the optimal solution of the model is given as follows:

$$c = 13{,}631.94, q_1 = 89.00929, q_2 = 123.0679, q_3 = 2497.848.$$

Worst Case IMEOQ (WIMEOQ):

$$\text{WIMEOQ1}: \text{ Min } \overline{c} = \frac{16{,}000}{q_1} + 3.5q_1 + \frac{18{,}540}{q_2} + 4q_2 + \frac{21{,}200}{q_3} + 5q_3$$

$$\text{s.t.}\frac{120{,}000}{q_1} + \frac{123{,}600}{q_2} + \frac{127{,}200}{q_3} \le 5000$$

$$3.5q_1 + 4q_2 + 5q_3 \le 15{,}000$$

$$8q_1 + 15q_2 + 27q_3 \ge 70{,}000$$

$$q_i > 0; i = 1, 2, 3, \ldots, n.$$

By using LINGO software, the optimal solution of the model is given as follows:

$$c = 13{,}631.94, q_1 = 89.00929, q_2 = 123.0679, q_3 = 2497.848.$$

And

$$\text{WIMEOQ2}: \text{ Min } \overline{c} = \frac{16{,}000}{q_1} + 3.5q_1 + \frac{18{,}540}{q_2} + 4q_2 + \frac{21{,}200}{q_3} + 5q_3$$

$$\text{s.t.}\frac{120{,}000}{q_1}+\frac{123{,}600}{q_2}+\frac{127{,}200}{q_3}\leq 5000$$

$$3.5q_1+4q_2+5q_3\leq 15{,}000$$

$$4q_1+11q_2+21q_3\leq 90{,}000$$

$$q_i>0;\, i=1,2,3,\ldots,n.$$

By using LINGO software, the optimal solution of the model is given as follows:

$$c=1677.921,\, q_1=75.80660,\, q_2=75.45966,\, q_3=71.49831.$$

5 Result Discussion

Model name	Cost value	
	Single item	Multi-item
Deterministic	579.0576	4248.712
Fuzzy	1594.242	10,377.09
Interval	473.2864	1677.921

6 Conclusion

In this paper, we discussed the EOQ models for a single item and multi-items in an uncertain environment in two cases. In case-1, the computational procedure for the defined EOQ model is carried out by using the signed-distance method and in case-2 the proposed model is discussed by using Best and Worst case (BWC) method. By using the concept of the signed-distance method we obtained the corresponding crisp model to our proposed fuzzy model. The corresponding deterministic model can be solved by using an optimization technique. But here we have used the LINGO software to obtain the optimal solutions. The defuzzification technique can also be applied to other fuzzy models. Also, the proposed (best and worst cases) technique is employed in case-2 and the worst case optimal value is rendered as the optimal solution of the proposed model.

References

1. K.A.M. Kotb, H.A. Fergany, Multi-item EOQ model with varying holding cost: a geometric programming approach. Int. Math. Forum **6**(23), 1135–1144 (2011)
2. T. Shaocheng, Interval number and fuzzy number linear programming. Fuzzy Sets Syst. **66**, 301–306 (1994)
3. M.O. Aboul-El-Ata, K.A.M. Kotb, Multi-item EOQ inventory model with varying holding cost under two restrictions: a geometric approach. Prod. Plan. Control. **8**(6), 608–611 (1997)
4. J.W. Chinneck, K. Ramadan, Linear programming with interval coefficients. J. Oper. Res. Soc. **51**(2), 209–220 (2002)
5. H.C. Chang, An application of fuzzy sets theory to the EOQ model with imperfect quality Items. Comput. Oper. Res. **31**(12), 2079–2092 (2004)
6. J.K. Sayed, L.A. Aziz, Fuzzy inventory model without shortage using signed distance method. Appl. Math. Inf. Sci. **1**(2) (2007)
7. D. Dutta, P. Kumar, Fuzzy inventory model without shortage using trapezoidal fuzzy number with sensitivity analysis. IOSR J. Math. **4**(3), 32–37 (2012). ISSN: 2278-5728
8. M. Allahdadi, H.M. Nehi, The optimum value bounds of the objective function in the interval linear programming problem. Chiang Mai J. Sci. **42**(2), 501–511 (2015)
9. A. Sahoo, J.K. Dash, optimal solution for a single period inventory model with fuzzy cost and demand as a fuzzy random variable. J. Intell. Fuzzy Syst. **28**, 1195–1203 (2015)
10. R.M. Rajalakshmi, G.M. Rosario, A fuzzy inventory model with allowable shortage using different fuzzy number. Int. J. Comput. Appl. Math. **12**(1) (2017). ISSN 1819-4966
11. A. Ren, Y. Wang, X. Xue, A novel approach based on the preference-based index for interval bilevel linear programming problem. J. Inequalities Appl. (2017)
12. H.A. Ashayerinasab, H.M. Nehi, M. Allahdadi, Solving the interval linear programming problem: a new algorithm for a general case. Expert Syst. Appl. **93**, 39–49 (2018)
13. M. Jayalakshmi, A new approach to solve fully interval linear programming problems. Int. J. Pure Appl. Math. **118**, 363–370 (2018)
14. M. Allahdadi, An improved three-step method for solving the interval linear programming problems. Yugoslav J. Oper. Res. **28**(4), 435–451 (2018)
15. A. Sahoo, J.K. Dash, Solving chance-constrained single-period inventory model with type-1 fuzzy set, in *Operation Research in the Development Sector* (2019), pp. 15–17

Free Poisson Elements Induced by Orthogonal Projections

Ilwoo Cho

Abstract In this paper, we construct-and-consider free (weighted-)Poisson elements in a topological noncommutative free $*$-probability space generated by countable-infinitely many mutually free, (weighted-)semicircular elements which are induced by mutually orthogonal countable-infinitely many projections in a fixed C^*-probability space.

Keywords Free probability · Weighted-semicircular elements · Semicircular elements · Free Poisson elements · Free weighted-Poisson elements

2010 Mathematics Subject Classification 05E15 · 11R56 · 46L54 · 47L30 · 47L55

1 Introduction

Let (A, ψ) be a C^*-probability space containing a family $\mathbf{Q} = \{q_j\}_{j \in \mathbb{Z}}$ of mutually orthogonal projections q_j's. In [8], we showed that such a family $\mathbf{Q}$ induces the corresponding free weighted-semicircular family $\{u_j\}_{j \in \mathbb{Z}}$ in a certain Banach $*$-probability space $(\mathfrak{L}_Q, \tau)$, and, under additional conditions, this free family $\{u_j\}_{j \in \mathbb{Z}}$ generates the free semicircular family $\{U_j\}_{j \in \mathbb{Z}}$ in $(\mathfrak{L}_Q, \tau)$. This free weighted-semicircular family $\{u_j\}_{j \in \mathbb{Z}}$ generates the free-probabilistic sub-structure $(\mathbb{L}_Q, \tau)$ of $(\mathfrak{L}_Q, \tau)$. The free probability on $\mathbb{L}_Q$ was studied in [6, 11, 12], and the corresponding free stochastic calculus was considered in [9].

There are various ways to construct semicircular elements. For example, see [1, 2, 17–21, 28, 30]. However, our construction of (weighted-)semicircular elements are different from those of the above earlier works; it is motivated by the construction(s)

I. Cho (✉)
Department of Mathematics and Statistics, Saint Ambrose University,
421 Ambrose Hall, 518 W. Locust St., Davenport, IA 52803, USA
e-mail: choilwoo@sau.edu

S. K. Paikray et al. (eds.), *New Trends in Applied Analysis and Computational Mathematics*, Advances in Intelligent Systems and Computing 1356,
https://doi.org/10.1007/978-981-16-1402-6_15

of the (weighted-)semicircular elements in the sense of [5, 7, 11]. For example, see [6, 8, 9, 11, 12]. In [5, 7, 11], (weighted-)semicircular elements are naturally constructed from *p-adic analyses* (and their globalization, the *Adelic analysis*) for all *primes* p, from $|\mathbb{Z}|$-many orthogonal projections induced by $|\mathbb{Z}|$-many measurable p-adic characteristic functions of boundaries of the p-adic number fields $\mathbb{Q}_p$, for all primes p (e.g., [26, 27]). Motivated by these, we mimic the construction of (weighted-)semicircular elements if a C^*-probability space (A, ψ) has mutually orthogonal $|\mathbb{Z}|$-many projections in [8]. For more about details and applications, see [6, 9, 11, 12].

The main purposes of this paper are (i) to construct *free Poisson elements* of the *Banach $*$-probability space* $(\mathbb{L}_Q, \tau)$, and to study corresponding free-distributional data in $\mathbb{L}_Q$, (ii) to construct *free weighted-Poisson elements* of $(\mathbb{L}_Q, \tau)$, similar to the construction of (i) from *weighted-semicircular elements* of $(\mathbb{L}_Q, \tau)$, and (iii) to study *free distributions* of such free (weighted-)Poisson elements in $(\mathbb{L}_Q, \tau)$.

Our main results provide concrete characterizations of free (weighted-)Poisson distributions by computing not only free cumulants, but also free moments of given free (weighted-)Poisson elements in $(\mathbb{L}_Q, \tau)$.

In earlier works (e.g., [23, 24, 26]), free Poisson elements, whose free distributions are the *(free) Poisson distributions*, have been studied. Here, different from them, we are interested in free Poisson elements in the Banach $*$-algebra $\mathbb{L}_Q$ generated by its maximal *free semicircular family* affected by the free probability of a fixed C^*-probability space (A, ψ).

Even though free Poisson elements, and free Poisson distributions, are well known and well characterized (e.g., see [15–17]), our free "weighted-Poisson" elements are newly introduced-and-studied. Moreover, their free distributions, the *free weighted-Poisson distributions*, would contain free-probabilistic information of a fixed C^*-probability space (A, ψ).

In Sect. 2, we briefly discuss about free probability theory, and the main free-probabilistic objects used in this paper. In Sects. 3 and 4, the (weighted-)semicircular elements introduced in [6, 8, 9, 11, 12] are re-considered. In Sects. 5 and 6, we construct the Banach $*$-probability space $(\mathbb{L}_Q, \tau)$ generated by the mutually free semicircular elements, and investigate free-probabilistic structures of $(\mathbb{L}_Q, \tau)$. And then, our free Poisson elements generated by the free generators of $\mathbb{L}_Q$ are studied in Sect. 7. The corresponding free Poisson distributions are characterized there. Different from Sect. 7, we study free weighted-Poisson elements of $\mathbb{L}_Q$, generated by the weighted-semicircular elements, in Sect. 8. Roughly speaking, it is shown that the free weighted-Poisson distributions are certain scalar multiples of free Poisson distributions obtained in Sect. 7. Interestingly, such scalar multiples contain C^*-probabilistic information of (A, ψ). In Sect. 9, free Poisson distributions induced by free sums of semicircular elements and those induced by other free Poisson elements are characterized in $\mathbb{L}_Q$.

2 Preliminaries

Free probability is operator-algebraic analogue of classical *measure theory* and *statistics*. The freeness on a noncommutative algebra for a fixed linear functional replaces the classical *independence* on a commutative set for a fixed measure in this noncommutative theory (e.g., [17, 22, 23, 30], [38]). It is widely applicable not only to pure mathematics (e.g., [3, 4, 20, 21, 24, 25, 28]), but also to related fields (e.g., [5–12]). Here, the combinatorial free probability theory is used (e.g., [17, 22, 23]).

By the *central limit theorem(s)*, studying semicircular elements, whose free distributions are *the semicircular law*, is one of the main subjects in free probability theory (e.g., [1, 2, 18–21, 28]). As application, *free Poisson elements*, whose free distributions are the *free Poisson distributions*, are obtained (e.g., [15–17]).

In this paper, we study free Poisson elements generated by the semicircular elements induced by the same Banach $*$-probabilistic settings of [6, 8, 9, 12]. Since free Poisson distributions are well known (e.g., see [17]), equivalently, since free Poisson elements are well characterized as free random variables, we also introduce-and-study free-Poisson-like elements, called *free weighted-Poisson elements* generated by the *weighted-semicircular elements* of [8, 12].

Our main results show that free weighted-Poisson distributions are characterized by free Poisson distributions under certain additional free-distributional information (See Sects. 7, 8, and 9, below).

3 Banach $*$-Algebras Induced by Projections

Let (B, φ_B) be a topological $*$-probability space (a *C^*-probability space*, or a *W^*-probability space*, or a *Banach $*$-probability space*, etc.), consisting of a topological $*$-algebra B (a C^*-algebra, resp., a W^*-algebra, resp., a Banach $*$-algebra, etc.), and a bounded linear functional φ_B on B. An operator a of B is called a *free random variable*, if it is regarded as an element of (B, φ_B). We say a free random variable a is *self-adjoint*, if it is self-adjoint as an operator, i.e., $a^* = a$ in B, where a^* is the *adjoint of a* (e.g., [14]).

Definition 1 A self-adjoint free random variable a is weighted-semicircular in (B, φ_B) with its weight $t_0 \in \mathbb{C}^\times = \mathbb{C} \setminus \{0\}$ (in short, t_0-semicircular), if a satisfies the free-cumulant computations,

$$k_n^B(a, \ldots, a) = \begin{cases} t_0 & \text{if } n = 2 \\ 0 & \text{otherwise,} \end{cases} \tag{3.1}$$

for all $n \in \mathbb{N}$, where $k_\bullet^B(\ldots)$ is the free cumulant on B in terms of φ_B under the Möbius inversion of [17, 22, 23].

If $t_0 = 1$ in (3.1), the corresponding 1-semicircular element a is said to be semicircular in (B, φ_B), i.e., a is semicircular in (B, φ_B), if

$$k_n^B(a, \ldots, a) = \begin{cases} 1 & \text{if } n = 2 \\ 0 & \text{otherwise,} \end{cases} \tag{3.2}$$

for all $n \in \mathbb{N}$.

By the Möbius inversion, one can characterize the weighted-semicircularity (3.1) with respect to free moments: a self-adjoint operator a is t_0-semicircular in (B, φ_B), if and only if

$$\varphi_B(a^n) = \omega_n \left(t_0^{\frac{n}{2}} c_{\frac{n}{2}} \right), \tag{3.3}$$

where

$$\omega_n \overset{def}{=} \begin{cases} 1 & \text{if } n \text{ is even} \\ 0 & \text{if } n \text{ is odd,} \end{cases}$$

for all $n \in \mathbb{N}$, and c_k are the kth *Catalan numbers*,

$$c_k \overset{def}{=} \frac{1}{k+1} \binom{2k}{k} = \frac{1}{k+1} \left(\frac{(2k)!}{k!(2k-k)!} \right) = \frac{(2k)!}{k!(k+1)!},$$

for all $k \in \mathbb{N}_0 = \mathbb{N} \cup \{0\}$. Similarly, a self-adjoint free random variable a is semicircular in (B, φ_B), if and only if

$$\varphi_B(a^n) = \omega_n c_{\frac{n}{2}}, \tag{3.4}$$

by (3.2) and (3.3), for all $n \in \mathbb{N}$.

Fix now a C^*-probability space (A, ψ), containing its $|\mathbb{Z}|$-many projections $\{q_j\}_{j \in \mathbb{Z}}$ in the C^*-algebra A, satisfying

$$q_j^* = q_j = q_j^2 \text{ in } A,$$

for all $j \in \mathbb{Z}$ (e.g., [14]). Assume moreover that the projections $\{q_j\}_{j \in \mathbb{Z}}$ are *mutually orthogonal* in A, in the sense that

$$q_i q_j = \delta_{i,j} q_j \text{ in } A, \text{ for all } i, j \in \mathbb{Z}, \tag{3.5}$$

where δ is the *Kronecker delta*. That is, we have a mutually orthogonal family of projections,

$$\mathbf{Q} = \{q_j : q_j \text{ satisfy } (3.5)\}_{j \in \mathbb{Z}} \text{ in } A. \tag{3.6}$$

Such a C^*-algebraic structure A containing a family $\mathbf{Q}$ of (3.6) does exist naturally, or artificially (e.g., see [5–10, 12]).

Define now the C^*-subalgebra Q of A generated by the family $\mathbf{Q}$ of (3.6), i.e.,

$$Q \overset{def}{=} C^* (\mathbf{Q}) \subseteq A. \tag{3.7}$$

Proposition 2 *Let Q be a C^*-subalgebra (3.7) of a fixed C^*-algebra A. Then*

$$Q \overset{*\text{-}iso}{=} \underset{j \in \mathbb{Z}}{\oplus} \left(\mathbb{C} \cdot q_j\right) \overset{*\text{-}iso}{=} \mathbb{C}^{\oplus |\mathbb{Z}|}, \ in\ A. \tag{3.8}$$

Proof The relation (3.8) is shown by the orthogonality (3.5) of $\mathbf{Q}$, and the definition (3.7) of Q in A. ■

Define linear functionals ψ_j on Q by

$$\psi_j\,(q_i) = \delta_{ij}\psi(q_j), \ \text{for all } i \in \mathbb{Z}, \tag{3.9}$$

for all $j \in \mathbb{Z}$, where ψ is the linear functional of (A, ψ). These linear functionals $\{\psi_j\}_{j\in\mathbb{Z}}$ of (3.9) are well defined on Q by (3.8).

Assumption In the rest of this paper, we assume that

$$\psi(q_j) \in \mathbb{C}^{\times} \overset{def}{=} \mathbb{C} \setminus \{0\}, \forall j \in \mathbb{Z}.$$

□

Definition 3 The C^*-probability spaces $\left(Q, \psi_j\right)$ are called the jth filters of Q in a given C^*-probability space (A, ψ), where Q is in the sense of (3.7), and ψ_j are the linear functionals of (3.9), for all $j \in \mathbb{Z}$.

Now, define bounded linear transformations $\mathbf{c}$ and $\mathbf{a}$ acting on the C^*-algebra Q, by morphisms satisfying

$$\mathbf{c}\left(q_j\right) = q_{j+1}, \ \text{and } \mathbf{a}\left(q_j\right) = q_{j-1}, \tag{3.10}$$

for all $j \in \mathbb{Z}$. By definition, the linear transformations $\mathbf{c}$ and $\mathbf{a}$of (3.10) are *Banach-space operators* in the *operator space* $B(Q)$ of [13], by regarding Q as a *Banach space* equipped with its C^*-*norm*. We call these operators the creation, and the annihilation, respectively.

Define now a new Banach-space operator $\mathbf{l}$ on Q by

$$\mathbf{l} = \mathbf{c} + \mathbf{a} \in B(Q). \tag{3.11}$$

Definition 4 We call the Banach-space operator $\mathbf{l} \in B(Q)$ of (3.11), the *radial operator on* Q.

From the radial operator $\mathbf{l}$ of (3.11), construct a closed subspace $\mathfrak{L}$ of $B(Q)$ by

$$\mathfrak{L} \overset{def}{=} \overline{\mathbb{C}[\{\mathbf{l}\}]}^{\|.\|}, \tag{3.12}$$

where $\|.\|$ is the operator-norm on the operator space $B(Q)$,

$$\|T\| = \sup\{\|Tq\|_Q : \|q\|_Q = 1\},$$

for all $T \in B(Q)$, where $\|.\|_Q$ is the C^**-norm on* Q (inherited from that on A), and where $\overline{X}^{\|.\|}$ mean the *operator-norm closures* of subsets X of the *operator space* $B(Q)$ (e.g., [13]). This subspace $\mathfrak{L}$ forms an algebra by (3.12), i.e., it is a Banach algebra $B(Q)$.

On the Banach algebra $\mathfrak{L}$ of (3.12), define a unary operation $(*)$ by

$$\left(\sum_{n=0}^{\infty} t_n \mathbf{l}^n\right)^* = \sum_{n=0}^{\infty} \overline{t_n} \mathbf{l}^n \text{ in } \mathfrak{L}, \tag{3.13}$$

where $\overline{z}$ are the *conjugates of* $z \in \mathbb{C}$. Then the operation (3.13) is a well-defined *adjoint on* $\mathfrak{L}$ (e.g., [8, 10]), and hence, all elements of $\mathfrak{L}$ are *adjointable* in $B(Q)$ (e.g., [13]). So, the algebra $\mathfrak{L}$ forms a *Banach* $*$*-algebra* in $B(Q)$ under (3.13).

Definition 5 We call the Banach $*$-algebra $\mathfrak{L}$ of (3.12), the *radial* (*Banach* $*$-*)algebra on* Q.

By taking the radial algebra $\mathfrak{L}$ on Q, define the *tensor product Banach* $*$*-algebra* $\mathfrak{L}_Q$ by

$$\mathfrak{L}_Q = \mathfrak{L} \otimes_{\mathbb{C}} Q, \tag{3.14}$$

where $\otimes_{\mathbb{C}}$ is the tensor product of Banach $*$-algebras.

Definition 6 The tensor product Banach $*$-algebra $\mathfrak{L}_Q$ of (3.14) is called the radial projection (Banach $*$-)algebra of Q.

4 Weighted-Semicircular Elements

In this section, we construct weighted-semicircular elements induced by the family $\mathbf{Q}$ of (3.6), in the radial projection algebra $\mathfrak{L}_Q$ of (3.14). Let (Q, ψ_j) be the jth filters of Q in (A, ψ), where ψ_j are the linear functionals (3.9), for all $j \in \mathbb{Z}$.

If u_j are the operators,

$$u_j \overset{def}{=} \mathbf{l} \otimes q_j \in \mathfrak{L}_Q, \text{ for all } j \in \mathbb{Z}, \tag{4.1}$$

then

$$u_j^n = (\mathbf{1} \otimes q_j)^n = \mathbf{l}^n \otimes q_j, \text{ for all } n \in \mathbb{N},$$

since $q_j^k = q_j$, for all $k \in \mathbb{N}$. Let's axiomatize

$$u_j^0 \overset{\text{axiom}}{=} \mathbf{l}^0 \otimes q_j = 1_Q \otimes q_j,$$

where 1_Q is the identity operator of $B(Q)$. Note that these operators $\{u_j\}_{j \in \mathbb{Z}}$ of (4.1) generate our radial projection algebra $\mathfrak{L}_Q$, by (3.8), (3.12) and (3.14).

Define a linear functional φ_j on $\mathfrak{L}_Q$ by a morphism satisfying that

$$\varphi_j (u_i^n) = \varphi_j (\mathbf{l}^n \otimes q_i) \overset{def}{=} \psi_j (\mathbf{l}^n (q_i)), \tag{4.2}$$

for all $n \in \mathbb{N}_0$, for all $i, j \in \mathbb{Z}$. The linear functionals $\{\varphi_j\}_{j \in \mathbb{Z}}$ of (4.2) are well defined on $\mathfrak{L}_Q$ by (3.14).

If $\mathbf{c}$ and $\mathbf{a}$ are the creation, respectively, the annihilation on Q, then

$$\mathbf{ca} = 1_Q = \mathbf{ac}, \tag{4.3}$$

since

$$\mathbf{ca}(q_j) = \mathbf{c}(\mathbf{a}(q_j)) = \mathbf{c}(q_{j-1}) = q_{j-1+1} = q_j,$$

and

$$\mathbf{ac}(q_j) = \mathbf{a}(\mathbf{c}(q_j)) = \mathbf{a}(q_{j+1}) = q_{j+1-1} = q_j,$$

for all $j \in \mathbb{Z}$. It implies that

$$\mathbf{c}^n \mathbf{a}^n = 1_Q = \mathbf{a}^n \mathbf{c}^n, \text{ for all } n \in \mathbb{N},$$

and

$$\mathbf{c}^{n_1} \mathbf{a}^{n_2} = \mathbf{a}^{n_2} \mathbf{c}^{n_1}, \text{ for all } n_1, n_2 \in \mathbb{N}, \tag{4.4}$$

by (4.3). Thus, we have

$$\mathbf{l}^n = (\mathbf{c} + \mathbf{a})^n = \sum_{k=0}^{n} \binom{n}{k} \mathbf{c}^k \mathbf{a}^{n-k} \text{ on } Q, \tag{4.5}$$

for all $n \in \mathbb{N}$, by (4.4), where

$$\binom{n}{k} = \frac{n!}{k!(n-k)!}, \forall k \leq n \in \mathbb{N}_0.$$

By (4.4) and (4.5), for any $n \in \mathbb{N}$,

$$\mathbf{l}^{2n-1} = \sum_{k=0}^{2n-1} \binom{2n-1}{k} \mathbf{c}^k \mathbf{a}^{n-k}. \tag{4.6}$$

Note that the formula (4.6) does not contain 1_Q-terms by (4.4). Meanwhile, for any $n \in \mathbb{N}$,

$$\mathbf{l}^{2n} = \sum_{k=0}^{2n} \binom{2n}{k} \mathbf{c}^k \mathbf{a}^{n-k} = \binom{2n}{n} \mathbf{c}^n \mathbf{a}^n + [\text{Rest terms}], \tag{4.7}$$

by (4.5), showing that l^{2n} contains $\binom{2n}{n}$-many 1_Q-terms.

Lemma 7 *Let l be the radial operator on Q. Then, for any $n \in \mathbb{N}$,*

$$l^{2n-1} \textit{ does not contain } 1_Q\textit{-terms in } \mathfrak{L}, \tag{4.8}$$

$$l^{2n} \textit{ contains } \binom{2n}{n} \cdot 1_Q \textit{ in } \mathfrak{L}, \textit{ as its summand.} \tag{4.9}$$

Proof The statements (4.8) and (4.9) are shown by (4.6) and (4.7). ■

Observe that

$$\varphi_j\left(u_j^{2n-1}\right) = \psi_j\left(\mathbf{l}^{2n-1}(q_j)\right) = 0, \tag{4.10}$$

for all $n \in \mathbb{N}$, by (3.9) and (4.8). Similarly,

$$\varphi_j\left(u_j^{2n}\right) = \psi_j\left(\mathbf{l}^{2n}(q_j)\right) = \psi_j\left(\binom{2n}{n} q_j + [\text{Rest terms}](q_j)\right)$$

by (4.7)

$$= \binom{2n}{n} \psi_j(q_j) = \binom{2n}{n} \psi(q_j),$$

by (3.9) and (4.9). That is,

$$\varphi_j\left(u_j^{2n}\right) = \binom{2n}{n} \psi(q_j), \tag{4.11}$$

for all $n \in \mathbb{N}$.

Lemma 8 *Fix $j \in \mathbb{Z}$, and let $u_k = \mathbf{1} \otimes q_k$ be the kth generating operators of the jth $*$-probability space $(\mathfrak{L}_Q, \varphi_j)$, for all $k \in \mathbb{Z}$. Then*

$$\varphi_j\left(u_k^n\right) = \delta_{j,k}\omega_n\left(\left(\frac{n}{2}+1\right)\psi(q_j)\right)c_{\frac{n}{2}}, \tag{4.12}$$

where ω_n are in the sense of (3.3) for all $n \in \mathbb{N}$, and c_m are the mth Catalan numbers for all $m \in \mathbb{N}$.

Proof Take the jth generating operator $u_j \in (\mathfrak{L}_Q, \varphi_j)$. Then

$$\varphi_j \left(u_j^{2n-1}\right) = 0,$$

by (4.8), and

$$\varphi_j \left(u_j^{2n}\right) = \binom{2n}{n} \psi\left(q_j\right) = \left(\frac{n+1}{n+1}\right) \binom{2n}{n} \psi\left(q_j\right)$$

$$= \left((n+1)\psi\left(q_j\right)\right) c_n,$$

for all $n \in \mathbb{N}$, by (4.9). Meanwhile, if $k \neq j$ in $\mathbb{Z}$, then, by (3.9) and (4.2),

$$\varphi_j \left(u_k^n\right) = 0, \ \forall n \in \mathbb{N}.$$

Therefore, the formula (4.12) holds. ■

Motivated by (4.12), define a bounded linear transformation,

$$E_{j,Q} : \mathfrak{L}_Q \to \mathfrak{L}_Q$$

by a morphism satisfying

$$E_{j,Q}\left(u_i^n\right) \overset{def}{=} \begin{cases} \frac{\psi(q_j)^{n-1}}{([\frac{n}{2}]+1)} u_j^n & \text{if } i = j \\ 0_{\mathfrak{L}_Q}, \text{ the zero operator of } \mathfrak{L}_Q & \text{otherwise,} \end{cases} \tag{4.13}$$

for all $n \in \mathbb{N}$, $i, j \in \mathbb{Z}$, where $[\frac{n}{2}]$ mean the *minimal integers* greater than or equal to $\frac{n}{2}$. These linear transformations $E_{j,Q}$ of (4.13) are well determined on $\mathfrak{L}_Q$, by the cyclicity (3.12) of the tensor factor $\mathfrak{L}$ of $\mathfrak{L}_Q$, and the structure theorem (3.8) of the other tensor factor Q of $\mathfrak{L}_Q$, for all $j \in \mathbb{Z}$.

With help of the linear transformations (4.13), define the linear functionals τ_j on $\mathfrak{L}_Q$ by

$$\tau_j \overset{def}{=} \varphi_j \circ E_{j,Q} \text{ on } \mathfrak{L}_Q, \text{ for all } j \in \mathbb{Z}, \tag{4.14}$$

where φ_j and $E_{j,Q}$ are in the sense of (4.2) and (4.13), respectively.

Definition 9 The Banach $*$-probability spaces,

$$\mathfrak{L}_Q(j) \overset{denote}{=} \left(\mathfrak{L}_Q, \tau_j\right), \tag{4.15}$$

are called the jth filtered (Banach-$*$-probability) spaces of $\mathfrak{L}_Q$, where τ_j are the linear functionals (4.14) on the radial projection algebra $\mathfrak{L}_Q$, for all $j \in \mathbb{Z}$.

On the jth filtered space $\mathfrak{L}_Q(j)$ of (4.15),

$$\tau_j\left(u_j^n\right) = \varphi_j\left(E_{j,Q}\left(u_j^n\right)\right)$$

$$= \varphi_j\left(\frac{\psi(q_j)^{n-1}}{([\frac{n}{2}]+1)}\left(u_j^n\right)\right) = \frac{\psi(q_j)^{n-1}}{([\frac{n}{2}]+1)}\varphi_j\left(u_j^n\right)$$

$$= \frac{\psi(q_j)^{n-1}}{([\frac{n}{2}]+1)}\omega_n\left(\left(\tfrac{n}{2}+1\right)\psi\left(q_j\right)\right)c_{\frac{n}{2}},$$

by (4.12), i.e.,

$$\tau_j\left(u_j^n\right) = \omega_n\psi(q_j)^n c_{\frac{n}{2}}, \tag{4.16}$$

for all $n \in \mathbb{N}$, for $j \in \mathbb{Z}$, where ω_n are in the sense of (3.3).

Lemma 10 *On the jth filtered space $\mathfrak{L}_Q(j)$, we have*

$$\tau_j\left(u_i^n\right) = \delta_{j,i}\left(\omega_n\psi(q_j)^n c_{\frac{n}{2}}\right), \tag{4.17}$$

for all $n \in \mathbb{N}$, for all $i \in \mathbb{Z}$.

Proof The formula (4.17) holds by (4.16), for all $n \in \mathbb{N}$. ■

By (4.17), we have the following result.

Theorem 11 *The "jth" generating operator $u_j \in \mathfrak{L}_Q(j)$ of (4.1) is $\psi(q_j)^2$-semicircular in the "jth" filtered space $\mathfrak{L}_Q(j)$.*

Proof Recall that all generating operators u_i are self-adjoint in $\mathfrak{L}_Q$, for all $i \in \mathbb{Z}$, since

$$u_i^* = (\mathbf{1} \otimes q_i)^* = \mathbf{1} \otimes q_i = u_i \text{ in } \mathfrak{L}_Q,$$

for all $i \in \mathbb{Z}$, by (3.13).

For $j \in \mathbb{Z}$, let $u_j = \mathbf{1} \otimes q_j$ be the jth generating operator (4.1) of the jth filtered space $\mathfrak{L}_Q(j)$. Then, by (4.17), we have that

$$\tau_j\left(u_j^n\right) = \omega_n\left(\psi\left(q_j\right)^2\right)^{\frac{n}{2}} c_{\frac{n}{2}}, \ \forall n \in \mathbb{N},$$

implying that, the self-adjoint element u_j is $\psi(q_j)^2$-semicircular in $\mathfrak{L}_Q(j)$ by (3.3). ■

The above theorem shows that the "jth" generating operator u_j is $\psi(q_j)^2$-semicircular in the "jth" filtered space $\mathfrak{L}_Q(j)$, for all $j \in \mathbb{Z}$. While, also by (4.17), it is not hard to verify the following result, too.

Theorem 12 *Let $u_i = \mathbf{1} \otimes u_i$ be the ith generating operators (4.1) of the jth filtered space $\mathfrak{L}_Q(j)$, for all $j \neq i \in \mathbb{Z}$. Then u_i have the zero free distribution in $\mathfrak{L}_Q(j)$.*

Proof Let $\mathfrak{L}_Q(j)$ be the jth filtered space for $j \in \mathbb{Z}$, and let $i \neq j$ in $\mathbb{Z}$. By the self-adjointness of a generating operator u_i, the free distribution of u_i is fully characterized by the free-momental sequence,

$$\left(\tau_j(u_i^n)\right)_{n=1}^{\infty} = (0, 0, 0, 0, \ldots),$$

identified with the zero sequence by (4.17). Therefore, the free distributions of u_i are the zero free distribution in $\mathfrak{L}_Q(j)$, whenever $i \neq j$ in $\mathbb{Z}$. ■

By the Möbius inversion of [17], if u_i are the ith generating operators of $\mathfrak{L}_Q(j)$, then

$$k_n^j(u_i, \ldots, u_i) = \delta_{j,i}\delta_{n,2} \tag{4.18}$$

for all $n \in \mathbb{N}$, and $i \in \mathbb{Z}$, by Theorems 11 and 12, where $k_\bullet^j(\ldots)$ is the free cumulant on $\mathfrak{L}_Q$ in terms of the linear functional τ_j, for all $j \in \mathbb{Z}$.

5 Semicircular Elements Induced by Q

Let $\mathfrak{L}_Q(j)$ be the jth filtered space (4.15) for $j \in \mathbb{Z}$. Then, the jth generating operator u_j of $\mathfrak{L}_Q$ is $\psi(q_j)^2$-semicircular in $\mathfrak{L}_Q(j)$, satisfying that

$$\tau_j\left(u_j^n\right) = \omega_n \psi(q_j)^n c_{\frac{n}{2}},$$

equivalently,

$$k_n^j\left(u_j, \ldots, u_j\right) = \begin{cases} \psi(q_j)^2 & \text{if } n = 2 \\ 0 & \text{otherwise,} \end{cases} \tag{5.1}$$

for all $n \in \mathbb{N}$, by (4.17) and (4.18).

From the $\psi\left(q_j\right)^2$-semicircular element $u_j \in \mathfrak{L}_Q(j)$, define a new free random variable U_j by

$$U_j \overset{def}{=} \frac{1}{\psi(q_j)} u_j \in \mathfrak{L}_Q(j), \tag{5.2}$$

for each $j \in \mathbb{Z}$. Recall that since it is automatically assumed that $\psi(q_k) \in \mathbb{C}^{\times}$, for all $k \in \mathbb{Z}$, the operators U_j of (5.2) are well defined in $\mathfrak{L}_Q(j)$.

Theorem 13 *Let $U_j = \frac{1}{\psi(q_j)} u_j$ be a free random variable (5.2) of $\mathfrak{L}_Q(j)$, for $j \in \mathbb{Z}$, where u_j is the jth generating operator of $\mathfrak{L}_Q$. If*

$$\psi(q_j) \in \mathbb{R}^\times = \mathbb{R} \setminus \{0\} \text{ in } \mathbb{C}^\times,$$

then U_j is semicircular in $\mathfrak{L}_Q(j)$.

Proof For $j \in \mathbb{Z}$, assume $\psi(q_j) \in \mathbb{R}^\times$ in $\mathbb{C}^\times$. Then

$$U_j^* = \left(\frac{1}{\psi(q_j)}u_j\right)^* = \frac{1}{\psi(q_j)}u_j = U_j,$$

by the self-adjointness of u_j in $\mathfrak{L}_Q$. Moreover, one has

$$\begin{aligned}\tau_j\left(U_j^n\right) &= \left(\tfrac{1}{\psi(q_j)}\right)^n \tau_j\left(u_j^n\right) \\ &= \left(\tfrac{1}{\psi(q_j)^n}\right)\left(\omega_n \psi(q_j)^n c_{\frac{n}{2}}\right) = \omega_n c_{\frac{n}{2}},\end{aligned} \tag{5.3}$$

for all $n \in \mathbb{N}$. So, this self-adjoint operator U_j is semicircular in $\mathfrak{L}_Q(j)$ by (3.4) and (5.3). ■

Assumption 5.1 (in short, **A 5.1**, from below) We assume that

$$\psi(q_j) \in \mathbb{R}^\times \text{ in } \mathbb{C}, \text{ for } q_j \in \mathbf{Q},$$

for all $j \in \mathbb{Z}$. □

6 The Free Filterization

Let (A, ψ) be a fixed C^*-probability space containing a family $\mathbf{Q} = \{q_k\}_{k\in\mathbb{Z}}$ of mutually orthogonal projections satisfying **A 5.1**, and let $\mathfrak{L}_Q(j)$ be the jth filtered spaces of Q, for all $j \in \mathbb{Z}$. From the system,

$$\{\mathfrak{L}_Q(j) : j \in \mathbb{Z}\},$$

define the *free product Banach $*$-probability space* $\mathfrak{L}_Q(\mathbb{Z})$ by

$$\begin{aligned}\mathfrak{L}_Q(\mathbb{Z}) &\overset{denote}{=} \left(\mathfrak{L}_Q(\mathbb{Z}),\ \tau\right) \\ &\overset{def}{=} \underset{j\in\mathbb{Z}}{\star} \mathfrak{L}_Q(j) = \left(\underset{j\in\mathbb{Z}}{\star} \mathfrak{L}_{Q,j},\ \underset{j\in\mathbb{Z}}{\star} \tau_j\right),\end{aligned} \tag{6.1}$$

with identity:

$$\mathfrak{L}_{Q,j} = \mathfrak{L}_Q, \text{ for all } j \in \mathbb{Z}.$$

That is, the Banach $*$-probability space $\mathfrak{L}_Q(\mathbb{Z})$ of (6.1) is generated by its free blocks, the jth filtered spaces $\left\{\mathfrak{L}_Q(j)\right\}_{j\in\mathbb{Z}}$ (e.g., [17, 30]).

Definition 14 Let $\mathfrak{L}_Q(\mathbb{Z})$ be the free product Banach $*$-probability space (6.1) of $\{\mathfrak{L}_Q(j)\}_{j\in\mathbb{Z}}$. Then, it is said to be the free filterization of $Q \subset (A, \psi)$.

Define now the subsets $\mathcal{X}$ and $\mathcal{S}$ of $\mathfrak{L}_Q(\mathbb{Z})$ by

$$\mathcal{X} = \{u_j \in \mathfrak{L}_Q(j) : j \in \mathbb{Z}\},$$

and

$$\mathcal{S} = \{U_j \in \mathfrak{L}_Q(j) : j \in \mathbb{Z}\}, \tag{6.2}$$

where $u_j \in \mathcal{X}$ and $U_j \in \mathcal{S}$ are in the sense of (4.1), respectively, (5.2) in $\mathfrak{L}_Q$.

A subset $\mathcal{Y}$ of a topological $*$-probability space (B, φ_B) is said to be a *free family*, if all elements of $\mathcal{Y}$ are mutually free from each other in (B, φ_B). A free family $\mathcal{Y}$ is called a *free (weighted-)semicircular family* in (B, φ_B), if the elements of $\mathcal{Y}$ are (weighted-)semicircular in (B, φ_B). (e.g., [8, 30]).

Theorem 15 *Let $\mathcal{X}$ and $\mathcal{S}$ be the subsets (6.2) in the free filterization $\mathfrak{L}_Q(\mathbb{Z})$ of (6.1).*

$$\textit{The family } \mathcal{X} \textit{ is a free weighted-semicircular family in } \mathfrak{L}_Q(\mathbb{Z}). \tag{6.3}$$

$$\textit{The family } \mathcal{S} \textit{ is a free semicircular family in } \mathfrak{L}_Q(\mathbb{Z}). \tag{6.4}$$

Proof All elements u_j of the family $\mathcal{X}$ of (6.2) are taken from mutually distinct free blocks $\mathfrak{L}_Q(j)$ of $\mathfrak{L}_Q(\mathbb{Z})$, for all $j \in \mathbb{Z}$, implies the mutual freeness of them in $\mathfrak{L}_Q(\mathbb{Z})$. That is, this family $\mathcal{X}$ is a free family in $\mathfrak{L}_Q(\mathbb{Z})$. Moreover, the nth powers u_j^n of $u_j \in \mathcal{X}$ are again contained in the free block $\mathfrak{L}_Q(j)$ as free reduced words with their lengths-1, for all $n \in \mathbb{N}$, for $j \in \mathbb{Z}$. Thus,

$$\tau\left(u_j^n\right) = \tau_j\left(u_j^n\right) = \omega_n\left(\psi\left(q_j\right)^2\right)^{\frac{n}{2}} c_{\frac{n}{2}},$$

for all $n \in \mathbb{N}$, by (5.1), for all $j \in \mathbb{Z}$. So, every element $u_j \in \mathcal{X}$ is $\psi(q_j)^2$-semicircular in $\mathfrak{L}_Q(\mathbb{Z})$, for all $j \in \mathbb{Z}$. Therefore, the statement (6.3) holds.

Similarly, one can verify that the family $\mathcal{S}$ of (6.2) is a free semicircular family in $\mathfrak{L}_Q(\mathbb{Z})$ (under **A 5.1**), showing that the statement (6.4) holds true. ■

We now restrict our interests to the Banach $*$-subalgebra $\mathbb{L}_Q$ of the free filterization $\mathfrak{L}_Q(\mathbb{Z})$, whose elements have possible non-zero free distributions in $\mathfrak{L}_Q(\mathbb{Z})$.

Definition 16 Define a Banach $*$-subalgebra $\mathbb{L}_Q$ of the free filterization $\mathfrak{L}_Q(\mathbb{Z})$ by

$$\mathbb{L}_Q \overset{def}{=} \overline{\mathbb{C}[\mathcal{X}]}, \tag{6.5}$$

where $\mathcal{X}$ is the free weighted-semicircular family (6.3) in $\mathfrak{L}_Q(\mathbb{Z})$, and $\overline{Y}$ are the Banach-topology closures of subsets Y of $\mathfrak{L}_Q(\mathbb{Z})$. Construct the Banach $*$-probability "subspace,"

$$\mathbb{L}_Q \overset{denote}{=} \left(\mathbb{L}_Q,\ \tau = \tau \mid_{\mathbb{L}_Q}\right), \tag{6.6}$$

of $\mathfrak{L}_Q(\mathbb{Z}) = \left(\mathfrak{L}_Q(\mathbb{Z}),\ \tau\right)$. We call $\mathbb{L}_Q$ of (6.5), or, of (6.6), the semicircular (free-sub-)filterization of $\mathfrak{L}_Q(\mathbb{Z})$.

Theorem 17 *Let $\mathbb{L}_Q$ be the semicircular filterization (6.5) of $\mathfrak{L}_Q(\mathbb{Z})$. Then*

$$\begin{aligned}\mathbb{L}_Q &\overset{def}{=} \overline{\mathbb{C}[\mathcal{X}]} = \overline{\mathbb{C}[\mathcal{S}]} \\ &\overset{*\text{-}iso}{=} \underset{j\in\mathbb{Z}}{\star} \overline{\mathbb{C}[\{u_j\}]} \overset{*\text{-}iso}{=} \overline{\mathbb{C}\left[\underset{j\in\mathbb{Z}}{\star}\{u_j\}\right]},\end{aligned} \tag{6.7}$$

in $\mathfrak{L}_Q(\mathbb{Z})$, where "$\overset{*\text{-}iso}{=}$" means "being Banach-$*$-isomorphic," and where $(\star)$ in the first $*$-isomorphic relation of (6.7) is the free-probabilistic free product of [17, 30], and $(\star)$ in the second $*$-isomorphic relation of (6.7) is the pure-algebraic free product inducing noncommutative free words in $\mathcal{X}$.

Proof The free weighted-semicircular family $\mathcal{X}$ of (6.3) can be re-written by

$$\mathcal{X} = \{\psi(q_j)U_j \in \mathfrak{L}_Q(j) : U_j \in \mathcal{S}\}$$

in $\mathfrak{L}_Q(\mathbb{Z})$, where $\mathcal{S}$ is the free semicircular family (6.4). Thus, one obtains the set-equality,

$$\overline{\mathbb{C}[\mathcal{X}]} = \overline{\mathbb{C}[\mathcal{S}]} \text{ in } \mathfrak{L}_Q(\mathbb{Z}).$$

By (6.5), the generating set $\mathcal{X}$ of $\mathbb{L}_Q$ is a free family $\mathcal{X}$, implying the first $*$-isomorphic relation of (6.7) in the free filterization $\mathfrak{L}_Q(\mathbb{Z})$ by (6.1). Since

$$\mathbb{L}_Q \overset{*\text{-iso}}{=} \underset{j\in\mathbb{Z}}{\star} \overline{\mathbb{C}[\{u_j\}]} \text{ in } \mathfrak{L}_Q(\mathbb{Z}),$$

every element T of $\mathbb{L}_Q$ is a limit of linear combinations of free reduced words in $\mathcal{X}$. Also, all (pure-algebraic) free words in $\mathcal{X}$ have their unique free-reduced-word forms under operator-multiplication on $\mathfrak{L}_Q(\mathbb{Z})$. Therefore, the second $*$-isomorphic relation of (6.7) holds, too. ∎

7 Free Poisson Elements of $\mathbb{L}_Q$

Let $(A,\ \psi)$ be a fixed C^*-probability space, containing the family $\mathbf{Q} = \{q_j\}_{j\in\mathbb{Z}}$ of mutually orthogonal projections, and let $\mathbb{L}_Q$ be the semicircular filterization (6.6) of the free filterization $\mathfrak{L}_Q(\mathbb{Z})$ of the C^*-subalgebra $Q = C^*(\mathbf{Q})$ of A. Throughout this section, we also assume **A 5.1**, and hence, the family $\mathcal{S}$ of (6.2) is a well-determined free semicircular family in $\mathfrak{L}_Q(\mathbb{Z})$, generating $\mathbb{L}_Q$ by (6.4) and (6.7).

7.1 Free Poisson Elements

Let (B, φ_B) be a topological $*$-probability space. If $x \in (B, \varphi_B)$ is a free random variable, then the free distribution of x is characterized by the *joint free moments of* $\{x, x^*\}$

$$\varphi_B\left(x^{r_1}x^{r_2}\dots x^{r_n}\right),$$

or, by the *joint free cumulants of* $\{x, x^*\}$,

$$k_n^B\left(x^{r_1}, x^{r_2}, \dots, x^{r_n}\right),$$

for all $(r_1, \dots, r_n) \in \{1, *\}^n$, for all $n \in \mathbb{N}$, where $k_\bullet^B(..)$ is the free cumulant on B in terms of φ_B under the Möbius inversion of [17, 22, 23]. And they provide equivalent free-distributional data of $x \in (B, \varphi_B)$. Thus, if x is a self-adjoint free random variable of (B, φ_B), satisfying $x = x^*$ in B, then the free distribution of x is characterized by the free-moment sequence,

$$\left(\varphi_B(x^n)\right)_{n=1}^{\infty} = \left(\varphi_B(x), \varphi_B(x^2), \varphi_B(x^3), \dots\right),$$

or, by the free-cumulant sequence,

$$\left(k_n^B(x, \dots, x)\right)_{n=1}^{\infty} = \left(k_1^B(x) = \varphi_B(x), k_2^B(x, x), \dots\right).$$

For example, every semicircular element $s \in (B, \varphi_B)$ has its free distribution, the semicircular law, characterized by the free-moment sequence,

$$(0, c_1, 0, c_2, 0, c_3, \dots),$$

or, by the free-cumulant sequence,

$$(0, 1, 0, 0, 0, \dots),$$

where c_k are the kth Catalan numbers for all $k \in \mathbb{N}$.

Notation From below, we will write "a free random variable $x \in (B, \varphi_B)$ has its free distribution $(\varphi_B(x^n))_{n=1}^{\infty}$," equivalently, "$x \in (B, \varphi_B)$ has its free distribution $\left(k_n^B(x, \dots, x)\right)_{n=1}^{\infty}$," if (i) x is self-adjoint in B, and (ii) the free distribution of x is characterized by

$$\left(\varphi_B(x^n)\right)_{n=1}^{\infty}, \text{ or, } \left(k_n^B(x, \dots, x)\right)_{n=1}^{\infty}.$$

□

Definition 18 Let $s \in (B, \varphi_B)$ be a semicircular element, and let a be a self-adjoint free random variable of (B, φ_B), having its free distribution $(\varphi_B(a^n))_{n=1}^{\infty}$. Assume that a and s are free in (B, φ_B). Then, a new free random variable

$$W_s^a = sas \in (B, \varphi_B) \tag{7.1.1}$$

is called the free Poisson element generated by s and a.

As in Chap. 12 of [17], the free random variable W_s^a of (7.1.1) is indeed free Poisson. By (7.1.1),

$$\left(W_s^a\right)^* = (sas)^* = s^* a^* s^* = sas = W_s^a \tag{7.1.2}$$

in B, and hence, it is a self-adjoint free random variable of (B, φ_B), too.

Let $\Omega = \{e_1, \ldots, e_n\}$ be a finite set with its cardinality $n \in \mathbb{N}$. Then, the *lattice* $NC(\Omega)$ *of noncrossing partitions of* Ω is well defined with its partial ordering $\leq$,

$$\theta_1 \leq \theta_2 \Longleftrightarrow \forall N \in \theta_1, \exists V \in \theta_2, \text{ s.t., } N \subseteq V,$$

where $\subseteq$ is the usual set inclusion, where "$S \in \theta$" means "S is a block of θ."

For example, if $\Omega_5 = \{1, 2, \ldots, 5\}$, and if

$$\theta_1 = \{(1, 4), (2, 3), (5)\},$$

and

$$\theta_2 = \{(1, 2, 3, 4), (5)\}$$

in $NC(\Omega_5)$, then $\theta_1 \leq \theta_2$.

Notation From below, if a given finite set Ω is a subset $\{1, 2, \ldots, n\}$ of $\mathbb{N}$, for some $n \in \mathbb{N}$, then we denote Ω by Ω_n. □

Also, under ($\leq$), the lattice $NC(\Omega)$ has its maximal element,

$$1_\Omega = \{(e_1, \ldots, e_n)\}, \text{ the 1-block partition,}$$

and its minimal element,

$$0_\Omega = \{(e_1), (e_2), \ldots, (e_n)\}, \text{ the } n \text{ -block partition.}$$

Now, let Ω be as above, and let $NC(\Omega)$ be the corresponding noncrossing-partition lattice. Suppose

$$\Omega = \Omega^1 \sqcup \Omega^2, \text{ with } \Omega^l \subset \Omega, \forall l = 1, 2.$$

where $\sqcup$ is the disjoint union, and let $NC\left(\Omega^l\right)$ be the noncrossing-partition lattices of Ω^l, for $l = 1, 2$.

Then, for $\theta_l \in NC(\Omega^l)$, for $l = 1, 2$, one can construct a noncrossing partition $\theta \in NC(\Omega)$,

$$\theta = \theta_1 \vee \theta_2 \in NC(\Omega),$$

as in [17, 22, 23]. For example, if

$$\theta_1 = \{(2, 5), (3)\} \text{ in } NC(\{2, 3, 5\}),$$

and

$$\theta_2 = \{(4), (1, 6, 7)\} \text{ in } NC(\{1, 4, 6, 7\}),$$

then we obtain

$$\theta_1 \vee \theta_2 = \{(1, 6, 7), (2, 5), (3), (4)\},$$

in $NC(\Omega_7)$.

Now, fix $n \in \mathbb{N}$, and let $NC(\Omega_{3n})$ be the noncrossing-partition lattice of

$$\Omega_{3n} = \{1, 2, \ldots, 3n\}.$$

Also, let

$$\Omega^1_{3n} = \{1, 3, 4, 6, 7, 9, 10, \ldots, 3n-3, 3n-2, 3n\},$$

and

$$\Omega^2_{3n} = \Omega_{3n} \setminus \Omega^1_{3n} = \{2, 5, 8, 11, \ldots, 3n-1\}, \tag{7.1.3}$$

satisfying

$$\Omega_{3n} = \Omega^1_{3n} \sqcup \Omega^2_{3n}.$$

And then take $\theta_o \in NC\left(\Omega^1_{3n}\right)$,

$$\theta_o = \{(1, 3n), (3, 4), (6, 7), \ldots, (3n-3, 3n-2)\},, \tag{7.1.4}$$

where Ω^1_{3n} is in the sense of (7.1.3).

Now, let's consider the free distribution of a free Poisson element $W^a_s = sas$ of (7.1.1) in (B, φ_B), in terms of its free cumulants. The following lemma is already shown (e.g., in p. 207 of [17]), but we provide a sketch of the proof for our future application.

Lemma 19 *Let $a \in (B, \varphi_B)$ be a self-adjoint free random variable having its free distribution $(\varphi_B(a^n))_{n=1}^{\infty}$, and let $s \in (B, \varphi_B)$ be a semicircular element. Assume that s and a are free in (B, φ_B), and let $W^a_s = sas$ be the corresponding free Poisson element (7.1.1). Then*

$$k^B_n\left(\underbrace{W^a_s,\ W^a_s, \ldots, W^a_s}_{n\text{-times}}\right) = \varphi_B(a^n), \tag{7.1.5}$$

for all $n \in \mathbb{N}$.

Proof By the semicircularity of s,

$$k_n^B(s, \ldots, s) = \begin{cases} 1 & \text{if } n = 2 \\ 0 & \text{otherwise,} \end{cases} \tag{7.1.6}$$

for all $n \in \mathbb{N}$. So,

$$k_n^B\left(W_s^a, \ldots, W_s^a\right) = k_n^B(sas,\ sas, \ldots, sas)$$

$$= \sum_{\theta \in NC(\Omega_{3n}),\ \theta \vee \pi_0 = 1_{3n}} k_\theta^B(sas, \ldots, sas)$$

where Ω_{3n} is in the sense of (7.1.3), and $k_\theta^B(\ldots)$ are the block-depending free cumulants, and where 1_{3n} is the maximal 1-block partition of $NC(\Omega_{3n})$, and

$$\pi_0 = \{(1, 2, 3), (4, 5, 6), \ldots, (3n-2, 3n-1, 3n)\}$$

(See [17, 22, 23] for details), and hence, it goes to

$$= \sum_{\theta \in NC(\Omega_{3n}^2),\ \theta_o \vee \theta \in NC(\Omega_{3n})} k_{\theta_o \vee \theta}^B(sas, \ldots, sas)$$

by the semicircularity (7.1.6) of $s \in (B, \varphi_B)$, where Ω_{3n}^2 is in the sense of (7.1.3), and θ_o is the partition (7.1.4)

$$= \sum_{\theta \in NC(\Omega_{3n}^2)} k_{\theta_o}^B\left(\underbrace{s,\ s,\ s, \ldots, s}_{2n\text{-times}}\right) k_\theta^B\left(\underbrace{a,\ a, \ldots, a}_{n\text{-times}}\right)$$

by the freeness of s and a in (B, φ_B) (Recall that s and a are free, if and only if all their "mixed" free cumulants vanish!)

$$= \sum_{\theta \in NC(\Omega_{3n}^2)} \left(\left(k_2^B(s,\ s)\right)^n\right) k_\theta^B(a,\ a, \ldots, a)$$

by (7.1.4)

$$= \sum_{\theta \in NC(\Omega_{3n}^2)} k_\theta^B\left(\underbrace{a,\ a, \ldots, a}_{n\text{-times}}\right)$$

by (7.1.6)

$$= \sum_{\pi \in NC(\Omega_n)} k_\pi^B(a, \ldots, a)$$

since the sub-lattice $NC\left(\Omega_{3n}^2\right)$ of $NC\left(\Omega_{3n}\right)$ is equivalent to the lattice $NC\left(\Omega_n\right)$

$$= \varphi_B(a^n)$$

by the Möbius inversion, for all $n \in \mathbb{N}$.

Therefore, the free-distributional data (7.1.5) is obtained. ∎

The free-cumulant formula (7.1.5) shows that the free distribution of a free Poisson element W_s^a is characterized by the free distribution $(\varphi_B(a^n))_{n=1}^{\infty}$ of a fixed self-adjoint free random variable a in (B, φ_B).

The following characterization of (7.1.5) is not widely used in earlier works, but it will be applied in our future works.

Proposition 20 *Let $W_s^a = sas \in (B, \varphi_B)$ be a free Poisson element (7.1.1). Then*

$$\varphi_B\left((W_s^a)^n\right) = \sum_{\theta \in NC(\Omega_n)} \left(\prod_{V \in \theta} \varphi_B\left(a^{|V|}\right)\right), \tag{7.1.7}$$

for all $n \in \mathbb{N}$, where $|V|$ are the cardinalities of blocks V (as sets).

Proof Observe that

$$\varphi_B\left((W_s^a)^n\right) = \sum_{\theta \in NC(\Omega_n)} k_\theta^B\left(W_s^a, \ldots, W_s^a\right)$$

by the Möbius inversion

$$= \sum_{\theta \in NC(\Omega_n)} \left(\prod_{V \in \theta} k_{|V|}^B\left(\underbrace{W_s^a, \ldots, W_s^a}_{|V|\text{-times}}\right)\right)$$

$$= \sum_{\theta \in NC(\Omega_n)} \left(\prod_{V \in \theta} \varphi_B\left(a^{|V|}\right)\right),$$

by (7.1.5), for all $n \in \mathbb{N}$. So, the free-distributional data (7.1.7) holds. ∎

It is not difficult to verify that the free-distributional data (7.1.7) is mathematically dual of (7.1.5).

7.2 Free Poisson Elements of $\mathbb{L}_Q$ Induced by $\mathcal{S}$

Let $\mathbb{L}_Q = \left(\mathbb{L}_Q, \tau\right)$ be the semicircular filterization (6.6) under **A 5.1**. Throughout this section, we fix a semicircular element $U_j = \frac{1}{\psi(q_j)} u_j$ of the free semicircular family $\mathcal{S}$ of (6.4), for $j \in \mathbb{Z}$. By Sect. 7.1, if $T \in \mathbb{L}_Q$ is a self-adjoint operator, and if T is free from U_j in $\mathbb{L}_Q$, then one obtains the corresponding free Poisson element,

$$W_j^T = U_j T U_j \text{ in } \mathbb{L}_Q,$$

satisfying that

$$k_n\left(W_j^T, \ldots, W_j^T\right) = \tau\left(T^n\right),$$

and

$$\tau\left(\left(W_j^T\right)^n\right) = \sum_{\theta \in NC(\Omega_n)} \left(\prod_{V \in \theta} \tau\left(T^{|V|}\right)\right), \tag{7.2.1}$$

by (7.1.5) and (7.1.7), for all $n \in \mathbb{N}$, where $k_\bullet(\ldots)$ is the free cumulant on $\mathbb{L}_Q$ in terms of the linear functional τ of (6.6).

By the structure theorem (6.7) of our semicircular filterization $\mathbb{L}_Q$, we are interested in the cases where fixed self-adjoint operators T in (7.2.1) are generated by semicircular elements $U_k \in \mathcal{S}$ in $\mathbb{L}_Q$, for $j \neq k \in \mathbb{Z}$.

Define free Poisson elements $W_j^k = W_j^{U_k}$ of (7.2.1) by

$$W_j^k \overset{def}{=} U_j U_k U_j \in \mathbb{L}_Q, \tag{7.2.2}$$

with $j \neq k$ in $\mathbb{Z}$.

By the freeness (6.7) on $\mathbb{L}_Q$, if $j \neq k$ in $\mathbb{Z}$, then two semicircular elements U_j and U_k are distinct in the free family $\mathcal{S}$, implying that they are free in $\mathbb{L}_Q$. So, the operator W_j^k is a well-defined free Poisson element by (7.1.1). By the self-adjointness (7.1.2) of W_j^k, the corresponding free distribution is characterized by the free-moment sequence, or the free-cumulant sequence. Consider that

$$k_n\left(W_j^k, \ldots, W_j^k\right) = \tau\left(U_k^n\right)$$

and

$$\tau\left(\left(W_j^k\right)^n\right) = \sum_{\theta \in NC(\Omega_n)} \left(\prod_{V \in \theta} \tau\left(U_k^{|V|}\right)\right), \tag{7.2.3}$$

by (7.2.1), for all $n \in \mathbb{N}$.

Proposition 21 *Let $W_j^k = U_j U_k U_j \in \mathbb{L}_Q$ be the free Poisson element (7.2.2) generated by two, free, semicircular elements U_k, $U_j \in \mathcal{S}$. Then*

$$k_n\left(W_j^k, \ldots, W_j^k\right) = \omega_n c_{\frac{n}{2}},$$

and

$$\tau\left(\left(W_j^k\right)^n\right) = \omega_n \left(\sum_{\theta \in NC_e(\Omega_n)} \left(\prod_{V \in \theta} \left(c_{\frac{|V|}{2}}\right)\right)\right), \tag{7.2.4}$$

for all $n \in \mathbb{N}$, where the subset,

$$NC_e(\Omega_n) = \{\pi \in NC(\Omega_n) : |V| \in 2\mathbb{N},\ \forall V \in \pi\},$$

of $NC_e(\Omega_n)$ is the set of all noncrossing partitions over Ω_n whose blocks have even cardinalities.

Proof The free-distributional data (7.2.4) is proven by (7.2.1) and (7.2.3). In particular, by (7.2.3), one has that

$$\tau\left(\left(W_j^k\right)^n\right) = \sum_{\theta \in NC(\Omega_n)} \left(\prod_{V \in \theta} \tau\left(U_k^{|V|}\right)\right),$$

and hence,

$$\tau\left(\left(W_j^k\right)^n\right) = \sum_{\theta \in NC(\Omega_n)} \left(\prod_{V \in \theta} \left(\omega_{|V|} c_{\frac{|V|}{2}}\right)\right),$$

by the semicircularity of $U_k \in \mathcal{S}$ in $\mathbb{L}_Q$. Thus,

$$\omega_{|V|} = 1, \Longleftrightarrow |V| \in 2\mathbb{N},\ \forall V \in \pi,$$

for all $\pi \in NC(\Omega_n)$, if and only if $\pi \in NC_e(\Omega_n)$ and $n \in 2\mathbb{N}$, because $NC_e(\Omega_n)$ is not empty in $NC(\Omega_n)$, if and only if n is even in $\mathbb{N}$.

Therefore, the free-moment formula in (7.2.4) holds as above. ■

Let $U_k \in \mathcal{S}$ be a semicircular element of $\mathbb{L}_Q$, for any $k \in \mathbb{Z}$, and let $N \in \mathbb{N}$. Observe that

$$k_n\left(U_k^N, \ldots, U_k^N\right) = \sum_{\pi \in NC(\Omega_n)} \left(\prod_{V \in \pi} \tau\left(U_k^{N|V|}\right)\right)$$

by the Möbius inversion

$$= \sum_{\pi \in NC(\Omega_n)} \left(\prod_{V \in \pi} \omega_{|V|N} c_{\frac{|V|N}{2}}\right)$$

by the semicircularity of $U_k \in \mathcal{S}$ in $\mathbb{L}_Q$

$$= \begin{cases} \sum\limits_{\pi \in NC(\Omega_n)} \left(\prod\limits_{V \in \pi} c_{\frac{|V|N}{2}}\right) & \text{if } N \text{ is even} \\ \omega_n \left(\sum\limits_{\theta \in NC_e(\Omega_n)} \left(\prod\limits_{V \in \theta} c_{\frac{|V|N}{2}}\right)\right) & \text{if } N \text{ is odd,} \end{cases}$$

where

$$NC_e(\Omega_n) \overset{def}{=} \{\theta \in NC(\Omega_n) : |V| \text{ is even}, \forall V \in \theta\}, \tag{7.2.5}$$

is in the sense of (7.2.4), for all $n \in \mathbb{N}$.

The first formula of (7.2.5) holds because if N is even, then $N \cdot k$ is even for all $k \in \mathbb{N}$; and the second formula of (7.2.5) holds, if N is odd, then $N \cdot k$ is even whenever k is even, meanwhile it is odd whenever k is odd, for all $k \in \mathbb{N}$.

Note that, if $N = 1$, then the formula (7.2.5) is identical to

$$\omega_n \left(\sum_{\theta \in NC_e(\Omega_n)} \left(\prod_{V \in \theta} c_{\frac{|V|N}{2}} \right) \right), \forall n \in \mathbb{N},$$

obtained in the second formula of (7.2.4).

Now, let $k \neq j$ in $\mathbb{Z}$, and take a semicircular element $U_k \in \mathcal{S}$ in $\mathbb{L}_Q$, and construct a self-adjoint operator $U_k^N \in \mathbb{L}_Q$, for $N \in \mathbb{N}$. Then, one obtains the corresponding free Poisson element,

$$W_j^{k,N} = U_j U_k^N U_j \in \mathbb{L}_Q, \text{ for } N \in \mathbb{N}. \tag{7.2.6}$$

Theorem 22 *Let $W_j^{k,N}$ be a free Poisson element (7.2.6) for $N \in \mathbb{N}$. Then*

$$k_n \left(W_j^{k,N}, \ldots, W_j^{k,N} \right) = \omega_{nN} c_{\frac{nN}{2}},$$

and

$$\tau \left(\left(W_j^{k,N} \right)^n \right) = \begin{cases} \sum_{\pi \in NC(\Omega_n)} \left(\prod_{B \in \theta} c_{\frac{N|B|}{2}} \right) & \text{if } N \text{ is even} \\ \omega_n \left(\sum_{\theta \in NC_e(\Omega_n)} \left(\prod_{V \in \theta} c_{\frac{N|V|}{2}} \right) \right) & \text{if } N \text{ is odd,} \end{cases} \tag{7.2.7}$$

for all $n \in \mathbb{N}$, where $NC_e(\Omega_n)$ are in the sense of (7.2.5).

Proof Observe first that

$$k_n \left(W_j^{k,N}, \ldots, W_j^{k,N} \right) = \tau \left(\left(U_k^N \right)^n \right)$$

$$= \tau \left(U_k^{nN} \right) = \omega_{nN} c_{\frac{nN}{2}},$$

by (7.2.1), for all $n \in \mathbb{N}$. Thus, the free-cumulant formula of (7.2.7) holds.

Consider now that

$$\tau\left(\left(W_j^{k,N}\right)^n\right) = k_n\left(U_k^N, \ldots, U_k^N\right)$$

by (7.1.7)

$$= \begin{cases} \sum\limits_{\pi \in NC(\Omega_n)} \left(\prod\limits_{V \in \theta} c_{\frac{N|V|}{2}}\right) & \text{if } N \text{ is even} \\ \\ \omega_n \left(\sum\limits_{\theta \in NC_e(\Omega_n)} \left(\prod\limits_{V \in \theta} c_{\frac{N|V|}{2}}\right)\right) & \text{if } N \text{ is odd,} \end{cases}$$

for all $n \in \mathbb{N}$, by (7.2.5). So, the free-moment formula of (7.2.7) holds. ■

The free-distributional data (7.2.7) generalizes (7.2.4).

7.3 Free Poisson Elements of $\mathbb{L}_Q$ Induced by $\mathcal{X}$

As in Sect. 7.2, let's fix a semicircular element $U_j \in \mathcal{S}$, and let $\mathcal{X}$ be the free weighted-semicircular family (6.3) in the semicircular filterization $\mathbb{L}_Q$. For any $k \neq j$ in $\mathbb{Z}$, we have the corresponding free Poisson elements,

$$Y_j^{k,N} = U_j u_k^N U_j \text{ in } \mathbb{L}_Q, \tag{7.3.1}$$

where $u_k \in \mathcal{X}$ is a $\psi(q_j)^2$-semicircular element of $\mathbb{L}_Q$, for all $N \in \mathbb{N}$. Note that

$$u_k = \psi(q_k)U_k \text{ in } \mathbb{L}_Q, \tag{7.3.2}$$

by the constructions of the free families $\mathcal{X}$ and $\mathcal{S}$ of (6.2) in $\mathbb{L}_Q$.

So, by the definition (7.3.1) of a free Poisson element $Y_j^{k,N}$,

$$Y_j^{k,N} = \psi(q_k)^N U_j U_k^N U_{p,j} = \psi(q_k)^N W_j^{k,N}, \tag{7.3.3}$$

in $\mathbb{L}_Q$, by (7.3.2), where $W_j^{k,N}$ is the free Poisson element (7.2.6).

Theorem 23 *Let $Y_j^{k,N}$ be a free Poisson element (7.3.1) in $\mathbb{L}_Q$, for $N \in \mathbb{N}$. Then*

$$k_n\left(Y_j^{k,N}, \ldots, Y_j^{k,N}\right) = \omega_{nN}\psi\left(q_k\right)^{nN} c_{\frac{nN}{2}},$$

and

$$\tau^{0}\left(\left(Y_{j}^{k,N}\right)^{n}\right)=\begin{cases}\psi(q_k)^{nN}\left(\sum\limits_{\pi\in NC(\Omega_n)}\left(\prod\limits_{B\in\theta}c_{\frac{N|B|}{2}}\right)\right) & \textit{if } N \textit{ is even}\\ \\ \omega_n\psi(q_k)^{nN}\left(\sum\limits_{\theta\in NC_e(\Omega_n)}\left(\prod\limits_{V\in\theta}c_{\frac{N|V|}{2}}\right)\right) & \textit{if } N \textit{ is odd,}\end{cases} \tag{7.3.4}$$

for all $n \in \mathbb{N}$.

Proof Consider that

$$k_n\left(Y_j^{k,N},\ldots,Y_j^{k,N}\right)=k_n\left(\psi(q_k)^N W_j^{k,N},\ldots,\psi(q_k)^N W_j^{k,N}\right)$$

by (7.3.3) and (7.2.7)

$$=\psi(q_k)^{nN}k_n\left(W_j^{k,N},\ldots,W_j^{k,N}\right)$$

by the bimodule-map property of free cumulants (e.g., [17, 22, 23])

$$=\psi(q_k)^{nN}\tau\left(U_k^{nN}\right)=\psi(q_k)^{nN}\left(\omega_{nN}c_{\frac{nN}{2}}\right)$$
$$=\omega_{nN}\psi(q_j)^{nN}c_{\frac{nN}{2}},$$

for all $n \in \mathbb{N}$.

Similarly, for any $n \in \mathbb{N}$, one obtains that

$$\tau\left(\left(Y_j^{k,N}\right)^n\right)=\tau\left(\psi(q_k)^{nN}\left(W_j^{k,N}\right)^n\right)$$

by (7.3.3)

$$=\psi(q_k)^{nN}\tau\left((W_j^{k,N})^n\right)$$

$$=\begin{cases}\psi(q_k)^{nN}\left(\sum\limits_{\pi\in NC(\Omega_n)}\left(\prod\limits_{B\in\theta}c_{\frac{N|B|}{2}}\right)\right) & \text{if } N \text{ is even}\\ \\ \omega_n\psi(q_k)^{nN}\left(\sum\limits_{\theta\in NC_e(\Omega_n)}\left(\prod\limits_{V\in\theta}c_{\frac{N|V|}{2}}\right)\right) & \text{if } N \text{ is odd,}\end{cases}$$

by (7.2.7). Therefore, the free-distributional data (7.3.4) hold. ∎

The above theorem shows that the free Poisson elements $Y_j^{k,N}$ of (7.3.1) induced by the free weighted-semicircular family $\mathcal{X}$ of (6.3) have their free Poisson distributions affected by the weights of fixed weighted-semicircular elements of $\mathcal{X}$ in $\mathbb{L}_Q$.

Corollary 24 *Let $Y_j^{k,N}$ and $W_j^{k,N}$ be the free Poisson elements (7.3.1), respectively, (7.2.6) in $\mathbb{L}_Q$. Then*

$$k_n\left(Y_j^{k,N}, \ldots, Y_j^{k,N}\right) = \psi(q_k)^{nN} k_n\left(W_j^{k,N}, \ldots, W_j^{k,N}\right),$$

and

$$\tau\left(\left(Y_j^{k,N}\right)^n\right) = \psi(q_k)^{nN} \tau\left(\left(W_j^{k,N}\right)^n\right), \tag{7.3.5}$$

for all $n \in \mathbb{N}$.

Proof The free-distributional data (7.3.5) is shown by (7.2.7) and (7.3.4). ∎

By (7.3.5), one obtains the following corollary, too.

Corollary 25 *Let $Y_j^{k,1}$ be in the sense of (7.3.1) (where $N = 1$). Then*

$$k_n\left(Y_j^{k,1}, \ldots, Y_j^{k,1}\right) = \psi(q_k)^n \left(\omega_n c_{\frac{n}{2}}\right),$$

and

$$\tau\left(\left(Y_j^{k,1}\right)^n\right) = \omega_n \psi(q_k)^n \left(\sum_{\theta \in NC_e(\Omega_n)} \left(\prod_{V \in \theta} c_{\frac{|V|}{2}}\right)\right) \tag{7.3.6}$$

for all $n \in \mathbb{N}$.

Proof The free-distributional data (7.3.6) holds by (7.3.5). ∎

8 Free Weighted-Poisson Elements of $\mathbb{L}_Q$

In this section, we consider free weighted-Poisson elements in $\mathbb{L}_Q$, different from Sect. 7. We are interested in free Poisson-like elements contained in our semicircular filterization $\mathbb{L}_Q$.

8.1 Free Weighted-Poisson Elements

Let (B, φ_B) be an arbitrary topological $*$-probability space, and let $x \in (B, \varphi_B)$ be a t_0-semicircular element for some $t_0 \in \mathbb{C}^{\times}$, satisfying

$$\varphi_B(x^n) = \omega_n t_0^{\frac{n}{2}} c_{\frac{n}{2}},$$

and

$$k_n^B(x, \ldots, x) = \begin{cases} t_0 & \text{if } n = 2 \\ 0 & \text{otherwise,} \end{cases} \tag{8.1.1}$$

for all $n \in \mathbb{N}$, where $k_n^B(\ldots)$ is the free cumulant on B in terms of φ_B.

Definition 26 Let $x \in (B, \varphi_B)$ be a t_0-semicircular element (8.1.1). Suppose $a \in (B, \varphi_B)$ is a self-adjoint free random variable, and assume that x and a are free in (B, φ_B). A free random variable

$$T_x^a = xax \in (B, \varphi_B) \tag{8.1.2}$$

is called a free weighted-Poisson element with its weight t_0 (in short, a free t_0-Poisson element) of (B, φ_B).

Let's consider free-distributional data of a free t_0-Poisson element T_x^a of (8.1.2).

Theorem 27 *Let $T_x^a = xax$ be a free t_0-Poisson element (8.1.2), where x is a fixed t_0-semicircular element (8.1.1) in (B, φ_B), and a has its free distribution $(\varphi_B(a^n))_{n=1}^{\infty}$. Then*

$$k_n^B\left(\underbrace{T_x^a, T_x^a, \ldots, T_x^a}_{n\text{-times}}\right) = t_0^n\left(\varphi_B(a^n)\right), \tag{8.1.3}$$

for all $n \in \mathbb{N}$.

Proof Observe that

$$k_n^B\left(T_x^a, \ldots, T_x^a\right) = k_n^B\left(xax, \; xax, \ldots, xax\right)$$
$$= \sum_{\theta \in NC(\Omega_{3n}),\ \theta \vee \pi_0 = 1_{3n}} k_\theta^B\left(xax, \ldots, xax\right),$$

where Ω_{3n} is in the sense of (7.1.3), and $k_\theta^B(\ldots)$ are the block-depending free cumulants of [17], and

$$\pi_0 = \{(1, 2, 3), (4, 5, 6), \ldots, (3n-2, 3n-1, 3n)\},$$

and hence, it goes to

$$= \sum_{\theta \in NC(\Omega_{3n}^2),\ \theta_o \vee \theta \in NC(\Omega_{3n})} k_{\theta_o \vee \theta}^B\left(xax, \ldots, xax\right)$$

by (7.1.6), where θ_o is in the sense of (7.1.4)

$$= \sum_{\theta \in NC(\Omega_{3n}^2)} k_{\theta_o}^B \left(\underbrace{x, \ x, \ x, \ldots, x}_{2n\text{-times}} \right) k_\theta^B \left(\underbrace{a, \ a, \ldots, a}_{n\text{-times}} \right)$$

$$= \sum_{\theta \in NC(\Omega_{3n}^2)} \left(\left(k_2^B(x, \ x) \right)^n \right) k_\theta^B \ (a, \ a, \ldots, a)$$

$$= t_0^n \left(\sum_{\theta \in NC(\Omega_{3n}^2)} k_\theta^B \ (a, \ a, \ldots, a) \right)$$

by (7.1.6), because x is a t_0-semicircular element satisfying (8.1.1)

$$= t_0^n \left(\sum_{\pi \in NC(\Omega_n)} k_\pi^B \ (a, \ldots, a) \right)$$

since the sub-lattice $NC\left(\Omega_{3n}^2\right)$ of $NC\left(\Omega_{3n}\right)$ is equivalent to the lattice $NC(\Omega_n)$

$$= t_0^n \left(\varphi_B(a^n) \right),$$

by the Möbius inversion, for all $n \in \mathbb{N}$. That is,

$$k_n^B \left(\underbrace{T_x^a, \ T_x^a, \ldots, T_x^a}_{n\text{-times}} \right) = t_0^n \left(\varphi_B(a^n) \right),$$

for all $n \in \mathbb{N}$. Therefore, the free-distributional data (8.1.3) holds. ■

The above theorem illustrates that, different from the free-Poisson case of Sect. 7, the free distributions of our free weighted-Poisson elements T_x^a of (8.1.2) are depending not only on the free distributions $(\varphi_B(a^n))_{n=1}^\infty$ of fixed self-adjoint free random variables a, but also on the weights of fixed weighted-semicircular elements x in (B, φ_B), by (8.1.3).

By the Möbius inversion, we obtain the following equivalent result of (8.1.3).

Theorem 28 *Let T_x^a be a free t_0-Poisson element (8.1.2) in (B, φ_B). Then*

$$\varphi_B\left((T_x^a)^n\right) = t_0^n \left(\sum_{\theta \in NC(\Omega_n)} \left(\prod_{V \in \theta} \varphi_B\left(a^{|V|}\right) \right) \right), \qquad (8.1.4)$$

for all $n \in \mathbb{N}$.

Proof Let $T_x^a \in (B, \varphi_B)$ be a free t_0-Poisson element (8.1.2). Then

$$\varphi_B\left((T_x^a)^n\right) = \sum_{\pi \in NC(\Omega_n)} k_\pi^B\left(T_x^a, \ldots, T_x^a\right)$$

$$= \sum_{\pi \in NC(\Omega_n)} \left(\prod_{V \in \pi} k_{|V|}^B\left(T_x^a, \ldots, T_x^a\right)\right)$$

by the Möbius inversion

$$= \sum_{\pi \in NC(\Omega_n)} \left(\prod_{V \in \pi} \left(t_0^{|V|} \varphi_B\left(a^{|V|}\right)\right)\right)$$

by (8.1.3)

$$= \sum_{\pi \in NC(\Omega_n)} \left(\prod_{V \in \pi} t_0^{|V|}\right)\left(\prod_{V \in \pi} \varphi_B\left(a^{|V|}\right)\right)$$

$$= \sum_{\pi \in NC(\Omega_n)} \left(t_0^{\sum_{V \in \pi} |V|}\right)\left(\prod_{V \in \pi} \varphi_B\left(a^{|V|}\right)\right)$$

$$= \sum_{\pi \in NC(\Omega_n)} \left(t_0^n\right)\left(\prod_{V \in \pi} \varphi_B\left(a^{|V|}\right)\right)$$

since

$$\sum_{V \in \pi} |V| = |\Omega_n| = n, \ \forall \pi \in NC\left(\Omega_n\right),$$

and hence, it goes to

$$= \left(t_0^n\right)\left(\sum_{\pi \in NC(\Omega_n)} \left(\prod_{V \in \pi} \varphi_B\left(a^{|V|}\right)\right)\right),$$

for all $n \in \mathbb{N}$. So, the free-momental data (8.1.4) is obtained. ■

The above free-distributional data (8.1.3) and (8.1.4) provide the equivalent free-distributional information of our free weighted-Poisson elements, which imply the free-distributional data (7.1.5) and (7.1.7) as special cases up to weights.

Theorem 29 *Let T_x^a be a free t_0-Poisson element (8.1.2) of (B, φ_B), and let W_s^a be a free Poisson element (7.1.1) of (B, φ_B), where s is an arbitrary semicircular element of (B, φ_B), which is free from the fixed self-adjoint free random variable $a \in (B, \varphi_B)$. Then*

$$k_n^B\left(T_x^a, \ldots, T_x^a\right) = t_0^n k_n^B\left(W_s^a, \ldots, W_s^a\right),$$

and

$$\varphi_B\left((T_x^a)^n\right) = t_0^n \varphi_B\left((W_s^a)^n\right), \tag{8.1.5}$$

for all $n \in \mathbb{N}$.

Proof Under hypothesis, one has that

$$k_n^B\left(T_x^a, \ldots, T_x^a\right) = t_0^n \varphi_B(a^n) = t_0^n k_n^B\left(W_s^a, \ldots, W_s^a\right),$$

for all $n \in \mathbb{N}$, by (7.1.5) and (8.1.3).

Similarly, we have

$$\varphi_B\left((T_x^a)^n\right) = t_0^n \left(\sum_{\pi \in NC(n)} \left(\prod_{V \in \pi} \varphi_B\left(a^{|V|}\right) \right) \right) = t_0^n \varphi_B\left((W_s^a)^n\right),$$

for all $n \in \mathbb{N}$, by (7.1.7) and (8.1.4).

Therefore, the relation (8.1.5) holds in (B, φ_B). ■

The above theorem not only characterizes the difference between the free Poisson distributions and the free weighted-Poisson distributions, but also shows how weights act free-probabilistically, by (8.1.5).

Corollary 30 *Let* $T = uau$ *be the free* t_0*-Poisson element of a topological* $*$*-probability space* (B_1, φ_{B_1})*, for* $t_0 \in \mathbb{C}^\times$*, where* u *is a* t_0*-semicircular element, free from a self-adjoint free random variable* a *in* (B_1, φ_{B_1})*. Suppose* $W = UxU$ *be a free Poisson element of a topological* $*$*-probability space* (B_2, φ_{B_2})*, where* U *is a semicircular element, free from a self-adjoint free random variable* x *in* (B_2, φ_{B_2})*. Assume that the free distribution of* a*, and that of* x *are identical in the sense that:*

$$\left(\varphi_{B_1}(a^n)\right)_{n=1}^{\infty} = \left(\varphi_{B_2}(x^n)\right)_{n=1}^{\infty}. \tag{8.1.6}$$

Then

$$k_n^{B_1}\left(\underbrace{T, T, \ldots, T}_{n\text{-times}}\right) = t_0^n k_n^{B_2}\left(\underbrace{W, W, \ldots, W}_{n\text{-times}}\right),$$

and

$$\varphi_{B_1}\left(T^n\right) = t_0^n \varphi_{B_2}(W^n), \tag{8.1.7}$$

for all $n \in \mathbb{N}$.

Proof By the assumption (8.1.6), the free-distributional data (8.1.7) holds by (8.1.5). ■

8.2 Free Weighted-Poisson Elements of $\mathbb{L}_Q$ Induced by $\mathcal{S}$

Let $\mathbb{L}_Q$ be the semicircular filterization, and let $\mathcal{S}$ be the free semicircular family (6.4) in $\mathbb{L}_Q$. Let's fix a $\psi(q_j)^2$-semicircular element $u_j \in \mathcal{X}$ in $\mathbb{L}_Q$, where $\mathcal{X}$ is the free weighted-semicircular family (6.3) of $\mathbb{L}_Q$. Then, as in (8.1.2), one can define the free $\psi(q_j)^2$-Poisson elements,

$$T_j^{k,N} = u_j U_k^N u_j \in \mathbb{L}_Q, \tag{8.2.1}$$

for $U_k \in \mathcal{S}$, and $N \in \mathbb{N}$, whenever $k \neq j$ in $\mathbb{Z}$.

By (8.1.3), (8.1.4) and (8.1.5), we obtain the following free-distributional data of free $\psi(q_j)^2$-Poisson elements (8.2.1) in $\mathbb{L}_Q$.

Theorem 31 *Let $T_j^{k,N} = u_j U_k^N u_j$ be a free $\psi(q_j)^2$-Poisson element (8.2.1) of $\mathbb{L}_Q$, for $N \in \mathbb{N}$. Then*

$$k_n\left(T_j^{k,N}, \ldots, T_j^{k,N}\right) = \psi(q_j)^{2n}\left(\omega_{nN} c_{\frac{nN}{2}}\right),$$

and

$$\tau\left(\left(T_j^{k,N}\right)^n\right) = \begin{cases} \psi(q_j)^{2n}\left(\sum\limits_{\pi \in NC(\Omega_n)}\left(\prod\limits_{V \in \pi} c_{\frac{N|V|}{2}}\right)\right) & \text{if } N \text{ is even} \\ \omega_n \psi(q_j)^{2n}\left(\sum\limits_{\theta \in NC_e(\Omega_n)}\left(\prod\limits_{V \in \theta} c_{\frac{N|V|}{2}}\right)\right) & \text{if } N \text{ is odd,} \end{cases} \tag{8.2.2}$$

for all $n \in \mathbb{N}$, where $NC_e(\Omega_n)$ are the subsets (7.2.4) of the lattices $NC(\Omega_n)$.

Proof Observe that, for any $n \in \mathbb{N}$,

$$k_n\left(T_j^{k,N}, \ldots, T_j^{k,N}\right) = \psi(q_j)^{2n}\tau\left(U_k^{nN}\right)$$

by (8.1.3)

$$= \psi(q_j)^{2n}\left(\omega_{nN} c_{\frac{nN}{2}}\right),$$

by the semicircularity of $U_k \in \mathcal{S}$ in $\mathbb{L}_Q$.

Consider now that

$$\tau\left(\left(T_j^{k,N}\right)^n\right) = \psi(q_j)^{2n}\left(\sum_{\pi \in NC(\Omega_n)}\left(\prod_{V \in \pi} \tau\left(U_k^{N|V|}\right)\right)\right)$$

by (8.1.4)

$$= \begin{cases} \psi(q_j)^{2n} \left(\sum\limits_{\pi \in NC(\Omega_n)} \left(\prod\limits_{V \in \pi} c_{\frac{N|V|}{2}} \right) \right) & \text{if } N \text{ is even} \\ \\ \omega_n \psi(q_j)^{2n} \left(\sum\limits_{\theta \in NC_e(\Omega_n)} \left(\prod\limits_{V \in \theta} c_{\frac{N|V|}{2}} \right) \right) & \text{if } N \text{ is odd,} \end{cases}$$

for all $n \in \mathbb{N}$.

Therefore, the free-distributional data (8.2.2) holds. ■

By the free-distributional data (7.2.7) and (8.2.2), we have the following special case of (8.1.5) on the semicircular filterization $\mathbb{L}_Q$, as in (8.1.7).

Corollary 32 *Let $T_j^{k,N} = u_j U_k^N u_j$ be a free $\psi(q_j)^2$-Poisson element (8.2.1), and let $W_i^{l,N} = U_i U_l^N U_i$ be a free Poisson element (7.2.6) in $\mathbb{L}_Q$, for $N \in \mathbb{N}$, where*

$$k \neq j \text{ and } l \neq i \text{ in } \mathbb{Z}.$$

Then

$$k_n\left(T_j^{k,N}, \ldots, T_j^{k,N}\right) = \psi(q_j)^{2n} k_n\left(W_i^{l,N}, \ldots, W_i^{l,N}\right),$$

and

$$\tau\left(\left(T_j^{k,N}\right)^n\right) = \psi(q_j)^{2n} \tau\left(\left(W_i^{l,N}\right)^n\right), \tag{8.2.3}$$

for all $n \in \mathbb{N}$.

Proof The proof of (8.2.3) is done by (7.2.7), (7.2.2), (8.1.5) and (8.2.2). ■

By (8.2.3), we obtain the following corollary.

Corollary 33 *Let $T_j^{k,1} = u_j U_k u_j$ be a free $\psi(q_j)^2$-Poisson element (8.2.1), where $N = 1$. Then*

$$k_n\left(T_j^{k,N}, \ldots, T_j^{k,N}\right) = \omega_n \psi(q_j)^{2n} c_{\frac{n}{2}},$$

and

$$\tau\left(\left(T_j^{k,N}\right)^n\right) = \omega_n \psi(q_j)^{2n} \left(\sum_{\theta \in NC_e(\Omega_n)} \left(\prod_{V \in \theta} c_{\frac{|V|}{2}} \right) \right), \tag{8.2.4}$$

for all $n \in \mathbb{N}$.

Proof The proof of (8.2.4) is done by (8.2.3). ■

8.3 Free Weighted-Poisson Elements of $\mathbb{L}_Q$ Induced by $\mathcal{X}$

Let $\mathbb{L}_Q$ be the semicircular filterization, and $\mathcal{X}$, the free weighted-semicircular family (6.3) of $\mathbb{L}_Q$, and let's fix $u_j \in \mathcal{X}$ in $\mathbb{L}_Q$. For any $k \neq j$ in $\mathbb{Z}$, define the corresponding free $\psi(q_j)^2$-Poisson elements,

$$X_j^{k,N} = u_j u_k^N u_j \in \mathbb{L}_Q, \tag{8.3.1}$$

for $N \in \mathbb{N}$.

Since

$$U_j = \frac{1}{\psi(q_j)} u_j \text{ in } \mathcal{S} \Longleftrightarrow u_j = \psi(q_j) U_j \text{ in } \mathcal{X},$$

in $\mathbb{L}_Q$, for all $j \in \mathbb{Z}$, our free $\psi(q_j)^2$-Poisson elements $X_j^{k,N}$ of (8.3.1) are also understood as

$$\begin{aligned} X_j^{k,N} &= \psi(q_j)^2 \psi(q_k)^N U_j U_k^N U_j \\ &= \left(\psi(q_j)^2 \psi(q_k)^N\right) W_j^{k,N}, \end{aligned} \tag{8.3.2}$$

in $\mathbb{L}_Q$, where $W_j^{k,N}$ are the free Poisson element (7.2.6) of $\mathbb{L}_Q$, for all $N \in \mathbb{N}$.

Theorem 34 *Let $X_j^{k,N} = u_j u_k^N u_j$ be a free $\psi(q_j)^2$-Poisson element (8.3.1) of $\mathbb{L}_Q$, for $N \in \mathbb{N}$. Then*

$$k_n\left(X_j^{k,N}, \ldots, X_j^{k,N}\right) = \beta_n\left(\omega_{nN} c_{\frac{nN}{2}}\right)$$

and

$$\tau\left(\left(X_j^{k,N}\right)^n\right) = \begin{cases} \beta_n\left(\sum\limits_{\pi \in NC(\Omega_n)} \left(\prod\limits_{V \in \pi} c_{\frac{N|V|}{2}}\right)\right) & \text{if } N \text{ is even} \\ \\ \omega_n \beta_n\left(\sum\limits_{\theta \in NC_e(\Omega_n)} \left(\prod\limits_{V \in \theta} c_{\frac{N|V|}{2}}\right)\right) & \text{if } N \text{ is odd,} \end{cases} \tag{8.3.3}$$

with

$$\beta_n = \left(\psi(q_j)^2 \psi(q_k)^N\right)^n = \psi(q_j)^{2n} \psi(q_k)^{nN},$$

in $\mathbb{R}^{\times}$, for all $n \in \mathbb{N}$.

Proof Consider that

$$\begin{aligned} & k_n\left(X_j^{k,N}, \ldots, X_j^{k,N}\right) \\ &\quad = \left(\psi(q_j)^{2n} \psi(q_k)^{nN}\right) k_n\left(W_j^{k,N}, \ldots, W_j^{k,N}\right) \end{aligned}$$

by (8.3.2)

$$= \left(\psi(q_j)^{2n}\psi(q_k)^{nN}\right)\left(\omega_{nN}c_{\frac{nN}{2}}\right),$$

for all $n \in \mathbb{N}$, by (7.2.7), where $W_j^{k,N}$ are in the sense of (7.2.6).

Observe now that

$$\tau\left(\left(X_j^{k,N}\right)^n\right) = \left(\psi(q_j)^{2n}\psi(q_k)^{nN}\right)\tau\left(\left(W_j^{k,N}\right)^n\right)$$

$$= \begin{cases} \left(\psi(q_j)^{2n}\psi(q_k)^{nN}\right)\left(\sum\limits_{\pi\in NC(\Omega_n)}\left(\prod\limits_{V\in\pi} c_{\frac{N|V|}{2}}\right)\right) & \text{if } N \text{ is even} \\ \omega_n\left(\psi(q_j)^{2n}\psi(q_k)^{nN}\right)\left(\sum\limits_{\theta\in NC_e(\Omega_n)}\left(\prod\limits_{V\in\theta} c_{\frac{N|V|}{2}}\right)\right) & \text{if } N \text{ is odd,} \end{cases}$$

by (7.2.7) and (8.3.2). Therefore, the free-distributional data (8.3.3) are obtained. ■

The above theorem characterizes free weighted-Poisson distributions on $\mathbb{L}_Q$ in terms of free Poisson distributions on $\mathbb{L}_Q$ up to weight.

Corollary 35 *Let $X_j^{k,N} = u_j u_k^N u_j$ be a free $\psi(q_j)^2$-Poisson element (8.3.1), and let $W_i^{l,N} = U_i U_l^N U_i$ be a free Poisson element (7.2.6) in $\mathbb{L}_Q$, for $N \in \mathbb{N}$, where $k \neq j$, and $l \neq i$ in $\mathbb{Z}$. Then*

$$k_n\left(X_j^{k,N},\ldots,X_j^{k,N}\right) = \beta_n k_n\left(W_j^{l,N},\ldots,W_j^{l,N}\right),$$

and

$$\tau\left(\left(T_j^{k,N}\right)^n\right) = \beta_n\tau\left(\left(W_j^{l,N}\right)^n\right), \tag{8.3.4}$$

where

$$\beta_n = \left(\psi(q_j)^2\psi(q_k)^N\right)^n \in \mathbb{R}^{\times},$$

for all $n \in \mathbb{N}$.

Proof The proof of (8.3.4) is done by (7.2.7) and (8.3.3). Also, see (8.1.5) and (8.1.7). ■

By (8.3.4), one obtains the following special case.

Corollary 36 *Let $X_j^{k,1} = u_j u_k u_j$ be a free $\psi(q_j)^2$-Poisson element (8.3.1), where $N = 1$. Then*

$$k_n\left(X_j^{k,N},\ldots,X_j^{k,N}\right) = \omega_n\beta_n^1 c_{\frac{n}{2}},$$

and

$$\tau\left(\left(T_j^{k,N}\right)^n\right) = \omega_n \beta_n^1 \left(\left(\sum_{\theta \in NC_e(\Omega_n)} \left(\prod_{V \in \theta} c_{\frac{|V|}{2}}\right)\right)\right), \tag{8.3.5}$$

where

$$\beta_n^1 = \left(\psi(q_j)^2 \psi(q_k)\right)^n \in \mathbb{R}^\times$$

for all $n \in \mathbb{N}$.

Proof The proof of (8.3.5) is done by (8.3.3) and (8.3.4). ∎

9 More About Free Poisson Elements of $\mathbb{L}_Q$

Throughout this section, we use same notations, concepts, and assumptions with those of Sects. 7 and 8. As we have seen, free weighted-Poisson distributions of the free random variables $X_j^{k,N} = u_j u_k^N u_j$ of (8.3.1), for $u_j, u_k \in \mathcal{X}$, are characterized by certain scalar multiples of free moments, or those of free cumulants, of the free Poisson distributions of the operators $W_j^{k,N} = U_j U_k^N U_j$ of (7.2.6), for $U_j, U_k \in \mathcal{S}$, in the semicircular filterization $\mathbb{L}_Q = (\mathbb{L}_Q, \tau)$, for all $N \in \mathbb{N}$. Thus, in this section, we concentrate on free Poisson elements of $\mathbb{L}_Q$ formed by $U_j T U_j$, where self-adjoint operators T are free from $U_j \in \mathcal{S}$ in $\mathbb{L}_Q$.

9.1 Free Poisson Elements Induced by Certain Free Sums in $\mathcal{S}$

In this section, fix $j \in \mathbb{Z}$, and the corresponding semicircular element $U_j \in \mathcal{S}$ in $\mathbb{L}_Q$.

Lemma 37 *Let $U_{k_l} \in \mathcal{S}$ be two semicircular elements of $\mathbb{L}_Q$, where k_1 and k_2 are distinct in $\mathbb{Z}$. Assume further that*

$$k_l \neq j \text{ in } \mathbb{Z}, \text{ for all } l = 1, 2.$$

Let U be the operator defined by

$$U = U_{k_1} + U_{k_2} \in \mathbb{L}_Q. \tag{9.1.1}$$

$$The\, operator\, W_j^U \overset{def}{=} U_j U U_j\, is\, a\, well-defined\, free\, Poisson\, element. \tag{9.1.2}$$

$$The\, operator\, U\, is\, 2-semicircular\, in\, \mathbb{L}_Q. \tag{9.1.3}$$

The free Poisson element W_j^U *of* (9.1.2) *satisfies that* (9.1.4)

$$k_n\left(W_j^U, \ldots, W_j^U\right) = \omega_n 2^{\frac{n}{2}} c_{\frac{n}{2}},$$

and

$$\tau\left(\left(W_j^U\right)^n\right) = 2\omega_n \left(\sum_{\theta \in NC_e(\Omega_n)} \left(\prod_{V \in \theta} c_{\frac{|V|}{2}}\right)\right),$$

for all $n \in \mathbb{N}$.

Proof Let $U \in \mathbb{L}_Q$ be in the sense of (9.1.1). Then this operator U is self-adjoint in $\mathbb{L}_Q$, by the self-adjointness of U_{k_1}, $U_{k_2} \in \mathcal{S}$. By the self-adjointness of U, the free Poisson element W_j^U is well defined in $\mathbb{L}_Q$, because U and U_j are free in $\mathbb{L}_Q$. Therefore, the statement (9.1.2) holds.

Now, let $U = U_{k_1} + U_{k_2}$ be in the sense of (9.1.1). Then

$$\begin{aligned} k_n(U, \ldots, U) &= k_n\left(U_{k_1} + U_{k_2}, \ldots, U_{k_1} + U_{k_2}\right) \\ &= k_n\left(U_{k_1}, \ldots, U_{k_2}\right) + k_n\left(U_{k_2}, \ldots, U_{k_2}\right) \end{aligned}$$

by the freeness of U_{k_1} and U_{k_2} in $\mathbb{L}_Q$ (i.e., all mixed free cumulants of U_{k_1} and U_{k_2} vanish. For example, see [17, 22, 23].)

$$= \begin{cases} 1 + 1 = 2 & \text{if } n = 2 \\ 0 + 0 = 0 & \text{otherwise,} \end{cases} \tag{9.1.5}$$

for all $n \in \mathbb{N}$, by the semicircularity of U_{k_1} and U_{k_2} in $\mathbb{L}_Q$.

By the Möbius inversion and (9.1.5), one also obtains that

$$\begin{aligned} \tau(U^n) &= \sum_{\pi \in NC(\Omega_n)} \left(\prod_{V \in \pi} k_{|V|}(U, \ldots, U)\right) \\ &= \omega_n \left(\sum_{\theta \in NC_2(\Omega_n)} \left(\prod_{V \in \theta} k_2(U, U)\right)\right) \end{aligned}$$

where

$$NC_2(\Omega_n) = \{\theta \in NC(\Omega_n) : V \in \theta \Longleftrightarrow |V| = 2\},$$

(Remark that $NC_2(\Omega_n)$ is empty whenever n is odd; so ω_n is applied.)

$$= \omega_n \left(\sum_{\theta \in NC_2(\Omega_n)} 2^{\frac{n}{2}} \right) = \omega_n 2^{\frac{n}{2}} \left| NC_2(\Omega_n) \right|$$

by (9.1.5), because every noncrossing partition θ of $NC_2(\Omega_n)$ contains $\left(\frac{n}{2}\right)$-many blocks, where $|NC_2(\Omega_n)|$ is the cardinality of the set $NC_2(\Omega_n)$

$$= \omega_n 2^{\frac{n}{2}} \left| NC\left(\Omega_{\frac{n}{2}}\right) \right| = \omega_n 2^{\frac{n}{2}} c_{\frac{n}{2}}, \tag{9.1.6}$$

since $NC_2(\Omega_n)$ is bijective (or equipotent) to $NC\left(\Omega_{\frac{n}{2}}\right)$, for all $n \in 2\mathbb{N}$, and since

$$|NC(\Omega_m)| = c_m, \ \forall m \in \mathbb{N}$$

(e.g., see [17, 22, 23]). Therefore, this self-adjoint free random variable U is 2-semicircular by (9.1.5) and (9.1.6). This proves the statement (9.1.3).

Now, let $W_j^U = U_j U U_j \in \mathbb{L}_Q$ be the free Poisson element (9.1.2) induced by U. Then

$$k_n\left(W_j^U, \ldots, W_j^U\right) = \tau\left(U^n\right) = \omega_n 2^{\frac{n}{2}} c_{\frac{n}{2}},$$

and

$$\tau\left(\left(W_j^U\right)^n\right) = 2\omega_n \left(\sum_{\theta \in NC_e(\Omega_n)} \left(\prod_{V \in \theta} c_{\frac{|V|}{2}} \right) \right), \tag{9.1.7}$$

by (7.1.5), (7.1.7) and (9.1.3), for all $n \in \mathbb{N}$. Therefore, the statement (9.1.4) is proven by (9.1.7). ■

By the above lemma, we obtain the following result.

Theorem 38 *Let k_l be mutually distinct in $\mathbb{Z}$, for $l = 1, \ldots, N$, for $N \in \mathbb{N} \setminus \{1\}$, and let $U_{k_l} \in \mathcal{S}$ be the corresponding mutually free semicircular elements in $\mathbb{L}_Q$. Define the operator T by*

$$T = \sum_{l=1}^{N} U_{k_l} \text{ in } \mathbb{L}_Q. \tag{9.1.8}$$

Suppose that $k_l \neq j$ in $\mathbb{Z}$, for all $l = 1, \ldots, N$. Then, the free Poisson element $W_j^T = U_j T U_j \in \mathbb{L}_Q$ satisfies that

$$k_n\left(W_j^T, \ldots, W_j^T\right) = \omega_n N^{\frac{n}{2}} c_{\frac{n}{2}},$$

and

$$\tau\left(\left(W_j^T\right)^n\right) = \omega_n N \left(\sum_{\theta \in NC_e(\Omega_n)} \left(\prod_{V \in \theta} c_{\frac{|V|}{2}} \right) \right), \tag{9.1.9}$$

for all $n \in \mathbb{N}$.

Proof Let T be in the sense of (9.1.8). Then, by the self-adjointness of mutually distinct semicircular elements $\{U_{k_l}\}_{l=1}^{N}$, this operator T is self-adjoint in $\mathbb{L}_Q$, and hence, the corresponding free Poisson element $W_j^T = U_j T U_j$ is well defined in $\mathbb{L}_Q$, since U_j and T are free in $\mathbb{L}_Q$. Consider that

$$k_n(T, \ldots, T) = k_n\left(\sum_{l=1}^{N} U_{k_l}, \ldots, \sum_{l=1}^{N} U_{k_l}\right)$$

$$= \sum_{l=1}^{N} k_n\left(U_{k_l}, \ldots, U_{k_l}\right)$$

by the bimodule-map property of $k_\bullet(\ldots)$, and by the freeness of $\{U_{k_l}\}_{l=1}^{N}$ (Note that all "mixed" free cumulants of U_{k_1}, ..., U_{k_N} vanish!)

$$= \begin{cases} N & \text{if } n = 2 \\ 0 & \text{otherwise,} \end{cases}$$

like in (9.1.5), for all $n \in \mathbb{N}$. Therefore, one obtains

$$\tau\left(T^n\right) = \omega_n N^{\frac{n}{2}} c_{\frac{n}{2}}, \text{ for all } n \in \mathbb{N},$$

by the Möbius inversion, as in (9.1.6). That is, this self-adjoint free random variable T is N-semicircular in $\mathbb{L}_Q$. Therefore, the free-distributional data (9.1.9) is obtained similar to the proof of (9.1.4). ■

The above theorem illustrates that the free Poisson elements,

$$U_j\left(\sum_{l=1}^{N} U_{k_l}\right)U_j = \sum_{l=1}^{N}\left(U_j U_{k_l} U_j\right),$$

satisfies the free distribution represented by (9.1.9) in $\mathbb{L}_Q$.

Let's compare the free Poisson distributions (7.2.5) and (9.1.9). The free Poisson distribution (7.2.5) says that the free Poisson element,

$$W_l = U_j U_{k_l} U_j,$$

has its free distribution,

$$\left(k_n(W_l, \ldots, W_l)\right)_{n=1}^{\infty} = \left(\omega_n c_{\frac{n}{2}}\right)_{n=1}^{\infty},$$

for all $l = 1$, ..., N, meanwhile, the free Poisson distribution (9.1.9) shows that the free Poisson element,

$$W = \sum_{l=1}^{N} W_l = U_j \left(\sum_{l=1}^{N} U_{k_l} \right) U_j,$$

has its free distribution

$$(k_n(W, \ldots, W))_{n=1}^{\infty} = \left(\omega_n N^{\frac{n}{2}} c_{\frac{n}{2}} \right)_{n=1}^{\infty}.$$

That is,

$$(k_n(W, \ldots, W))_{n=1}^{\infty} = \left(N^{\frac{n}{2}} k_n(W_l, \ldots, W_l) \right)_{n=1}^{\infty}, \tag{9.1.10}$$

for all $l = 1, \ldots, N$. Similarly, one has

$$\left(\tau(W^n) \right)_{n=1}^{\infty} = \left(N^{\frac{n}{2}} \tau(W_l^n) \right)_{n=1}^{\infty}, \tag{9.1.11}$$

for all $l = 1, \ldots, N$, in $\mathbb{L}_Q$.

Let (B_l, φ_l) be topological $*$-probability spaces, and let $a_l \in (B_l, \varphi_l)$ be self-adjoint free random variables for $l = 1, 2$. Recall that $a_1 \in (B_1, \varphi_1)$ is identically free-distributed to $a_2 \in (B_2, \varphi_2)$, if their free distributions are identically same in the sense that

$$\left(\varphi_1(a_1^n) \right)_{n=1}^{\infty} = \left(\varphi_2(a_2^n) \right)_{n=1}^{\infty},$$

or

$$\left(k_n^{B_1}(a_1, \ldots, a_1) \right)_{n=1}^{\infty} = \left(k_n^{B_2}(a_2, \ldots, a_2) \right)_{n=1}^{\infty}.$$

For example, all semicircular elements are identically free-distributed from each other.

By (9.1.10), (9.1.11) and the identically free-distributedness, we obtain the following generalized result.

Theorem 39 *Let $s_1, \ldots, s_N$ be mutually free, semicircular elements in a topological $*$-probability space (B, φ_B) for $N \in \mathbb{N}$, and let s_0 be a semicircular element of (B, φ_B). Assume further that s_l are free from s_0 in (B, φ_B), for all $l = 1, \ldots, N$. Then the free Poisson elements,*

$$x_l = s_0 s_l s_0 \in (B, \varphi_B)$$

have their free distributions,

$$\left(k_n^B(x_l, \ldots, x_l) \right)_{n=1}^{\infty} = \left(\omega_n c_{\frac{n}{2}} \right)_{n=1}^{\infty},$$

equivalently,

$$\left(\varphi_B(x_l^n)\right)_{n=1}^{\infty} = \left(\omega_n\left(\sum_{\theta \in NC_e(\Omega_n)}\left(\prod_{V \in \theta} c_{\frac{|V|}{2}}\right)\right)\right)_{n=1}^{\infty}, \tag{9.1.12}$$

for all $l = 1, \ldots, N$, *where* $k_n^B(\ldots)$ *is the free cumulant on* B *in terms of* φ_B. *Also, the free Poisson element,*

$$x_0 = s_0\left(\sum_{l=1}^{N} s_l\right)s_0 = \sum_{l=1}^{N} s_0 s_l s_0 \in (B, \varphi_B),$$

has its free distribution,

$$\left(k_n^B(x_0, \ldots, x_0)\right)_{n=1}^{\infty} = \left(N^{\frac{n}{2}} k_n^B(x_l, \ldots, x_l)\right)_{n=1}^{\infty},$$

equivalently,

$$\left(\varphi_B(x_0^n)\right)_{n=1}^{\infty} = \left(N^{\frac{n}{2}} \varphi_B(x_l^n)\right)_{n=1}^{\infty}, \tag{9.1.13}$$

for any $l = 1, \ldots, N$.

Proof The proofs of (9.1.12) and (9.1.13) are done by (7.2.5), (9.1.9), (9.1.10) and (9.1.11), with help of the identically free-distributedness (3.2) and (3.4) of semicircular elements. ∎

9.2 Free Poisson Elements of $\mathbb{L}_Q$ Induced by Free Poisson Elements

As in Sect. 9.1, fix $j \in \mathbb{Z}$, and the corresponding semicircular element $U_j \in \mathcal{S}$ in the semicircular filterization $\mathbb{L}_Q$. Also, let's fix the mutually distinct integers $k_1 \neq k_2 \in \mathbb{Z}$, and the corresponding semicircular elements $U_{k_1}, U_{k_2} \in \mathcal{S}$ in $\mathbb{L}_Q$. By the distinctness of k_1 and k_2, the semicircular elements U_{k_1} and U_{k_2} are free in $\mathbb{L}_Q$, and hence, the free Poisson element

$$W_{k_1}^{k_2, N} = U_{k_1} U_{k_2}^N U_{k_1} \tag{9.2.1}$$

is well defined in $\mathbb{L}_Q$, for any $N \in \mathbb{N}$, as in (7.2.6).

Recall that this free Poisson element $W_{k_1}^{k_2, N}$ of (9.2.1) satisfies the free Poisson distribution,

$$k_n\left(W_{k_1}^{k_2, N}, \ldots, W_{k_1}^{k_2, N}\right) = \omega_n N c_{\frac{nN}{2}},$$

and

$$\tau\left(\left(W_{k_1}^{k_2,N}\right)^n\right) = \begin{cases} \sum\limits_{\pi\in NC(\Omega_n)} \left(\prod\limits_{V\in\pi} c_{\frac{N|V|}{2}}\right) & \text{if } N \text{ is even} \\ \omega_n \left(\sum\limits_{\theta\in NC_e(\Omega_n)} \left(\prod\limits_{B\in\theta} c_{\frac{N|B|}{2}}\right)\right) & \text{if } N \text{ is odd,} \end{cases} \tag{9.2.2}$$

for all $n \in \mathbb{N}$, by (7.2.7).

Theorem 40 *Let $W_N \overset{denote}{=} W_{k_1}^{k_2,N}$ be a free Poisson element (9.2.1) in $\mathbb{L}_Q$. Assume that $k_l \neq j$ in $\mathbb{Z}$, for all $l = 1, 2$. Define a free random variable T_N by*

$$T_N = U_j W_N U_j \text{ in } \mathbb{L}_Q. \tag{9.2.3}$$

$$The\ operator\ T_N\ of\ (9.2.3)\ is\ a\ free\ Poisson\ element. \tag{9.2.4}$$

$$The\ operator\ T_N\ has\ its\ free\ Poisson\ distribution\ with \tag{9.2.5}$$

$$k_n\left(T_N, \ldots, T_N\right) = \begin{cases} \sum\limits_{\pi\in NC(\Omega_n)} \left(\prod\limits_{V\in\pi} c_{\frac{N|V|}{2}}\right) & \textit{if } N \textit{ is even} \\ \omega_n \left(\sum\limits_{\theta\in NC_e(\Omega_n)} \left(\prod\limits_{B\in\theta} c_{\frac{N|B|}{2}}\right)\right) & \textit{if } N \textit{ is odd,} \end{cases}$$

and

$$\tau\left(T_N^n\right) = \begin{cases} \sum\limits_{\pi\in NC(\Omega_n)} \left(\prod\limits_{V\in\pi} \left(\sum\limits_{\pi\in NC(\Omega_{|V|})} \left(\prod\limits_{V\in\pi} c_{\frac{N|V|}{2}}\right)\right)\right) & \textit{if } N \textit{ is even} \\ \omega_n \left(\sum\limits_{\pi\in NC_e(\Omega_n)} \left(\prod\limits_{V\in\pi} \left(\sum\limits_{\theta\in NC_e(\Omega_{|V|})} \left(\prod\limits_{B\in\theta} c_{\frac{N|B|}{2}}\right)\right)\right)\right) & \textit{if } N \textit{ is odd,} \end{cases}$$

for all $n \in \mathbb{N}$.

Proof Let T_N be a free random variable (9.2.3) of $\mathbb{L}_Q$. By assumption, the three semicircular elements U_{k_1}, U_{k_2}, and U_j are mutually free from each other in the semicircular filterization $\mathbb{L}_Q$. Thus, the subsets $\{U_{k_1}, U_{k_2}\}$ and $\{U_j\}$ are free in $\mathbb{L}_Q$, too. It shows that the free Poisson element W_N of (9.2.1) and the semicircular element U_j are free in $\mathbb{L}_Q$. So, by the self-adjointness of W_N, the operator T_N of (9.2.3) is a well-defined free Poisson element generated by U_j and W_N. It shows the statement (9.2.4) holds.

By (9.2.4), one can verify that the operator T_N satisfies

$$k_n\left(T_N, \ldots, T_N\right) = \tau\left(W_N^n\right),$$

and

$$\tau\left(T_N^n\right) = \sum_{\pi \in NC(\Omega_n)} \left(\prod_{V \in \pi} \tau\left(W_N^{|V|}\right)\right), \tag{9.2.6}$$

by (7.1.5), respectively, by (7.1.7).

Thus, one has

$$k_n\left(T_N, \ldots, T_N\right) = \begin{cases} \sum_{\pi \in NC(\Omega_n)} \left(\prod_{V \in \pi} c_{\frac{N|V|}{2}}\right) & \text{if } N \text{ is even} \\ \omega_n \left(\sum_{\theta \in NC_e(\Omega_n)} \left(\prod_{B \in \theta} c_{\frac{N|B|}{2}}\right)\right) & \text{if } N \text{ is odd,} \end{cases}$$

and

$$\tau\left(T_N^n\right) = \begin{cases} \sum_{\pi \in NC(\Omega_n)} \left(\prod_{V \in \pi} \left(\sum_{\pi \in NC(\Omega_{|V|})} \left(\prod_{V \in \pi} c_{\frac{N|V|}{2}}\right)\right)\right) & \text{if } N \text{ is even} \\ \sum_{\pi \in NC(\Omega_n)} \left(\prod_{V \in \pi} \omega_{|V|} \left(\sum_{\theta \in NC_e(\Omega_{|V|})} \left(\prod_{B \in \theta} c_{\frac{N|B|}{2}}\right)\right)\right) & \text{if } N \text{ is odd,} \end{cases} \tag{9.2.7}$$

for all $n \in \mathbb{N}$, by (9.2.2) and (9.2.6).

Therefore, the free-distributional data in (9.2.5) hold true, by (9.2.7). ■

By (9.2.4) and (9.2.5), we obtain the following corollary.

Corollary 41 *Let $W_1 = W_{k_1}^{k_2,1}$ be a free Poisson element (9.2.1) of $\mathbb{L}_Q$, and let T_1 be the free Poisson element (9.2.3) induced by W_1 in $\mathbb{L}_Q$. Then*

$$k_n\left(T_1, \ldots, T_1\right) = \omega_n \left(\sum_{\theta \in NC_e(\Omega_n)} \left(\prod_{V \in \theta} k_{|V|}\left(W_1, \ldots, W_1\right)\right)\right),$$

and

$$\tau\left(T_1^n\right) = \omega_n \left(\sum_{\pi \in NC_e(\Omega_n)} \left(\prod_{V \in \pi} \tau\left(W_1^{|V|}\right)\right)\right), \tag{9.2.8}$$

for all $n \in \mathbb{N}$.

Proof Let $W_1 = U_{k_1} U_{k_2} U_{k_1} \in \mathbb{L}_Q$ be a free Poisson element (9.2.1). Then, the free Poisson element $T_1 = U_j W_1 U_j$ is well defined in $\mathbb{L}_Q$ by (9.2.4); and the operator T_1 satisfies that

$$k_n(T_1, \ldots, T_1) = \omega_n \left(\sum_{\theta \in NC_e(\Omega_n)} \left(\prod_{V \in \theta} c_{\frac{|V|}{2}} \right) \right)$$

$$= \omega_n \left(\sum_{\theta \in NC_e(\Omega_n)} \left(\prod_{V \in \theta} k_{|V|}(W_1, \ldots, W_1) \right) \right),$$

and

$$\tau\left(T_1^n\right) = \omega_n \left(\sum_{\pi \in NC_e(\Omega_n)} \left(\prod_{V \in \pi} \left(\sum_{\theta \in NC_e(\Omega_{|V|})} \left(\prod_{B \in \theta} c_{\frac{N|B|}{2}} \right) \right) \right) \right)$$
$$= \omega_n \left(\sum_{\pi \in NC_e(\Omega_n)} \left(\prod_{V \in \pi} \tau\left(W_1^{|V|}\right) \right) \right), \tag{9.2.9}$$

for all $n \in \mathbb{N}$, by (9.2.2) and (9.2.5). Therefore, the free Poisson distribution (9.2.8) is proven by (9.2.9). ■

The above corollary shows the relation between two free Poisson distributions induced by the free Poisson elements,

$$U_{k_1} U_{k_2} U_{k_1}, \text{ and } U_j \left(U_{k_1} U_{k_2} U_{k_1} \right) U_j,$$

in the semicircular filterization $\mathbb{L}_Q$.

More general to (9.2.8), we obtain the following recurrence relation.

Theorem 42 *Let $\mathcal{S}$ be the free semicircular family (6.4) in the semicircular filterization $\mathbb{L}_Q$. Let i_k be mutually distinct in $\mathbb{Z}$, and let $U_{(k)} = U_{i_k} \in \mathcal{S}$ be the corresponding semicircular elements of $\mathbb{L}_Q$, for all $k \in \mathbb{N}_0$. Define the free Poisson elements $\{W_{(l)}\}_{l=1}^{\infty}$ of $\mathbb{L}_Q$ by*

$$W_{(1)} = U_{(0)} U_{(1)} U_{(0)}, \; W_{(2)} = U_{(2)} W_{(1)} U_{(2)}, \; \ldots,$$

and

$$W_{(l+1)} = U_{(l+1)} W_{(l)} U_{(l+1)}, \text{ for all } l \in \mathbb{N}. \tag{9.2.10}$$

$$The\, operators\; W_{(l)}\; of\; (9.2.10)\; are\; free\; Poisson\; elements. \tag{9.2.11}$$

$$The\; free\; Poisson\; distributions\; of\; W_{(l+1)}\; satisfy\; that \tag{9.2.12}$$

$$k_n\left(W_{(l+1)}, \ldots, W_{(l+1)}\right) = \omega_n \left(\sum_{\pi \in NC_e(\Omega_n)} \left(\prod_{V \in \pi} k_{|V|}(W_{(l)}, \ldots, W_{(l)}) \right) \right),$$

and

$$\tau\left(W_{(l+1)}^n\right) = \omega_n \left(\sum_{\theta \in NC_e(\Omega_n)} \left(\prod_{V \in \theta} \tau\left(W_{(l)}^{|V|}\right) \right) \right),$$

with

$$k_n\left(W_{(1)}, \ldots, W_{(1)}\right) = \omega_n c_{\frac{n}{2}},$$

and

$$\tau\left(W_{(1)}^n\right) = \omega_n \left(\sum_{\pi \in NC_e(\Omega_n)} \left(\prod_{V \in \pi} c_{\frac{|V|}{2}} \right) \right),$$

for all $n, l \in \mathbb{N}$.

Proof By the definition (9.2.10) of the operators $\{W_{(l)}\}_{l=1}^{\infty}$, and by the freeness on the sub-family $\{U_{(m)}\}_{m=0}^{\infty}$ of $\mathcal{S}$ in $\mathbb{L}_Q$, all operators $W_{(l)}$ are well-defined free Poisson elements of $\mathbb{L}_Q$, for all $l \in \mathbb{N}$. Indeed, the given semicircular elements $U_{(l+1)}$ and the operators $W_{(l)}$ are free from each other in $\mathbb{L}_Q$, and hence, $W_{(l+1)} = U_{l+1} W_{(l)} U_{l+1}$ are free Poisson elements by the self-adjointness of $W_{(l)}$ in $\mathbb{L}_Q$, for all $l \in \mathbb{N}$. Therefore, the statement (9.2.11) holds.

The free Poisson distributions (9.2.12) of $W_{(l+1)}$ are obtained by induction on (9.2.8). ■

References

1. M. Ahsanullah, Some inferences on semicircular distribution. J. Stat. Theory Appl. **15**(3), 207–213 (2016)
2. H. Bercovici, D. Voiculescu, Superconvergence to the central limit and failure of the cramer theorem for free random variables. Probab. Theory Relat. Fields **103**(2), 215–222 (1995)
3. M. Bozejko, W. Ejsmont, T. Hasebe, Noncommutative probability of type *D*. Int. J. Math. **28**(2), 1750010, 30 (2017)
4. M. Bozheuiko, E.V. Litvinov, I.V. Rodionova, An extended anyon fock space and non-commutative Meixner-type orthogonal polynomials in the infinite-dimensional case. Uspekhi Math. Nauk. **70**(5), 75–120 (2015)
5. I. Cho, Semicircular families in free product banach $*$-algebras Induced by p-adic number fields over primes p. Complex Anal. Oper. Theory **11**(3), 507–565 (2017)
6. I. Cho, Acting semicircular elements induced by orthogonal projections on von Neumann algebras. Mathematics **5**, 74 (2017). https://doi.org/10.3390/math5040074
7. I. Cho, Semicircular-like and semicircular laws on banach $*$-probability spaces induced by dynamical systems of the finite adele ring. Adv. Oper. Theory (2018). Special Issue: Trends in Operators on Banach Spaces (dedicated to Prof. S. Banach). To Appear. https://doi.org/10.15352/aot.1802-1317
8. I. Cho, Semicircular-like laws and the semicircular law induced by orthogonal projections. Complex Anal. Oper. Theory (2018). To Appear
9. I. Cho, Free stochastic integrals for weighted-semicircular motion induced by orthogonal projections, in *Applied Mathematical Analysis: Theory, Methods, and Applications*. Monograph Series (Taylor & Fransis, 2019). To Appear
10. I. Cho, P.E.T. Jorgensen, Semicircular elements induced by p-adic number fields. Opuscula Math. **35**(5), 665–703 (2017)
11. I. Cho, P.E.T. Jorgensen, Banach $*$-algebras generated by semicircular elements induced by certain orthogonal projections. Opuscula Math. **38**(4), 501–535 (2018)

12. I. Cho, P.E.T. Jorgensen, Semicircular elements induced by projections on separable hilbert spaces, in *Operator Theory: Advances & Applications*. Monograph Series (Birkhauser, Basel, 2019). To Appear
13. A. Connes, *Noncommutative Geometry*. (Academic, San Diego, CA, 1994). ISBN: 0-12-185860-X
14. P. R. Halmos, *Hilbert Space Problem Books*. Graduate Texts in Mathematics, vol. 19 (Springer, 1982). ISBN: 978-0387906850
15. I. Kaygorodov, I. Shestakov, Free generic poisson fields and algebras. Commun. Alg. **46**(4) (2018). https://doi.org/10.1080/00927872.2017.1358269
16. L. Makar-Limanov, I. Shestakov, Polynomials and poisson dependence in free poisson algebras and free poisson fields. J. Alg. **349**(1), 372–379 (2012)
17. A. Nica, R. Speicher, Lectures on the Combinatorics of Free Probability, 1st edn. London Mathematical Society Lecture Note Series, vol. 335 (Cambridge University Press, 2006). ISBN-13:978-0521858526
18. I. Nourdin, G. Peccati, R. Speicher, Multi-dimensional semicircular limits on the free Wigner Chaos. Progr. Probab. **67**, 211–221 (2013)
19. V. Pata, The central limit theorem for free additive convolution. J. Funct. Anal. **140**(2), 359–380 (1996)
20. F. Radulescu, Random matrices, amalgamated free products and subfactors of the C^*-algebra of a free group of nonsingular index. Invent. Math. **115**, 347–389 (1994)
21. F. Radulescu, Free group factors and hecke operators, in *Proceedings of 24-th Conference in Operator Theory*, notes taken by N. Ozawa. Theta Advanced Series in Mathematics (Theta Foundation, 2014)
22. R. Speicher, Combinatorial theory of the free product with amalgamation and operator-valued free probability theory. Am. Math. Soc. Mem. **132**(627) (1998)
23. R. Speicher, A conceptual proof of a basic result in the combinatorial approach to freeness. Infin. Dimens. Anal. Quantum Probab. Relat. Top. **3**, 213–222 (2000)
24. R. Speicher, U. Haagerup, Brown's spectrial distribution measure for R-diagonal elements in finite Von Neumann algebras. J. Funct. Anal. **176**(2), 331–367 (2000)
25. R. Speicher, T. Kemp, Strong Haagerup inequalities for free R-diagonal elements. J. Funct. Anal. **251**(1), 141–173 (2007)
26. V.S. Vladimirov, p-adic quantum mechanics. Commun. Math. Phys. **123**(4), 659–676 (1989)
27. V.S. Vladimirov, I.V. Volovich, E.I. Zelenov, *p-Adic Analysis and Mathematical Physics*. Series on Soviet & East European Mathematics, vol. 1 (World Scientific, 1994). ISBN: 978-981-02-0880-6
28. D. Voiculescu, Free probability and the Von Neumann algebras of free groups. Rep. Math. Phys. **55**(1), 127–133 (2005)
29. D. Voiculescu, Aspects of free analysis. Jpn. J. Math. **3**(2), 163–183 (2008)
30. D. Voiculescu, K. Dykemma, A. Nica, *Free Random Variables*. C.R.M. Monograph Series, vol. 1 (American Mathematical Society, 1992). ISBN-13: 978–0821811405

On Approximation of Functions in the Generalized Zygmund Class Using $(E, r)(N, q_n)$ Mean Associated with Conjugate Fourier Series

B. P. Padhy, Susanta Kumar Paikray, Anwesha Mishra, and U. K. Misra

Abstract In the past few decades, a number of researchers have studied the error estimation of functions in various function spaces and obtained some useful results by using various summability techniques due to their wide applicability in science and engineering. The present study aims to establish a result on degree of approximation of conjugate Fourier series of functions in the generalized Zygmund class by using Euler–Nörlund product mean which generalizes several known results.

Keywords Degree of approximation · Generalized Zygmund class · Fourier series · Conjugate Fourier series · (E, r)-summability mean · (N, q_n)-summability mean · $(E, r)(N, q_n)$-summability mean

AMS Classification No 41A25 · 42B05 · 42B08

B. P. Padhy · A. Mishra
Department of Mathematics, Kalinga Institute of Industrial Technology, Deemed to be University, Bhubaneswar 751024, Odisha, India
e-mail: birupakhya.padhyfma@kiit.ac.in

A. Mishra
e-mail: m.anwesha17@gmail.com

S. K. Paikray (✉)
Department of Mathematics, Veer Surendra Sai University of Technology, Burla 768018, Odisha, India
e-mail: skpaikray_math@vssut.ac.in

U. K. Misra
Department of Mathematics, National Institute of Science and Technology, Pallur Hills, Golanthara, Berhampur 761008, Odisha, India
e-mail: umakanta_misra@yahoo.com

S. K. Paikray et al. (eds.), *New Trends in Applied Analysis and Computational Mathematics*, Advances in Intelligent Systems and Computing 1356,
https://doi.org/10.1007/978-981-16-1402-6_16

1 Introduction

The approximation of functions in several function spaces such as Hölder, Zygmund, Lipschitz, and Besov spaces using different summability techniques of trigonometric Fourier series and conjugate Fourier series have been received as a growing interest of many researchers in the last decades. It has been observed that for a 2π periodic function h, the error estimation of h and its conjugate $\overline{h}$ belonging to Lipschitz class in Hölder space using single summability means have been studied by several researchers (for details see [3, 7–11, 16–18, 20, 21]).

Working in this direction, Lal and Shireen [4], Leindler [6], Móricz [12], Móricz and Nemeth [13] and Pradhan et al. [19] have established some useful results concerning Zygmund class. Very recently, Das et al. [1], Nigam [14], and Singh et al. [22] proved some results concerning approximation of functions in the generalized Zygmund class by using different summability means.

Motivated essentially by the above mentioned results, here in this article, we propose to investigate on the degree of approximation of a function in the generalized Zygmund class $Z_l{}^{(m)}$ $(l \geq 1)$ based on $(E, r)(N, q_n)$ product mean of the conjugate Fourier series.

2 Definitions and Notations

For $l \geq 1$, suppose

$$L_l[0, 2\pi] = \Big\{h : [0, 2\pi] \subset \mathbb{R} : \int_0^{2\pi} |h(x)|^l dx < \infty\Big\}.$$

Let

$$\sum_{n=0}^{\infty} u_n(x) = \frac{a_0}{2} + \sum_{n=1}^{\infty}\Big(a_n \cos nx + b_n \sin nx\Big) \tag{2.1}$$

be the Fourier series associated with $h(x)$ and

$$\sum_{n=0}^{\infty} \overline{u}_n(x) = \sum_{n=1}^{\infty}\Big(a_n \cos nx - b_n \sin nx\Big) \tag{2.2}$$

be the conjugate Fourier series of (2.1), where a_0, a_n, b_n have their usual meaning.

Assuming $\overline{S_k}\Big(h; x\Big)$ as the kth partial sum of conjugate Fourier series, we can write

$$\overline{S_k}\left(h; x\right) - \overline{h}(x) - = \frac{1}{2\pi} \int_0^{\pi} \Psi(x; v) \frac{\cos\left(k + \frac{1}{2}\right)v - \cos\frac{v}{2}}{\sin\frac{v}{2}} dv,$$

where

$$\overline{h}(x) = -\frac{1}{2\pi} \int_0^{\pi} \Psi(x; v) \cot\frac{v}{2} dv.$$

We define

$$\|h\|_l = \left(\frac{1}{2\pi} \int_0^{2\pi} |h(x)|^l dx\right)^{\frac{1}{l}}, 1 \leq l < \infty$$

and

$$\|h\|_l = ess \sup_{0 \leq x \leq 2\pi} |h(x)|, l = \infty.$$

Suppose the Zygmund modulus of continuity of $h(x)$ be

$$m(h; r) = \sup_{0 \leq r, x \in R} |h(x + v) - h(x - v).| \text{(see [23])}$$

Let **B** represent the Banach space of all 2π periodic functions which are continuous and defined over $[0, 2\pi]$ under the supremum norm. Clearly,

$$Z_{(\alpha)} = \left\{h \in \mathbf{B} : |h(x + v) - h(x - v)| = O\left(|v|^{\alpha}\right) 0 < \alpha \leq 1\right\}$$

is a Banach space under the norm $\|.\|_{(\alpha)}$ defined by

$$\|h\|_{(\alpha)} = \sup_{0 \leq x \leq 2\pi} |h(x)| + \sup_{x, t \neq 0} \frac{|h(x + v) - h(x - v)|}{|v|^{\alpha}}.$$

For $h \in L_l[0, 2\pi]$, $(l \geq 1)$, the integral Zygmund modulus of continuity is defined by

$$m_l(h; r) = \sup_{0 < v \leq r} \left\{\frac{1}{2\pi} \int_0^{2\pi} |h(x + v) - h(x - v)|^l dx\right\}^{\frac{1}{l}}$$

and for $h \in \mathbf{B}$, $l = \infty$,

$$m_{\infty}(h; r) = \sup_{0 < v \leq r} \max_{x} |h(x + v) - h(x - v)|.$$

Clearly,

$$m_l(h; r) \to 0 \quad \text{as} \quad l \to 0.$$

Let us define the space

$$Z_{(\alpha),\, l} = \left\{h \in L_l[0, 2\pi] : \left(\int_0^{2\pi} |h(x+v) - h(x-v)|^l dx\right)^{\frac{1}{l}},\right.$$

which is a Banach space under the norm $\|.\|_{(\alpha),\, l}$ for $0 < \alpha \leq 1$ and $l \geq 1$. Clearly,

$$\|h\|_{(\alpha),\, l} = \|h\|_l + \sup_{v \neq 0} \frac{\|h(.+v) - h(.-v)\|_l}{|v|^\alpha}.$$

Let

$$Z^{(m)} = \left\{h \in \mathbf{B} : |h(x+v) - h(x-v)| = O\Big(m(v)\Big)\right\},$$

where m is a Zygmund modulus of continuity satisfying

(a) $m(0) = 0$
(b) $m(v_1 + v_2) \leq m(v_1) + m(v_2)$.

Suppose $m : [0, 2\pi] \to \mathbf{R}$, a function with $m(v) > 0$ for $0 \leq v < 2\pi$ and

$$\lim_{v \to 0^+} m(v) = m(0) = 0.$$

Define

$$Z_l^{(m)} = \left\{h \in L_l : 1 \leq l < \infty, \sup_{v \neq 0} \frac{\|h(.+v) - h(.-v)\|_l}{m(v)} < \infty\right\},$$

where

$$\|h\|_l^{(m)} = \|h\|_l + \sup_{v \neq 0} \frac{\|h(.+v) - h(.-v)\|_l}{m(v)}, l \geq 1.$$

Clearly, $\|.\|_l^{(m)}$ is a norm $Z_l^{(m)}$. Also, $Z_l^{(m)}$ is complete since L_l, $(l \geq 1)$ is complete. So, $Z_l^{(m)}$ is a Banach space under $\|.\|_l^{(m)}$.

If we assume $m(v)$ and $\mu(v)$ as the Zygmund moduli of continuity such that $\left(\frac{m(v)}{\mu(v)}\right)$ is positive and non-decreasing, then

$$\|h\|_l^{(\mu)} \leq \max.\left(1, \frac{m(2\pi)}{\mu(2\pi)}\right) \|h\|_l^{(m)} \leq \infty. \tag{2.3}$$

Clearly,

$$Z_l^{(m)} \subseteq Z_l^{(\mu)} \subseteq L_l, \quad (l \geq 1).$$

Let $\sum u_n$ be an infinite series with sequence of partial sums $\{s_n\}$. Let $\{q_k\}$ represent the sequence of non-negative integers such that

$$Q_n = \sum_{k=0}^{n} q_k \to \infty \quad \text{as} \quad n \to \infty.$$

Let

$$\tau_n^N = \frac{1}{Q_n} \sum_{k=0}^{n} q_{n-k} s_k, \quad n = 0, 1, 2, \ldots$$

represent the (N, q_n) mean of $\{s_n\}$ generated by the sequence $\{q_n\}$.

By (N, q_n) method, the series $\sum u_n$ is said to be summable to s if

$$\lim_{n\to\infty} \tau_n^N \to s.$$

We know, (N, q_n) method is regular [1].

The (E, r) transform of $\{s_n\}$ is given by

$$E_n^r = \frac{1}{(1+r)^n} \sum_{k=0}^{n} C(n, k)\ r^{n-k} s_k. \tag{2.4}$$

If $E_n^r \to s$ as $n \to \infty$ then $\sum u_n$ is summable to s by (E, r) summability. Also, (E, r) method is regular [1].

The $(E, r)(N, q_n)$ transform of $\{s_n\}$ is given by

$$\tau_n^{E_r, N} = \frac{1}{(1+r)^n} \sum_{k=0}^{n} C(n, k) \left\{ \frac{1}{Q_k} \sum_{\nu=0}^{k} q_{k-\nu} s_\nu \right\} \tag{2.5}$$

The series $\sum u_n$ is summable to s by the $(E, r)(N, q_n)$ transform if $\tau_n^{E_r, N} \to s$ as $n \to \infty$.

The following notations are used in the rest part of our paper:

$$\Psi(x, v) = h(x+v) - h(x-v);$$

$$\overline{K}_n^{E_r, N}(v) = \frac{1}{2\pi(1+r)^n} \sum_{k=0}^{n} C(n, k) \left\{ \frac{1}{Q_k} \sum_{\nu=0}^{k} q_{k-\nu} \frac{\cos\left(\nu + \frac{1}{2}\right)v - \cos\left(\frac{v}{2}\right)}{\sin\left(\frac{v}{2}\right)} \right\}.$$

3 Known Result

Using Hausdorff mean, Nigam [14] proved the following theorem.

Theorem 3.1 *Error approximation of a conjugate function $\overline{h}$ of a 2π periodic function $h \in Z_l^{(m)}, l \geq 1$, using $H = \left(\theta_{j,\alpha}\right)$ of conjugate Fourier series (2.2) is given by*

$$\|\overline{M}_j^H(h; .) - \overline{h}(.)\|_l^{(m)} = O\Big(\frac{1}{j+1}\int_{\frac{1}{j+1}}^{\pi} \frac{m(v)}{v^2\,\mu(v)}dv\Big),$$

where $m(v)$ and $\mu(v)$ are Zygmund moduli of continuity, provided

$$\int_0^{\pi} \frac{m(v)}{v\mu(v)}dv = O\Big(\frac{m(\eta)}{\mu(\eta)}\Big),\ 0 < \eta < \pi.$$

4 Main Result

Theorem 4.1 *Let $h : [0, 2\pi] \to \boldsymbol{R}$ be a periodic function (with period 2π) belonging to $Z_l^{(m)}$, $(l \geq 1)$, which is integrable in the sense of Lebesgue. Then the degree of approximation of $\overline{h}$ by using $(E, r)(N, q_n)$ mean of (2.2) is given by*

$$E_n(\overline{h}) = \inf_n \|\psi_n(.)\|_l^{\mu} = O\Big(\int_{\frac{1}{(n+1)}}^{\pi} \frac{m(v)}{v\,\mu(v)}dv\Big), \tag{4.1}$$

where $m(v)$ and $\mu(v)$ are the Zygmund moduli of continuity and $\frac{m(v)}{v\,\mu(v)}$ is positive and non-decreasing.

We require the below-mentioned lemmas to prove our main theorem.

Lemma 4.2 *Let*

$$|\overline{K}_n^{E_r,N}| = O(n)\ \ \texttt{for}\ 0 \leq v \leq \frac{1}{(n+1)}.$$

Lemma 4.3 *Let*

$$|\overline{K}_n^{E_r,N}| = O\Big(\frac{1}{v}\Big)\ \ \texttt{for}\ \frac{1}{(n+1)} \leq v \leq \pi.$$

Lemma 4.4 (see [4]) *Let Let $h \in Z_l^{(m)}$ then for $0 < v \leq \pi$,*

(i) $\|\Psi(., v)\|_l = O\Big(m(v)\Big)$,

(ii) $\|\Psi(.+y, v) - \Psi(.-y, v)\|_l = O\Big(m(v)\Big)$ *or* $O\Big(m(y)\Big)$.

(iii) If $m(v)$ *and* $\mu(v)$ *are as defined in Theorem 1, then*

$$\|\Psi(.+y, v) - \Psi(.-y, v)\|_l = O\Big(\mu(y)\frac{m(v)}{\mu(v)}\Big),$$

where $\Psi(x, v) = h(x+v) - h(x-v)$.

5 Proofs of the Theorem and Lemmas

Proof of Lemma 4.2. For $0 \le v \le \frac{1}{n+1}$ and $\sin nv \le n \ \sin v$, we have

$$|\overline{K}_n^{E_r,N}(v)| = \frac{1}{2\pi(1+r)^n}\left|\sum_{k=0}^{n} C(n,k)\ r^{n-k}\left\{\frac{1}{Q_k}\sum_{\nu=0}^{k} q_{k-\nu}\ \frac{\cos\left(\nu+\frac{1}{2}\right)v - \cos\frac{v}{2}}{\sin\frac{v}{2}}\right\}\right|$$

$$= \frac{1}{2\pi(1+r)^n}\left|\sum_{k=0}^{n} C(n,k)\ r^{n-k}\right.$$

$$\left.\times\left\{\frac{1}{Q_k}\sum_{\nu=0}^{k} q_{k-\nu}\ \frac{(\cos(\nu v).\cos\frac{v}{2} - \sin(\nu v).\sin\frac{v}{2} - \cos\frac{v}{2}}{\sin\frac{v}{2}}\right\}\right|$$

$$= \frac{1}{2\pi(1+r)^n}\left|\sum_{k=0}^{n} C(n,k)\ r^{n-k}\left\{\frac{1}{Q_k}\sum_{\nu=0}^{k} q_{k-\nu}\Big(\frac{2\sin^2\nu\frac{v}{2}.\cos\frac{v}{2}}{\sin\frac{v}{2}} + \sin(\nu v)\Big)\right\}\right|$$

$$\le \frac{1}{2\pi(1+r)^n}\left|\sum_{k=0}^{n} C(n,k)\ r^{n-k}\left\{\frac{1}{Q_k}\sum_{\nu=0}^{k} q_{k-\nu}\Big(O(\nu) + O(\nu)\Big)\right\}\right|$$

$$\le \frac{1}{\pi(1+r)^n}\left|\sum_{k=0}^{n} C(n,k)\ r^{n-k}\left\{\frac{O(k)}{Q_k}\sum_{\nu=0}^{k} q_{k-\nu}\right\}\right|$$

$$= O(n). \tag{5.1}$$

□

Proof of Lemma 4.3. By Jordan's lemma, we have

$$sin\left(\frac{v}{2}\right) \ge \frac{v}{\pi},\ sin\ nv \le 1\ \left(\frac{1}{n+1} \le v \le \pi\right).$$

Now,

$$
\begin{aligned}
|\overline{K}_n^{E_r,N}(v)| &= \frac{1}{2\pi(1+r)^n}\Big|\sum_{k=0}^{n} C(n,k)\ r^{n-k}\Big\{\frac{1}{Q_k}\sum_{\nu=0}^{k} q_{k-\nu}\ \frac{cos\left(\nu+\frac{1}{2}\right)v - cos\frac{v}{2}}{sin\frac{v}{2}}\Big\}\Big| \\
&\leq \frac{1}{2\pi(1+r)^n}\Big|\sum_{k=0}^{n} C(n,k)\ r^{n-k}\Big\{\frac{1}{Q_k}\sum_{\nu=0}^{k} \frac{\pi}{v}\ q_{k-\nu}\Big\}\Big| \\
&= \frac{1}{2v(1+r)^n}\Big|\sum_{k=0}^{n} C(n,k)\ r^{n-k}\Big\{\frac{1}{Q_k}\sum_{\nu=0}^{k} q_{k-\nu}\Big\}\Big| \\
&= \frac{1}{2v(1+r)^n}\Big|\sum_{k=0}^{n} C(n,k)\ r^{n-k}\Big| \\
&= O\Big(\frac{1}{v}\Big).
\end{aligned}
\tag{5.2}
$$

□

Proof of the Theorem 4.1. Let $\overline{S_k}\Big(h; x\Big)$ denote the kth partial sum of conjugate Fourier series and is given by

$$\overline{S_k}\Big(h; x\Big) - \overline{h}(x) - = \frac{1}{2\pi}\int_0^{\pi} \Psi(x; v)\frac{\cos\left(k+\frac{1}{2}\right)v - \cos\frac{v}{2}}{\sin\frac{v}{2}}\,dv.$$

Then its (N, q_n) transform is given by

$$
\begin{aligned}
&\frac{1}{Q_n}\sum_{k=0}^{n} q_{n-k}\left\{\overline{S_k}(h; x) - \overline{h}(x)\right\} \\
&= \frac{1}{2\pi}\int_0^{\pi} \Psi(x; v)\frac{1}{Q_n}\sum_{k=0}^{n} q_{n-k}\frac{\cos\left(k+\frac{1}{2}\right)v - \cos\frac{v}{2}}{\sin\frac{v}{2}}\,dv.
\end{aligned}
$$

Denoting the $(E, r)(N, q_n)$ transform of $\overline{S_k}\Big(h; x\Big)$ by $\overline{\tau}_n^{E_r,N}$, we have

$$
\begin{aligned}
&\overline{\tau}_n^{E_r,N} - \overline{h}(x) \\
&= \frac{1}{2\pi(1+r)^n}\int_0^{\pi} \Psi(x; v)\sum_{k-0}^{n} C(n,k) r^{n-k}\left\{\frac{1}{Q_k}\sum_{\nu=0}^{k} q_{k-\nu}\frac{\cos\left(\nu+\frac{1}{2}\right)v - \cos\frac{v}{2}}{\sin\frac{v}{2}}\,dv\right\} \\
&= \int_0^{\pi} \Psi(x; v)\overline{k}_n^{E_r,N}(v)\,dv \\
&= \psi_n(x) \quad \text{(say)}.
\end{aligned}
$$

Then,

$$\psi_n(x+y) - \psi_n(x-y) = \int_0^{\pi} \{\Psi(x+y;v) - \Psi(x-y;v)\} \overline{K}_n^{E_r,N}(v)\, dv.$$

Using Minkowski's inequality, we have

$$\begin{aligned}
\|\psi_n(.+y) - \psi_n(.-y)\|_l &= \left\{ \frac{1}{2\pi} \int_0^{2\pi} |\psi_n(x+y) - \psi_n(x-y)|^l\, dx \right\}^{\frac{1}{l}} \\
&= \frac{1}{2\pi} \left[\int_0^{2\pi} \left| \int_0^{\pi} \{\Psi(x+y;v) - \Psi(x-y;v)\} \overline{K}_n^{E_r,N}(v)\, dv \right|^l dx \right]^{\frac{1}{l}} \\
&\leq \int_0^{\pi} \left\{ \frac{1}{2\pi} \int_0^{2\pi} |\Psi(x+y;v) - \Psi(x-y;v)|^l\, dx \right\}^{\frac{1}{l}} |\overline{K}_n^{E_r,N}|\, dv \\
&= \int_0^{\pi} \|\Psi(.+y;v) - \Psi(.-y;v)\|_l\, |\overline{K}_n^{E_r,N}(v)|\, dv \\
&= \int_0^{\frac{1}{n+1}} \|\Psi(.+y;v) - \Psi(.-y;v)\|_l\, |\overline{K}_n^{E_r,N}(v)|\, dv \\
&\quad + \int_{\frac{1}{n+1}}^{\pi} \|\Psi(.+y;v) - \Psi(.-y;v)\|_l\, |\overline{K}_n^{E_r,N}(v)|\, dv \\
&= I_1 + I_2 \quad \text{(say)}.
\end{aligned} \tag{5.3}$$

Next, by using Lemmas 4.2, 4.4 and monotonicity of $\left(\frac{m(v)}{\mu(v)}\right)$ with respect to 'v', we have

$$\begin{aligned}
I_1 &= \int_0^{\frac{1}{n+1}} \|\Psi(.+y,v) - \Psi(.-y,v)\|_l\, |\overline{K}_n^{E_r,N}(v)|\, dv \\
&= \int_0^{\frac{1}{n+1}} O\Big(\mu(y)\frac{m(v)}{\mu(v)}\Big) O(n)\, dv.
\end{aligned}$$

By using the second mean value theorem of integral, we have

$$\begin{aligned}
I_1 &\leq O\Bigg(n\, \mu(y)\, \frac{m\left(\frac{1}{n+1}\right)}{\mu\left(\frac{1}{n+1}\right)} \int_0^{\frac{1}{n+1}} dv\Bigg) \\
&= O\Bigg(\frac{n}{n+1}\, \mu(y)\, \frac{m\left(\frac{1}{n+1}\right)}{\mu\left(\frac{1}{n+1}\right)}\Bigg) \\
&= O\Bigg(\mu(y)\, \frac{m\left(\frac{1}{n+1}\right)}{\mu\left(\frac{1}{n+1}\right)}\Bigg).
\end{aligned} \tag{5.4}$$

Again, by using Lemmas 4.3 and 4.4, we get

$$\begin{aligned} I_2 &= \int_{\frac{1}{n+1}}^{\pi} \|\Psi(.+y,v) - \Psi(.-y,v)\|_l |\overline{\mathcal{K}}_n^{E_r,N}(v)|\,dv \\ &\leq \int_{\frac{1}{n+1}}^{\pi} O\Big(\mu(y)\frac{m(v)}{\mu(v)}\Big)\frac{1}{v}\,dv \\ &= O\Big(\mu(y)\int_{\frac{1}{n+1}}^{\pi} \frac{m(v)}{v\,\mu(v)}\,dv\Big). \end{aligned} \tag{5.5}$$

Further, by (5.3)–(5.5), we have

$$\begin{aligned} &\|\psi_n(.+y) - \psi_n(.-y)\|_l \\ &= O\Big(\mu(y)\,\frac{m\Big(\frac{1}{n+1}\Big)}{\mu\Big(\frac{1}{n+1}\Big)}\Big) + O\Big(\mu(y)\int_{\frac{1}{n+1}}^{\pi} \frac{m(v)}{v\,\mu(v)}\,dv\Big). \end{aligned}$$

Therefore, we have

$$\begin{aligned} &\sup_{y\neq 0}\frac{\|\psi_n(.+y) - \psi_n(.-y)\|_l}{\mu(y)}. \\ &= O\Big(\frac{m\Big(\frac{1}{n+1}\Big)}{\mu\Big(\frac{1}{n+1}\Big)}\Big) + O\Big(\int_{\frac{1}{n+1}}^{\pi} \frac{m(v)}{v\,\mu(v)}\,dv\Big). \end{aligned} \tag{5.6}$$

Now, since, $h \in Z_l^{(m)}$ and $\Psi(x;v) = h(x+v) - h(x-v)$, by Minkowski's inequality, we have

$$\begin{aligned} \|\Psi(x,v)\|_l &= \|h(x+v) - h(x-v)\|_l \\ &= O\big(m(v)\big). \end{aligned} \tag{5.7}$$

Next, by using Lemmas 4.2, 4.3 and (5.7),

$$\begin{aligned} \|\psi_n(.)\|_l &\leq \Big(\int_0^{\frac{1}{n+1}} + \int_{\frac{1}{n+1}}^{\pi}\Big)\,\|\Psi(.,v)\|_l|\;\overline{\mathcal{K}}_n^{E_r,N}(v)|\,dv \\ &= O\Big(n\int_0^{\frac{1}{n+1}} m(v)\,dv\Big) + O\Big(\int_{\frac{1}{n+1}}^{\pi} \frac{m(v)}{v}dv\Big) \\ &= O\Big(m\Big(\frac{1}{n+1}\Big)\Big) + O\Big(\int_{\frac{1}{n+1}}^{\pi} \frac{m(v)}{v}dv\Big). \end{aligned} \tag{5.8}$$

From (5.6) and (5.8), we have

$$\|\psi_n(.)\|_l^\mu = \|\psi_n(.)\|_l + \sup_{y\neq 0}\frac{\|\psi_n(.+y)-\psi_n(.-y)\|_l}{\mu(y)}$$

$$= O\Big(m\Big(\frac{1}{n+1}\Big)\Big) + O\Big(\int_{\frac{1}{n+1}}^{\pi}\frac{m(v)}{v}dv\Big) + O\Bigg(\frac{m\Big(\frac{1}{n+1}\Big)}{\mu\Big(\frac{1}{n+1}\Big)}\Bigg) + O\Big(\int_{\frac{1}{n+1}}^{\pi}\frac{m(v)}{v\,\mu(v)}\,dv\Big)$$

$$= \sum_{j=1}^{4} G_j.$$

In view of monotonicity of $\mu(v)$ for $0 < v \leq \pi$, we have

$$m(v) = \frac{m(v)}{\mu(v)}\cdot\mu(v) \leq \mu(\pi).\frac{m(v)}{\mu(v)} = O\Big(\frac{m(v)}{\mu(v)}\Big).$$

Therefore,

$$G_1 = O(G_3).$$

Again, by using monotonicity of $\mu(v)$,

$$G_2 = \int_{\frac{1}{n+1}}^{\pi}\frac{m(v)}{v}dv = \int_{\frac{1}{n+1}}^{\pi}\frac{m(v)}{v\,\mu(v)}\,\mu(v)dv \leq \mu(\pi)\int_{\frac{1}{n+1}}^{\pi}\frac{m(v)}{v\,\mu(v)}dv = O(G_4).$$

Since, $\frac{m(v)}{\mu(v)}$ is positive and increasing, we have

$$G_4 = \int_{\frac{1}{n+1}}^{\pi}\frac{m(v)}{v\,\mu(v)}dv = \frac{m\Big(\frac{1}{n+1}\Big)}{\mu\Big(\frac{1}{n+1}\Big)}\int_{\frac{1}{n+1}}^{\pi}\frac{dv}{v} \geq \frac{m\Big(\frac{1}{n+1}\Big)}{\mu\Big(\frac{1}{n+1}\Big)}.$$

Clearly,

$$G_3 = O(G_4).$$

Thus,

$$\|\psi_n(.)\|_l^\mu = O(G_4) = O\Big(\int_{\frac{1}{n+1}}^{\pi}\frac{m(v)}{v\,\mu(v)}dv\Big).$$

Hence,

$$E_n(\overline{h}) = \inf_n \|\psi_n(.)\|_l^\mu = O\Big(\int_{\frac{1}{n+1}}^{\pi}\frac{m(v)}{v\,\mu(v)}dv\Big).$$

This completes the proof of the Theorem 4.1. □

6 Concluding Remarks and Observations

In this concluding section of our investigation, we present several remarks and observations concerning the result which we have proved in this article. In fact, our theorem effectively extends and improves some of the previously existing results in this direction.

Remark 6.1 If the $(E, r)(N, q_n)$ mean is replaced by $(E, 1)(\overline{N}, q_n)$ mean in our main theorem then the degree of approximation of a function $\overline{h} \in Z_l^{(m)}$, $(l \geq 1)$ by $(E, 1)(\overline{N}, q_n)$ mean

$$\tau_n^{E,\overline{N}} = \frac{1}{(2)^n}\sum_{k=0}^{n} C(n,k)\Big\{\frac{1}{Q_k}\sum_{\nu=0}^{k} q_\nu s_\nu\Big\}$$

of (2.2) is given by (see [19])

$$E_n(\overline{h}) = O\Big(\int_{\frac{1}{(n+1)}}^{\pi} \frac{m(v)}{v\,\mu(v)} dv\Big).$$

Thus, our theorem extends the result of Pradhan et al. [19]

Remark 6.2 If the $(E, r)(N, q_n)$ mean is replaced by $(E, 1)(C, 1)$ mean in our main theorem then the degree of approximation of a function $\overline{h} \in Z_l^{(m)}$, $(l \geq 1)$ by $(E, 1)(C, 1)$ mean

$$\tau_n^{E,C} = \frac{1}{(2)^n}\sum_{k=0}^{n} C(n,k)\Big\{\frac{1}{k+1}\sum_{\nu=0}^{k} s_\nu\Big\}$$

of (2.2) is given by (see [15])

$$E_n(\overline{h}) = O\Big(\int_{\frac{1}{(n+1)}}^{\pi} \frac{m(v)}{v\,\mu(v)} dv\Big).$$

Thus, our theorem extends the result of Nigam et al. [15].

Remark 6.3 If the $(E, r)(N, q_n)$ mean is replaced by the Euler–Hausdorff mean in our main theorem then the degree of approximation of a function $\overline{h} \in Z_l^{(m)}$, $(l \geq 1)$ by Euler–Hausdorff mean (see [5])

$$\tau_n^{E_r,H} = \frac{1}{(1+r)^n}\sum_{k=0}^{n} C(n,k)\, r^{n-k}\sum_{\nu=0}^{k} h_{k,\nu} s_\nu$$

of (2.2) is given by

$$E_n(\overline{h}) = O\Big(\frac{1}{n+1}\int_{\frac{1}{(n+1)}}^{\pi} \frac{m(v)}{v^2\,\mu(v)}dv\Big).$$

Thus, our theorem extends the result of Lal and Mishra [5].

References

1. A.A. Das, S.K. Paikray, T. Pradhan, Approximation of signals in the weighted Zygmund class via Euler-Hausdorff product summability mean of Fourier series. J. Indian Math. Soc. **87**, 22–36 (2020)
2. G.H. Hardy, *Divergent Series*, 1st edn. (Oxford University Press, 1949)
3. B.B. Jena, L.N. Mishra, S.K. Paikray, U.K. Misra, Approximation of signals by general matrix summability with effects of Gibbs phenomenon. Bol. Soc. Parana. Mat. **38**, 141–158 (2020)
4. S. Lal, Shireen, Best approximation of functions of generalized Zygmund class by matrix-Euler summability mean of Fourier series. Bull. Math. Anal. Appl. **5**, 1–13 (2013)
5. S. Lal, A. Mishra, Euler-Hausdorff matrix summability operator and trigonometric approximation of the conjugate of a function belonging to generalized Lipschitz class. J. Inequal. Appl. (2013). Article ID: 59
6. L. Leindler, Strong approximation and generalized Zygmund class. Acta. Sci. Math. **43**, 301–309 (1981)
7. V.N. Mishra, L.N. Mishra, Trigonometric approximation of signals (functions) in L^p, $(p \geq 1)$-norm. Int. J. Contemp. Math. Sci. **7**, 909–918 (2012)
8. V.N. Mishra, K. Khatri, L.N. Mishra, Approximation of functions belonging to $Lip\Big(\xi(t), r\Big)$ class by $(N, p_n)(E, q)$-summability of conjugate series of Fourier series. J. Inequal. Appl. (2012). Article ID-296
9. V.N. Mishra, K. Khatri, L.N. Mishra, Product $(N, p_n)(C, 1)$-summability of a sequence of a sequence of Fourier coefficients. Math. Sci. (2012). Article ID: 38
10. L.N. Mishra, V.N. Mishra, K. Khatri, Deepmala, On the trigonometric approximation of signals belonging to generalized weighted Lipschitz class $W\Big(L^r, \xi(t)\Big)$, $(r \geq 1)$-class by matrix (C^1, N_p) operator of conjugate series of its Fourier series. Appl. Math. Comput. **237**, 252–263 (2014)
11. V.N. Mishra, K. Khatri, L.N. Mishra, Deepmala, Trigonometric approximation of periodic signals belonging to generalized weighted Lipschitz $W\Big(L_r, \xi(t)\Big)$, $(r \geq 1)$-class by Norlund Euler $(N, p_n)(E, q)$ operator of conjugate series of its Fourier series. J. Class. Anal. **5**, 91–105 (2014)
12. F. Moricz, Enlarged Lipschitz and Zygmund classes of functions and Fourier transforms. East. J. Approx. **16**, 259–271 (2010)
13. F. Moricz, J. Nemeth, Generalized Zygmund classes of functions and strong approximation of Fourier series. Acta Sci. Math. **73**, 637–647 (2007)
14. H.K. Nigam, On approximation in generalized Zygmund class. Demonstr. Math. **52**, 370–387 (2019)
15. H.K. Nigam, K. Sharma, On $(E, 1)(C, 1)$ summability of Fourier series and its conjugate series. Int. J. Pure. Appl. Math. **82**, 365–375 (2013)
16. B.P. Padhy, P.K. Das, M. Misra, P. Samanta, U.K. Misra, Trigonometric Fourier approximation of the conjugate series of a function of generalized Lipschitz class by product summability, in

Advances in Intelligent Systems and Computing, vol. 410 (2016). https://doi.org/10.1007/978-81-322-2734-2-19
17. T. Pradhan, S.K. Paikray, U.K. Misra, Approximation of signals belonging to generalized Lipschitz class using $(\overline{N}; p_n; q_n)(E; s)$-summability mean of Fourier series. Cogent Math. **1250343**, 1–9 (2016)
18. T. Pradhan, B.B. Jena, S.K. Paikray, H. Dutta, U.K. Misra, On approximation of the rate of convergence of Fourier series in the generalized Holder metric by deferred N örlund mean. Afr. Mat. **30**, 1119–1131 (2019)
19. T. Pradhan, S.K. Paikray, A.A. Das, H. Dutta, On approximation of signals in the generalized Zygmund class via $(E; 1)(\overline{N}; p_n)$ summability means of conjugate Fourier series. Proyecciones J. Math. **38**, 981–998 (2019)
20. P. Parida, S.K. Paikray, M. Das, U.K. Misra, Degree of approximation by product $(N; p_n; q_n)(E; q)$ summability Fourier series of a signal belonging to $Lip(\alpha; r)$-class. TWMS J. Appl. Eng. Math. **9**, 901–908 (2019)
21. P. Parida, S.K. Paikray, H. Dutta, On approximation of signals in $Lip(\alpha, r)$-class using the product $(\overline{N}, p_n, q_n)(E, s)$- summability means of conjugate Fourier series. Nonlinear Stud. **27**, 1–9 (2020)
22. M.V. Singh, M.L. Mittal, B.E. Rhoades, Approximation of functions in the generalized Zygmund class using Hausdorff means. J. Inequal. Appl. **101** (2017). https://doi.org/10.1186/s13600-017-1361-8
23. A. Zygmund, *Trigonometric Series*, vol. 1, 2nd revised edn. (Cambridge University Press, Cambridge, 1939)

Electrification Effect of Nanoparticles on Nanofluid Flow over a Continuous Stretching Sheet

Kamala Kumar Pradhan, Ashok Misra, and Saroj Kumar Mishra

Abstract The main concern here is to study nanofluid flow past a continuous stretching sheet including electrification of nanoparticles along with Brownian diffusion and thermophoresis to show the impact on the enhancement of thermal conductivity and cooling process. The governing equations of the flow field are derived from revised Buongiorno's model including the electrification of nanoparticles and the formulated equations are reduced to dimensionless local similarity equations using similarity variables. The locally similar equations are solved numerically by shooting approach using MATLAB built-in-package bvp4c. Electrification effect of nanoparticles on dimensionless velocity, normalised temperature, dimensionless nanoparticle concentration as well as the non-dimensional heat and mass transfer coefficients is investigated through table and graphs. It is found that the electrification of nanoparticles is a possible mechanism for heat transfer enhancement of base fluids.

Keywords Nanofluids · BVP4C function · Electrification of particle · Viscous dissipation

1 Introduction

Flow and heat transfer over a continuous stretching sheet play vital roles in several industrial applications like production of polymer and plastic sheet, aerodynamic extraction, manufacturing of paper, glass blowing, steel spinning, plastic film

K. K. Pradhan (✉) · A. Misra · S. K. Mishra
School of Applied Science, Centurion University of Technology and Management, Paralakhemundi, Gajapati, Odisha, India
e-mail: mr.kml1411@gmail.com

A. Misra
e-mail: amisra1972@gmail.com

S. K. Mishra
e-mail: s1_mishra@yahoo.com

S. K. Paikray et al. (eds.), *New Trends in Applied Analysis and Computational Mathematics*, Advances in Intelligent Systems and Computing 1356,
https://doi.org/10.1007/978-981-16-1402-6_17

drawing, etc. The quality of their final products as per the requirement strictly depends on the rate of cooling, rate of stretching/shrinking and the process of stretching. To overcome these expectations, though some fluids like water, ethylene glycol and engine oil have been used to enhance the rate of heat transfer, these are having limited capabilities in terms of thermal properties. It imposes a restriction in thermal applications. Since the solid metals are having much higher thermal conductivity compared to the liquids, it can be expected that the fluids with the suspension of solid particles may increase its thermal conductivity. Taking a view on this, Choi and Eastman [4] has coined the term nanofluid and Masuda et al. [11] has observed its thermal conductivity enhancement phenomena. The first investigation of flow over continuous stretching sheet has been studied by Sakiadis [18]. After this the flow over continuous stretching sheet by taking force convection, impulsive motion, suction or injection effects and temperature-dependent viscosity has been studied by Kumari and Nath [10], Ali and Al-Yousef [1], Chen [3] and Elbashbeshy and Bazid [5], respectively.

Viscous dissipation behaves like an energy source. Generally, it is a process of converting the mechanical energy of downward flowing fluid into thermal. The first idea of viscous dissipation was first given by Zulkifi et al. [25] and the experiment using this effect was carried out by Pal and Mondal [15]. Later, many researchers have extended their works and took this effect in their experiments like Vajravelu et al. [21] and Partha et al. [16] and Yirga and Shankar [23]. They have found that viscous dissipation develops the velocity and enhances the thermal conductivity.

Pal and Mondal [15] and Gangadhar [7] took an experiment on flow of electrically conducting fluid in presence of magnetic field over stretching surface. It was found that the increase of magnetic field parameter produces a force called Lorentz's force which produces more resistance to the fluid. Hence, the velocity of fluid decreases as per the increase of magnetic parameter. But in case of electrically non-conducting fluids, a triboelectrification of particle in base fluid comes into the picture due to the particle–particle collision and particle–wall interaction. It produces an effective drag force on ions and gives negligible magnetic effect [20]. Homogeneity of the fluid has been assumed by several authors like Khan et al. [9] and Khan and Pop [8]. But when a nanoparticle migration phenomenon occurs, the non-homogeneity of nanofluid is well justified [22].

From the above literatures, it is observed that no investigation has been done to study the flow and heat transfer characteristic of nanofluids with electrification effect of nanoparticle. So, the present experiment investigates the influence of electrification of nanoparticles in presence of viscous dissipation using Buongiorno's two-component non-homogeneous model over a continuous stretching sheet using electrically non-conducting fluids.

2 Mathematical Formulations

A two-dimensional steady boundary layer flow over a stretching sheet immersed in a nanofluid is considered. Let T be assumed as the temperature inside the boundary layer and T_∞, T_w are the ambient temperature and wall temperature, respectively. It is also assumed that U is the free stream velocity and $U_w(x) = \varepsilon U(x)$ be the velocity on the plate where ε is the plate velocity parameter. C is taken as the nanoparticle volume fraction and C_w, C_∞ are the nanoparticle volume fraction at the surface and ambient nanoparticle volume fraction, respectively. u and v are the velocity components along the x and y directions, respectively. At the constant pressure, μ_f is the dynamic viscosity, ν_f is the kinematic viscosity, ρ_f is the density, k_f is the thermal conductivity and C_{pf} is the specific heat capacity. D_B and D_T are the Brownian diffusion coefficient and thermophoresis diffusion coefficient of the base fluid, respectively, τ the ratio of the effective heat capacity of the nanoparticle material and the heat capacity of the base fluid. The geometrical model with the coordinate system of this problem is shown in Fig. 1.

The governing equations by using above assumptions and the effect of Brownian Diffusion, thermophorosis along with electrification and viscous dissipation using the Buongiorno model [2] are given by

Continuity equation for fluid:

$$\frac{\partial u}{\partial x} + \frac{\partial v}{\partial y} = 0 \tag{1}$$

Continuity equation for nanoparticle:

Fig. 1 Physical model of the coordinate system

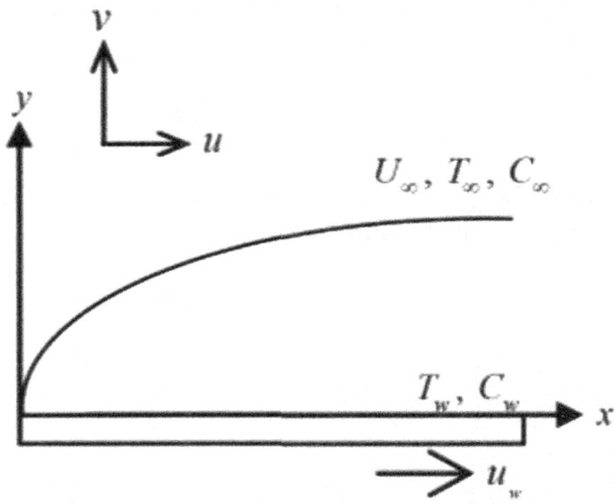

$$u\frac{\partial C}{\partial x}+v\frac{\partial C}{\partial y}=D_B\frac{\partial^2 C}{\partial y^2}+\frac{D_T}{T}\frac{\partial^2 T}{\partial y^2}+\left(\frac{q}{m}\right)\frac{1}{F}\left(\frac{\partial(\mathrm{CE_x})}{\partial x}+\frac{\partial(\mathrm{CE_y})}{\partial y}\right) \tag{2}$$

Momentum equation in the X-direction:

$$\rho_{nf}\left[u\frac{\partial u}{\partial x}+v\frac{\partial u}{\partial y}\right]=-\frac{\partial p}{\partial x}+\mu_{nf}\frac{\partial^2 u}{\partial y^2}+C\rho_s\left(\frac{q}{m}\right)E_x \tag{3}$$

Momentum equation in Y-direction:

$$\frac{\partial p}{\partial y}=O(\delta) \tag{4}$$

Energy equation

$$(\rho c)_{nf}\left[u\frac{\partial T}{\partial x}+v\frac{\partial T}{\partial y}\right]=k_{nf}\left[\frac{\partial^2 T}{\partial y^2}\right]+\rho_s c_s D_B\frac{\partial C}{\partial y}\frac{\partial T}{\partial y}+\frac{\rho_s c_s D_T}{T}\left(\frac{\partial T}{\partial y}\right)^2$$
$$+\left(\frac{q}{m}\right)\frac{Cc_s\rho_s}{F}\left(E_x\frac{\partial T}{\partial x}+E_y\frac{\partial T}{\partial y}\right)+\mu_{nf}\left(\frac{\partial u}{\partial y}\right)^2 \tag{5}$$

The E-field is given by the following equation:

$$\frac{\partial \mathrm{E_x}}{\partial x}+\frac{\partial \mathrm{E_y}}{\partial y}=\frac{\rho_s}{\epsilon_0}\frac{q}{m} \tag{6}$$

where ϵ_0 is the permittivity.

Following Soo [19] neglecting the change in electric field in the x-direction, the transverse electric field is given by

$$\frac{\partial \mathrm{E_y}}{\partial y}=\frac{\rho_s}{\epsilon_0}\frac{q}{m} \tag{7}$$

and E_x in potential flow outside the boundary layer which is assumed almost negligible with the boundary conditions

$$\left.\begin{array}{c} y=0, u=U_w=\varepsilon U, v=0, T=T_w, C=C_w \\ y=\infty, u=0, T=T_\infty, C=C_\infty \end{array}\right\} \tag{8}$$

3 Similarity Transformation

Introducing the set of similarity transformations.

$$\eta = y\sqrt{\frac{U}{2\nu_f x}}, \theta = \frac{T - T_\infty}{T_w - T_\infty}, u = Uf', v = -\sqrt{\frac{\nu_f U}{2x}}(f - \eta f'),$$
$$\psi = \sqrt{2\nu_f x U}\, f(\eta), S = \frac{C - C_\infty}{C_w - C_\infty}$$

Equation (1) is satisfied and Eqs. (2), (3) and (5) are also transformed into

$$S'' + Scf S' + \frac{Nt}{Nb}\theta'' - MSc\eta S' + 2Sc\frac{N_F}{N_{Re}}(S + \eta S' + N_c) = 0 \quad (9)$$

$$\varphi_1 f''' + ff'' + 2\frac{MN_b}{N_F}Le\varphi_2 s = 0 \quad (10)$$

$$\theta'' + \frac{Pr}{\varphi_3\varphi_4} f\theta' + \frac{Pr}{\varphi_4}N_b\theta' s' + \frac{Pr}{\varphi_4}N_t(\theta')^2 + 2\frac{Pr}{\varphi_4}\frac{N_F}{N_{Re}}\eta\theta' C_\infty\tau - \frac{1}{2}\tau\frac{Pr}{\varphi_4}M\eta\theta'$$
$$+ \varphi_5\frac{PrEc}{\varphi_4}(f'')^2 = 0, \quad (11)$$

respectively, by considering $\left(\frac{T_w - T_\infty}{T_\infty}\right) \ll 1$

where $M = \left(\frac{q}{m}\right)\frac{1}{FU}E_x$, $N_F = \frac{U}{Fx}$, $\frac{1}{N_{Re}} = \left(\left(\frac{q}{m}\right)^2\frac{\rho_s x^2}{U^2\epsilon_0}\right)$, $Nb = \frac{\tau D_B(C_w - C_\infty)}{\nu_f}$, $Nt = \frac{\tau D_T(T_w - T_\infty)}{\nu_f T_\infty}$, $Pr = \frac{\nu_f}{\alpha_f}$, $Sc = \frac{\nu_f}{D_B}$, $\tau = \frac{\rho_s c_s}{\rho_f c_f}$, $N_c = \frac{C_\infty}{C_w - C_\infty}$.

Also, by using the Maxwell model [12] for thermal conductivity, we get the thermophysical constants $\varphi_1, \varphi_2, \varphi_3, \varphi_4, \varphi_5$ as

$$\varphi_1 = \frac{1}{(1-C)^{2.5}}\frac{1}{\left[(1-C) + C\frac{\rho_s}{\rho_f}\right]}, \varphi_2 = \frac{\rho_s}{\rho_f}\frac{1}{\left[(1-C) + C\frac{\rho_s}{\rho_f}\right]}, \varphi_3 = \frac{1}{(1-C) + C\tau},$$
$$\varphi_4 = \frac{k_s + 2k_f - 2C(k_f - k_s)}{k_s + 2k_f + C(k_f - k_s)}, \varphi_5 = \frac{1}{(1-C)^{2.5}}$$

subjected to the boundary conditions

$$\left.\begin{array}{l} \eta = 0, f(0) = 0, f'(0) = \varepsilon, \theta(0) = 1, S(0) = 1 \\ \eta = \infty, \theta(\infty) = 0, S(\infty) = 0, f'(\infty) = 1 \end{array}\right\} \quad (12)$$

4 The Quantities of Physical Interest

The physical quantities of the problem are the skin friction coefficient $C_f(f''(0))$, Nusselt number N_{ux} $(-\theta'(0))$ and Sherwood's number S_{hx} $(-S'(0))$ which are given by C_f, N_{ux} and S_{hx}

$$\left.\begin{aligned} C_f &= \frac{\tau_w}{\rho u_e^2} \\ N_{ux} &= \frac{x q_w}{k(T_w - T_\infty)} \\ S_{hx} &= \frac{x j_w}{D_B(C_w - C_\infty)} \end{aligned}\right\} \quad (13)$$

Here ρ is the density of the fluid. τ_w, q_w and j_w are the surface heat flux, surface heat flux and surface mass flux and can be calculated as

$$\left.\begin{aligned} \tau_w &= \mu\left(\frac{\partial u}{\partial y}\right)_{y=0} \\ q_w &= -k\left(\frac{\partial T}{\partial y}\right)_{y=0} \\ j_w &= -D_B\left(\frac{\partial C}{\partial y}\right)_{y=0} \end{aligned}\right\} \quad (14)$$

By using the similarity variables, the above expression (14) will be

$$\left.\begin{aligned} C_f(2Re_x)^{\frac{1}{2}} &= f''(0) \\ N_{ux}\left(\frac{Re_x}{2}\right)^{-\frac{1}{2}} &= -\theta'(0) \\ S_{hx}\left(\frac{Re_x}{2}\right)^{-\frac{1}{2}} &= -S'(0) \end{aligned}\right\} \quad (15)$$

Here, $Re_x = \frac{U_\infty x}{\nu}$ is called local Reynold's number and $C_f(2Re_x)^{\frac{1}{2}}$, $N_{ux}\left(\frac{Re_x}{2}\right)^{-\frac{1}{2}}$ and $S_{hx}\left(\frac{Re_x}{2}\right)^{-\frac{1}{2}}$ are called reduced skin friction, reduced Nusselt Number and reduced Sherwoods number, respectively.

5 Results and Discussion

From Eqs. 9 to 11, we see that the parameters $M, N_F, \frac{1}{N_{Re}}$ are functions of x, and hence these equations do not possess self-similar solutions (the solutions for which velocity and temperature profiles remain geometrically similar despite the growth of the boundary layer with distance x from the leading edge). Since the present approach preserves the mass and momentum conservation, the flow dynamics is valid within the boundary layer [6]. Thus, the solution of Eqs. (9)–(11) subject to the boundary conditions (12) for fixed value of $M, N_F, \frac{1}{N_{Re}}$ would be locally similar. The local similarity solution means the dimensionless quantities $M, N_F, \frac{1}{N_{Re}}$ are determined at any x-station and the upstream history of the flow will be ignored, except as far as it influences the similarities variables. Here, in this problem, we treat them as constants in the numerical solutions and the results are only locally similar. The locally similar Eqs. (9)–(11) have been solved numerically using bvp4c function in the MATLAB software.

In this investigation, we have considered copper (Cu) nanoparticles with pure water as a base working fluid. The nanoparticle volume fraction (φ) is taken as 0.01.

Table 1 Thermophysical properties of base fluid and nanofluid [14]

Physical properties	Base fluid	Nanoparticles
	Water	Cu
C_p(J/kgK)	4179	385
ρ(kg/m^3)	997.1	8933
k(W/mK)	0.613	400

The thermophysical properties of base fluid and nanofluid are shown in given in Table 1.

To obtain the accurateness and validity of our numerical results, the values of $-\theta^{'}(0)/\sqrt{2}$ are obtained and compared with the published results of Rosca et al. [17], Mohamed et al. [13] and Zulkifi et al. [24] in Tables 2. It gives a favourable agreement.

Table 3 describes the computed results of different flow quantities on dimensional velocity ($f''(0)$), normalised temperature ($-\theta'(0)$) and non-dimensional nanoparticle concentration ($-S'(0)$).

Table 2 Comparison values of $-\theta^{'}(0)/\sqrt{2}$ with previous published results for different values of Pr, when $\varepsilon = Nb = N_F = Nt = Ec = Le = M = 0$

Pr	Roşca and Pop [17]	Mohamed et al. [13]	Zulkifi et al. [25]	Present
0.7	0.29268	0.29268	0.292680	0.292595
0.8	0.30691	0.306917	0.306917	0.306715
1	0.33205	0.332057	0.332057	0.332056
5	0.57668	0.576689	0.576689	0.576680
10	0.72814	0.728141	0.728141	0.728132

Table 3 .

ε	φ	Sc	Pr	N_F	Nc	N_{Re}	Ec	Nt	Nb	M	$f''(0)$	$-\theta'(0)$	$-S'(0)$
										0.1	0.48911	1.48471	−1.5088
0.5	0.01	0.1	6.2	0.1	0.1	2.0	0.1	0.1	0.1	0.5	1.05734	1.43415	−1.4614
										1.0	1.70386	1.20401	−1.2357
										1.5	2.27202	0.86595	−0.9026
										2.0	2.97838	−2.1856	2.19435
									0.1		0.48911	1.48475	−1.5088
0.5	0.01	0.1	6.2	0.1	0.1	2.0	0.1	0.1	1.0	0.1	1.16799	2.25458	−0.7732
									2.0		1.86584	5.32477	−0.7468
									3.0		2.52482	7.35009	−0.7421
								0.1			0.48911	1.48475	−1.5088
0.5	0.01	0.1	6.2	0.1	0.1	2.0	0.1	0.2	0.1	0.1	0.55239	1.32204	−2.2084
								0.3			0.60883	1.16771	−2.6089

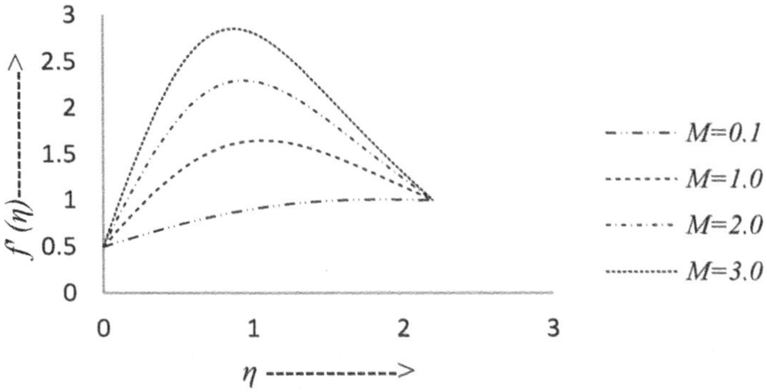

Fig. 2 Variation of Non-dimensional Velocity $f'(\eta)$ with M

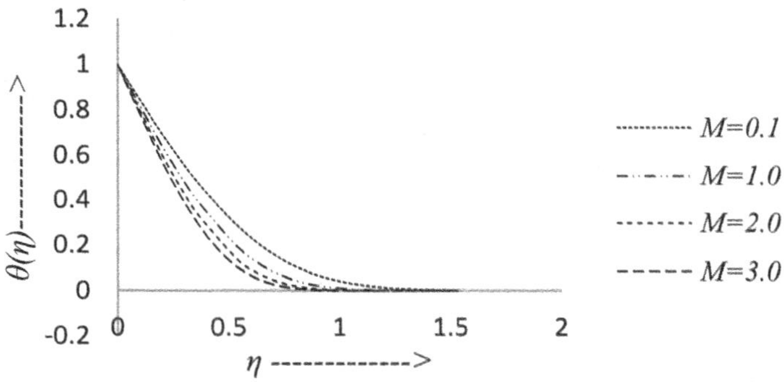

Fig. 3 Variation of Non-dimensional Temperature $\theta(\eta)$ with M

5.1 Impact of Electrification

The influence of electrification on velocity, temperature and concentration is shown in Figs. 2, 3 and 4, respectively. It is found that the velocity increases and the temperature decreases as per the increment of electrification parameter. This is due to the electrification of particle within the boundary layer that produces a drag force which accelerates the resistive-type force known as Lorentz force which accelerates the flow.

5.2 Impact of Brownian Diffusion

Figures 5, 6 and 7 illustrate the effect of Brownian diffusion on dimensionless velocity, temperature and concentration, respectively. From Fig. 5, it is found that the dimensionless velocity increases with the increase of Nb and Fig. 6 shows the

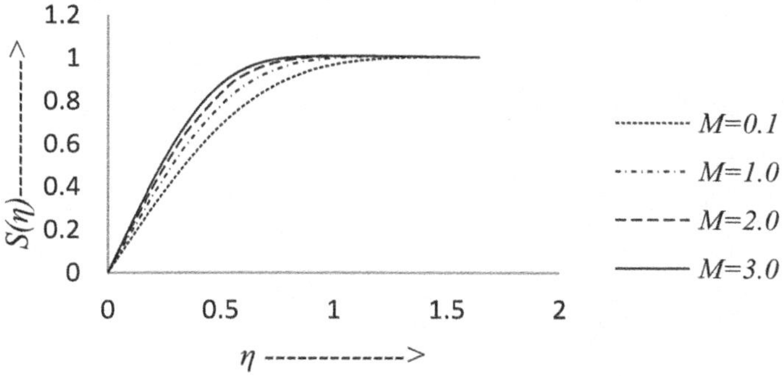

Fig. 4 Variation of Non-dimensional Concentration $S(\eta)$ with M

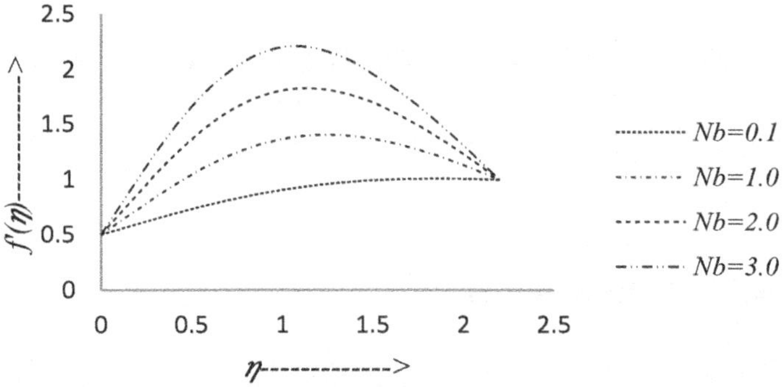

Fig. 5 Variation of Non-dimensional Velocity $f'(\eta)$ with Nb

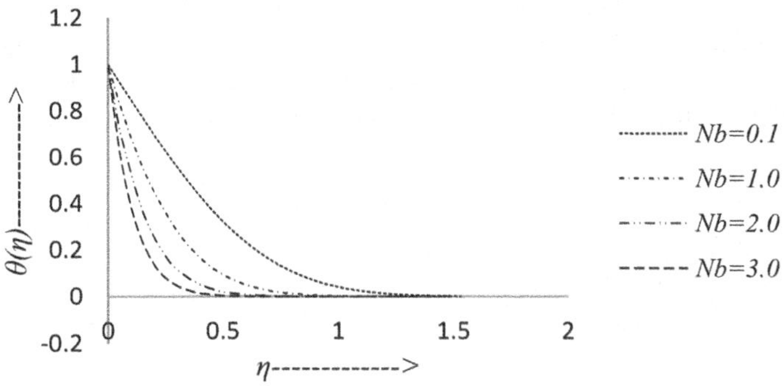

Fig. 6 Variation of Non-dimensional Temperature $\theta(\eta)$ with Nb

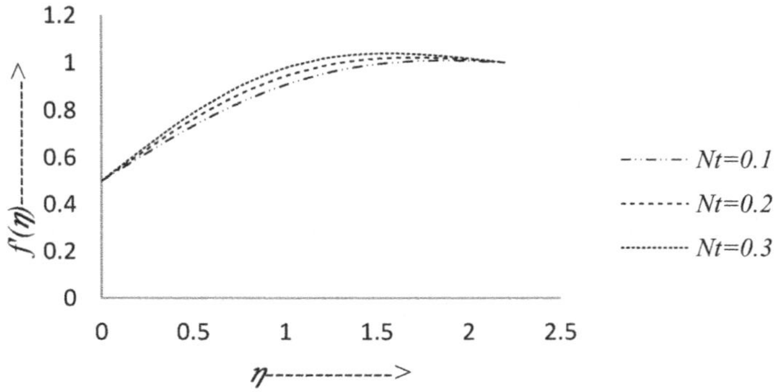

Fig. 7 Variation of Non-dimensional Velocity $f'(\eta)$ with Nt

increase of dimensionless temperature with the increase of Nb. It is observed from Fig. 7 that there is an increase in concentration with Brownian diffusion parameter Nb. For the above observation, the value of Sc is taken as 0.1, i.e. the Brownian motion is significant.

5.3 Impact of Thermophoresis

The effect of Nt on normalised velocity, temperature and concentration profile is presented in Figs. 7, 8 and 9, respectively. The figures demonstrate that normalised velocity and concentration decrease when Nt increases. From Fig. 9, it is found that the normalised temperature θ increases when Nt increases. In thermophoresis, the particle from the heated region is transferred to the cold region [25]. Hence, it causes the nanofluid temperature to be increased due to a large number of nanoparticles transferred from the hot region which enhance the fluid temperature.

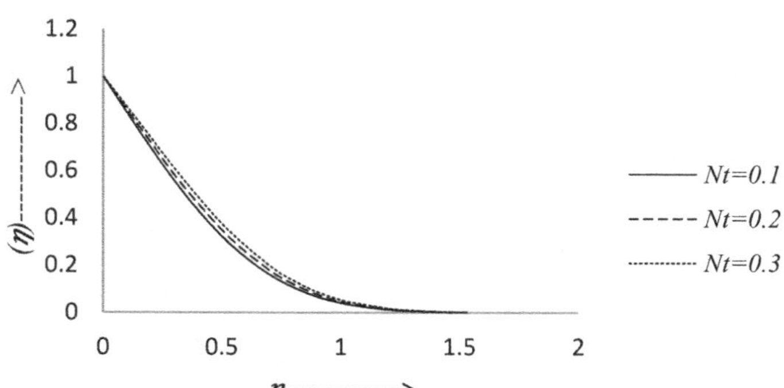

Fig. 8 Variation of Non-dimensional Temperature $\theta(\eta)$ with Nt

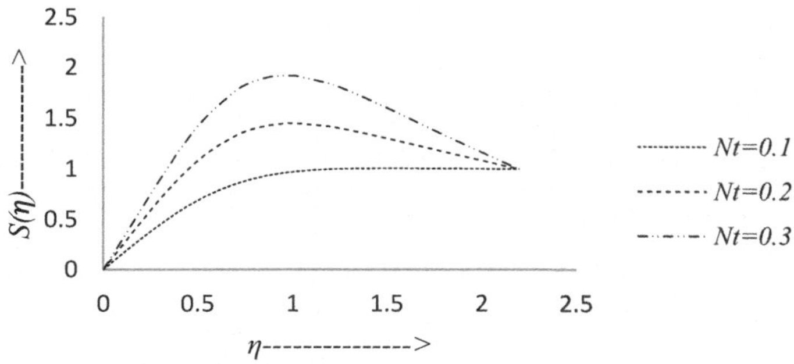

Fig. 9 Variation of Non-dimensional Concentration $S(\eta)$ with Nt

6 Conclusions

The flow and heat transfer over a continuous stretching sheet play a vital role in the field of engineering and technology due to its large number of applications. With this view, an investigation has been made to study the effects of electrification on heat and mass transfer analysis of nanofluid Buongiorno's model. The effect of various flows of parameters was discussed through graphs and tables.

i. The increment of electrification parameter (M) causes the reduction of surface temperature and increases the velocity of fluid flow. Hence, the fluid temperature and concentration are enhanced due to the effect of Lorentz's force.

ii. The Brownian motion causes the increase of temperature distribution and decrease of nanoparticle concentration, because the larger Nb corresponds to stronger random motion of nanoparticles within a fluid. So the fluid temperature and its boundary layer thickness are enhanced.

iii. Due to the increasing of thermophoresis, the boundary layer thickness gets thicker and a large number of nanoparticles are transferred from the hotter region to cold region [13]. It results in the enhancement of temperature of base fluid. Hence, due to thermophoretic effect, the fluid temperature increases and the nanoparticle concentration increases.

iv. The negative values of the concentration towards the edge of the boundary layer means that the concentration of the nanoparticles would remain zero as the negative values of the concentration of the nanoparticles are not allowed. It represents that the nanoparticles are swept from the region near the edge of the boundary layer towards the sheet.

v. Positive value of $-\theta'(0)$ represents that the flow of heat occurs from higher temperature region to lower as per Fourier's law of heat conduction. Here, due to the plate temperature $T_w > T_\infty$. the heat flows from plate to fluid and the fluid temperature gets increased. This may be due to the effect of strong Brownian diffusion which overcomes the Lorentz force.

References

1. M. Ali, F. Al-Yousef, Laminar mixed convection from a continuously moving vertical surface with suction or injection. Heat Mass Transf. **33**(4), 301–306 (1998)
2. J. Buongiorno, Convective transport in nanofluids. ASME J. Heat Transf. **128**, 240–250 (2006)
3. C.H. Chen, Forced convection over a continuous sheet with suction or injection moving in a flowing fluid. Acta Mech. **138**(1–2), 1–11 (1999)
4. S.U.S. Choi, J.A. Eastman, Enhancing thermal conductivity of fluids with nanoparticles, in *Proceedings of the ASME International Mechanical Engineering Congress and Exposition*, San Francisco, CA, USA, November 1995, vol. 12–17, pp. 99–105 (1995)
5. E.M.A. Elbashbeshy, M.A.A. Bazid, The effect of temperature dependent viscosity on heat transfer over a continuous moving surface. J. Phys. D Appl. Phys. **33**(21), 2716 (2000)
6. T. Fang, C.F. Lee, A moving wall boundary layer flow of slightly rarefied gas free stream over a moving flat plate. Appl. Math. Lett. **18**, 487–495
7. K. Gangadhar, Radiation, heat generation and viscous dissipation effects on MHD boundary layer flow for the Blasius and Sakiadis flows with a convective surface boundary condition. J. Appl. Fluid Mech. **8**(3), 559–570 (2015)
8. W.A. Khan, I. Pop, Boundary-layer flow of a nanofluid past a stretching sheet. Int. J. Heat Mass Transf. **53**, 2477–2483 (2010)
9. U. Khan, N. Ahmed, S.I.U. Khan, S.T. Mohyud-din, Thermodiffusion effects on MHD stagnation point flow towards a stretching sheet in a nanofluid. Propuls. Power Res. **3**(3), 151–158 (2014)
10. M. Kumari, G. Nath, Boundary layer development on a continuous moving surface with a parallel free stream due to impulsive motion. Heat Mass Transf. **31**(4), 283–289 (2014)
11. H.A. Masuda, K. Ebata, N. Hishinuma, Alteration of thermal conductivity and viscosity of liquid by dispersing ultra-fine particles. Netsu Bussei **7**, 227–233 (1993)
12. J. Maxwell, C. James, *A Treatise on Electricity and Magnetism*, 2nd edn. (Clarendon Press, Oxford, UK, 1873)
13. M.K.A. Mohamed, N.A.Z. Noar, M.Z. Salleh, A. Ishak, Mathematical model of boundary layer flow over a moving plate in a nanofluid with viscous dissipation. J. Appl. Fluid Mech. **9**(5), 2369–2377 (2016)
14. H.F. Oztop, E. Abu-Nada, Numerical study of natural convection in partially heated rectangular enclosures filled with nanofluids. Int. J. Heat Fluid Flow **29**, 1326–1336 (2008)
15. D. Pal, H. Mondal, Soret-Dufour effects on hydromagnetic non-darcy convective- radiative heat and mass transfer over a stretching sheet in porous medium with viscous dissipation and Ohmic heating. J. Appl. Fluid Mech. **7**(3), 513–523 (2014)
16. M.K. Partha, P. Murthy, G.P. Rajasekhar, Effect of viscous dissipation on the mixed convection heat transfer from an exponentially stretching surface. Heat Mass Transf. **41**(4), 360–366 (2005)
17. N.C. Roşca, I. Pop, Unsteady boundary layer flow of a nanofluid past a moving surface in an external uniform free stream using Buongiorno's model. Comput. Fluids **95**, 49–55 (2014)
18. B.C. Sakiadis, Boundary-layer behavior on continuous solid surfaces: I. Boundary-layer equations for two-dimensional and axisymmetric flow. Am. Inst. Chem. Eng.(AIChE) J. **7**(1), 26–28 (1961)
19. S.L. Soo, Particulates and Continuum - multiphase. Fluid Dyn. 273. SBIN 9780471970767
20. S.L. Soo, Effect of electrification on the dynamics of a particulate system. I and EC Fund **3**, 75–80 (1964)
21. K. Vajravelu, K.V. Prasad, J. Lee, C. Lee, I. Pop, R.A. Van Gorder, Convective heat transfer in the flow of viscous Ag–water and Cu–water nanofluids over a stretching surface. Int. J. Therm. Sci. **50**(5), 843–851 (2011)
22. D. Wen, L. Zhangi, Y. He, Flow and migration of nanoparticle in a single channel. Heat Mass Transf. **45**, 1061–1067 (2009)
23. Y. Yirga, B. Shankar, Effects of thermal radiation and viscous dissipation on magnetohydrodynamic stagnation point flow and heat transfer of nanofluid towards a stretching sheet. J. Nanofluids **2**(4), 283–291 (2013)

24. S.M. Zokri, N.S. Arifin, M.Z. Salleh, A.R.M. Kasim, N.F. Mohammad, W.N.S.W. Yusoff, MHD Jeffrey nanofluid past a stretching sheet with viscous dissipation effect. J. Phys.: Conf. Ser. **890**(1), 012002 (2017). IOP Publishing
25. S.N. Zulkifi, N.M. Sarif, Md.Z. Salleh, Numerical solution of boundary layer flow over a moving plate in a nanofluid with viscous dissipation- a revised model. J. Adv. Res. Fluid Mech. Thermal Sci. **56**, 287–295 (2019)

Determinantal Formulas and Recurrent Relations for Bi-Periodic Fibonacci and Lucas Polynomials

Bai-Ni Guo, Emrah Polatlı, and Feng Qi

Abstract In the paper, the authors find closed formulas and recurrent relations for bi-periodic Fibonacci polynomials and for bi-periodic Lucas polynomials in terms of the Hessenberg determinants. Consequently, the authors derive closed formulas and recurrent relations for the Fibonacci, Lucas, bi-periodic Fibonacci, and bi-periodic Lucas numbers in terms of the Hessenberg determinants.

Keywords Determinantal formula · Recurrent relation · Hessenberg determinant · Bi-periodic Fibonacci polynomial · Bi-periodic Lucas polynomial · Bi-periodic Fibonacci number · Bi-periodic Lucas number · Fibonacci number · Lucas number

1991 Mathematics Subject Classification Primary 11B39 · Secondary 11B83 · 11C20 · 11Y55

1 Introduction and Preliminaries

The Fibonacci and Lucas sequences are two important integer sequences which have been studied by many amateurs and professional mathematicians for centuries. There

---

Dedicated to Dr. Professor Silvestru Sever Dragomir at Victoria University in Australia.

B.-N. Guo
School of Mathematics and Informatics, Henan Polytechnic University, Jiaozuo 454010, Henan, China
e-mail: bai.ni.guo@gmail.com

E. Polatlı
Department Of Mathematics, Zonguldak Bülent Ecevit University, 67100 Zonguldak, Turkey
e-mail: emrahpolatli@gmail.com

F. Qi (✉)
School of Mathematical Sciences, Tianjin Polytechnic University, Tianjin 300387, China
e-mail: qifeng618@gmail.com
URL: https://qifeng618.wordpress.com

S. K. Paikray et al. (eds.), *New Trends in Applied Analysis and Computational Mathematics*, Advances in Intelligent Systems and Computing 1356,
https://doi.org/10.1007/978-981-16-1402-6_18

are numerous studies related to these sequences in the literature. They are defined, respectively, by recurrence relations

$$F_{n+2} = F_{n+1} + F_n \quad \text{and} \quad L_{n+2} = L_{n+1} + L_n \tag{1.1}$$

for $n \geq 0$, where $F_0 = 0$, $F_1 = 1$, $L_0 = 2$, and $L_1 = 1$. The values of the Fibonacci numbers F_n and the Lucas numbers L_n for $1 \leq n \leq 14$ are

$$1, \quad 1, \quad 2, \quad 3, \quad 5, \quad 8, \quad 13, \quad 21, \quad 34, \quad 55, \quad 89, \quad 144, \quad 233, \quad 377$$

and

$$1, \quad 3, \quad 4, \quad 7, \quad 11, \quad 18, \quad 29, \quad 47, \quad 76, \quad 123, \quad 199, \quad 322, \quad 521, \quad 843$$

respectively. See the monograph [8] and closely related references therein.

There are many generalizations for the Fibonacci sequence in the literature. An interesting one of them was given in [5] by

$$q_0 = 0, \quad q_1 = 1, \quad q_n = \begin{cases} aq_{n-1} + q_{n-2}, & \text{if } n \text{ is even} \\ bq_{n-1} + q_{n-2}, & \text{if } n \text{ is odd} \end{cases} \tag{1.2}$$

for $n \geq 2$, where a and b are two nonzero real numbers. Various identities and other results for the sequence $\{q_n\}_{n=0}^{\infty}$ were given in [5]. What we are interested in this paper is the generating function

$$\frac{x(1 + ax - x^2)}{1 - (ab + 2)x^2 + x^4} = \sum_{n=0}^{\infty} q_n x^n = x + ax^2 + (ab + 1)x^3 + a(ab + 2)x^4 + \cdots$$

derived in [5].

In [1], the Lucas sequence was generalized by

$$\ell_0 = 2, \quad \ell_1 = a, \quad \ell_n = \begin{cases} a\ell_{n-1} + \ell_{n-2}, & \text{if } n \text{ is odd} \\ b\ell_{n-1} + \ell_{n-2}, & \text{if } n \text{ is even} \end{cases} \tag{1.3}$$

for $n \geq 2$, where a and b are two nonzero real numbers. One of main results in [1] is the generating function

$$\begin{aligned} \frac{2 + ax - (ab + 2)x^2 + ax^3}{1 - (ab + 2)x^2 + x^4} &= \sum_{n=0}^{\infty} \ell_n x^n \\ &= 2 + ax + (ab + 2)x^2 + a(ab + 3)x^3 + \cdots \end{aligned}$$

In [37], bi-periodic Fibonacci and Lucas polynomials are defined, respectively, as

$$q_0(x) = 0, \quad q_1(x) = 1, \quad q_n(x) = \begin{cases} axq_{n-1}(x) + q_{n-2}(x), & \text{if } n \text{ is even} \\ bxq_{n-1}(x) + q_{n-2}(x), & \text{if } n \text{ is odd} \end{cases} \tag{1.4}$$

and

$$\ell_0(x) = 2, \quad \ell_1(x) = ax, \quad \ell_n(x) = \begin{cases} bx\ell_{n-1}(x) + \ell_{n-2}(x), & \text{if } n \text{ is even} \\ ax\ell_{n-1}(x) + \ell_{n-2}(x), & \text{if } n \text{ is odd} \end{cases} \tag{1.5}$$

for $n \ge 2$, where a and b are any two nonzero real numbers. The generating functions

$$Q(t;x) = \frac{t + axt^2 - t^3}{1 - (2 + abx^2)t^2 + t^4} = \sum_{n=0}^{\infty} q_n(x)t^n$$
$$= t - axt^2 + (abx^2 + 1)t^3 - ax(abx^2 + 2)t^4 + (a^2b^2x^4 + 3abx^2 + 1)t^5 + \cdots$$

and

$$L(t;x) = \frac{2 + axt - (2 + abx^2)t^2 + axt^3}{1 - (2 + abx^2)t^2 + t^4} = \sum_{n=0}^{\infty} \ell_n(x)t^n$$
$$= 2 + axt + (abx^2 + 2)t^2 + ax(abx^2 + 3)t^3 + (a^2b^2x^4 + 4abx^2 + 2)t^4 + \cdots$$

were also discovered in [37].

Inheriting from terminologies for $q_n(x)$ and $\ell_n(x)$, we call q_n and ℓ_n defined in (1.2) and (1.3) bi-periodic Fibonacci numbers and bi-periodic Lucas numbers, respectively.

A determinant $|A| = |a_{ij}|_{n\times n}$ is called a lower (or an upper, respectively) Hessenberg determinant if and only if $a_{ij} = 0$ for all pairs (i, j) such that $i + 1 < j$ (or $j + 1 < i$, respectively). For more details, please refer to [6, 9] and closely related references therein.

In the paper, we will find closed formulas and recurrent relations for bi-periodic Fibonacci polynomials $q_n(x)$ and for bi-periodic Lucas polynomials $\ell_n(x)$ in terms of some Hessenberg determinants. Consequently, we will derive closed formulas and recurrent relations for bi-periodic Fibonacci numbers q_n, for bi-periodic Lucas numbers ℓ_n, for the Fibonacci numbers F_n, and for the Lucas numbers L_n in terms of some Hessenberg determinants.

2 Determinantal Formulas

Our main results can be stated and proved as follows:

Theorem 2.1 *For $n \ge 0$, bi-periodic Fibonacci polynomials $q_n(x)$ can be formulated determinantally as*

$$q_n(x) = \frac{(-1)^n}{n!} \begin{vmatrix} 0 & 1 & 0 & 0 & \cdots & 0 & 0 & 0 & 0 \\ 1 & 0 & 1 & 0 & \cdots & 0 & 0 & 0 & 0 \\ 2ax & -2\binom{2}{0}(2+abx^2) & 0 & 1 & \cdots & 0 & 0 & 0 & 0 \\ -6 & 0 & -2\binom{3}{1}(2+abx^2) & 0 & \cdots & 0 & 0 & 0 & 0 \\ 0 & 24\binom{4}{0} & 0 & -2\binom{4}{2}(2+abx^2) & \cdots & 0 & 0 & 0 & 0 \\ 0 & 0 & 24\binom{5}{1} & 0 & \cdots & 0 & 0 & 0 & 0 \\ 0 & 0 & 0 & 24\binom{6}{2} & \cdots & 0 & 0 & 0 & 0 \\ \vdots & \vdots & \vdots & \vdots & \ddots & \vdots & \vdots & \vdots & \vdots \\ 0 & 0 & 0 & 0 & \cdots & -2\binom{n-2}{n-4}(2+abx^2) & 0 & 1 & 0 \\ 0 & 0 & 0 & 0 & \cdots & 0 & -2\binom{n-1}{n-3}(2+abx^2) & 0 & 1 \\ 0 & 0 & 0 & 0 & \cdots & 24\binom{n}{n-4} & 0 & -2\binom{n}{n-2}(2+abx^2) & 0 \end{vmatrix}.$$

Proof In [10, Sect. 2.2, p. 849], [14, p. 94], and [36, Lemma 2.1], Exercise 5) in [2, p. 40] was reformulated as the following conclusion. Let $u(t)$ and $v(t) \ne 0$ be two differentiable functions. Let $U_{(n+1)\times 1}(t)$ be an $(n+1)\times 1$ matrix whose elements $u_{k,1}(t) = u^{(k-1)}(t)$ for $1 \le k \le n+1$, let $V_{(n+1)\times n}(t)$ be an $(n+1)\times n$ matrix whose elements $v_{i,j}(t) = \binom{i-1}{j-1} v^{(i-j)}(t)$ for $1 \le i \le n+1$ and $1 \le j \le n$, and let $|W_{(n+1)\times(n+1)}(t)|$ denote the determinant of the $(n+1)\times(n+1)$ matrix $W_{(n+1)\times(n+1)}(t) = \big(U_{(n+1)\times 1}(t) \ V_{(n+1)\times n}(t)\big)$. Then the nth derivative of the ratio $\frac{u(t)}{v(t)}$ can be computed by

$$\frac{\mathrm{d}^n}{\mathrm{d}x^n}\left[\frac{u(t)}{v(t)}\right]=(-1)^n\frac{\left|W_{(n+1)\times(n+1)}(t)\right|}{v^{n+1}(t)}. \tag{2.1}$$

Letting $u(t)=t+axt^2-t^3$ and $v(t)=1-\left(2+abx^2\right)t^2+t^4$ in (2.1) yields

$$\frac{\partial^n Q(t;x)}{\partial t^n}=\frac{(-1)^n}{[1-(2+abx^2)t^2+t^4]^{n+1}}\begin{vmatrix}
t+axt^2-t^3 & 1-(2+abx^2)t^2+t^4 & 0 & 0 & \cdots & 0 & 0 & 0 & 0\\
1+2axt-3t^2 & \binom{1}{0}[4t^3-2(2+abx^2)t] & 1-(2+abx^2)t^2+t^4 & 0 & \cdots & 0 & 0 & 0 & 0\\
2ax-6t & \binom{2}{0}[12t^2-2(2+abx^2)] & \binom{2}{1}[4t^3-2(2+abx^2)t] & 1-(2+abx^2)t^2+t^4 & \cdots & 0 & 0 & 0 & 0\\
-6 & 24\binom{3}{0}t & \binom{3}{1}[12t^2-2(2+abx^2)] & \binom{3}{2}[4t^3-2(2+abx^2)t] & \cdots & 0 & 0 & 0 & 0\\
0 & 24\binom{4}{0} & 24\binom{4}{1}t & \binom{4}{2}[12t^2-2(2+abx^2)] & \cdots & 0 & 0 & 0 & 0\\
0 & 0 & 24\binom{5}{1} & 24\binom{5}{2}t & \cdots & 0 & 0 & 0 & 0\\
0 & 0 & 0 & 24\binom{6}{2} & \cdots & 0 & 0 & 0 & 0\\
\vdots & \vdots & \vdots & \vdots & \ddots & \vdots & \vdots & \vdots & \vdots\\
0 & 0 & 0 & 0 & \cdots & \binom{n-2}{n-4}[12t^2-2(2+abx^2)] & \binom{n-2}{n-3}[4t^3-2(2+abx^2)t] & 1-(2+abx^2)t^2+t^4 & 0\\
0 & 0 & 0 & 0 & \cdots & 24\binom{n-1}{n-4}t & \binom{n-1}{n-3}[12t^2-2(2+abx^2)] & \binom{n-1}{n-2}[4t^3-2(2+abx^2)t] & 1-(2+abx^2)t^2+t^4\\
0 & 0 & 0 & 0 & \cdots & 24\binom{n}{n-4} & 24\binom{n}{n-3}t & \binom{n}{n-2}[12t^2-2(2+abx^2)] & \binom{n}{n-1}[4t^3-2(2+abx^2)t]
\end{vmatrix}.$$

Letting $t\to 0$ for $n\in\mathbb{N}$ in the above equation, we obtain

$$
\lim_{t\to 0}\frac{\partial^n Q(t;x)}{\partial t^n}=(-1)^n
\begin{vmatrix}
0 & 1 & 0 & 0 & \cdots & 0 & 0 & 0 & 0\\
1 & 0 & 1 & 0 & \cdots & 0 & 0 & 0 & 0\\
2ax & -2\binom{2}{0}(2+abx^2) & 0 & 1 & \cdots & 0 & 0 & 0 & 0\\
-6 & 0 & -2\binom{3}{1}(2+abx^2) & 0 & \cdots & 0 & 0 & 0 & 0\\
0 & 24\binom{4}{0} & 0 & -2\binom{4}{2}(2+abx^2) & \cdots & 0 & 0 & 0 & 0\\
0 & 0 & 24\binom{5}{1} & 0 & \cdots & 0 & 0 & 0 & 0\\
0 & 0 & 0 & 24\binom{6}{2} & \cdots & 0 & 0 & 0 & 0\\
\vdots & \vdots & \vdots & \vdots & \ddots & \vdots & \vdots & \vdots & \vdots\\
0 & 0 & 0 & 0 & \cdots & -2\binom{n-2}{n-4}(2+abx^2) & 0 & 1 & 0\\
0 & 0 & 0 & 0 & \cdots & 0 & -2\binom{n-1}{n-3}(2+abx^2) & 0 & 1\\
0 & 0 & 0 & 0 & \cdots & 24\binom{n}{n-4} & 0 & -2\binom{n}{n-2}(2+abx^2) & 0
\end{vmatrix}.
$$

The proof of Theorem 2.1 is thus complete. □

Corollary 2.1 *The bi-periodic Fibonacci numbers q_n and the Fibonacci numbers F_n for $n \ge 0$ can be determinantally formulated by*

$$
q_n=\frac{(-1)^n}{n!}
\begin{vmatrix}
0 & 1 & 0 & 0 & \cdots & 0 & 0 & 0 & 0\\
1 & 0 & 1 & 0 & \cdots & 0 & 0 & 0 & 0\\
2a & -2\binom{2}{0}(2+ab) & 0 & 1 & \cdots & 0 & 0 & 0 & 0\\
-6 & 0 & -2\binom{3}{1}(2+ab) & 0 & \cdots & 0 & 0 & 0 & 0\\
0 & 24\binom{4}{0} & 0 & -2\binom{4}{2}(2+ab) & \cdots & 0 & 0 & 0 & 0\\
0 & 0 & 24\binom{5}{1} & 0 & \cdots & 0 & 0 & 0 & 0\\
0 & 0 & 0 & 24\binom{6}{2} & \cdots & 0 & 0 & 0 & 0\\
\vdots & \vdots & \vdots & \vdots & \ddots & \vdots & \vdots & \vdots & \vdots\\
0 & 0 & 0 & 0 & \cdots & -2\binom{n-2}{n-4}(2+ab) & 0 & 1 & 0\\
0 & 0 & 0 & 0 & \cdots & 0 & -2\binom{n-1}{n-3}(2+ab) & 0 & 1\\
0 & 0 & 0 & 0 & \cdots & 24\binom{n}{n-4} & 0 & -2\binom{n}{n-2}(2+ab) & 0
\end{vmatrix}.
$$

and

$$F_n = \frac{(-1)^n}{n!}\begin{vmatrix} 0 & 1 & 0 & 0 & \cdots & 0 & 0 & 0 & 0 \\ 1 & 0 & 1 & 0 & \cdots & 0 & 0 & 0 & 0 \\ 2 & -6\binom{2}{0} & 0 & 1 & \cdots & 0 & 0 & 0 & 0 \\ -6 & 0 & -6\binom{3}{1} & 0 & \cdots & 0 & 0 & 0 & 0 \\ 0 & 24\binom{4}{0} & 0 & -6\binom{4}{2} & \cdots & 0 & 0 & 0 & 0 \\ 0 & 0 & 24\binom{5}{1} & 0 & \cdots & 0 & 0 & 0 & 0 \\ 0 & 0 & 0 & 24\binom{6}{2} & \cdots & 0 & 0 & 0 & 0 \\ \vdots & \vdots & \vdots & \vdots & \ddots & \vdots & \vdots & \vdots & \vdots \\ 0 & 0 & 0 & 0 & \cdots & -6\binom{n-2}{n-4} & 0 & 1 & 0 \\ 0 & 0 & 0 & 0 & \cdots & 0 & -6\binom{n-1}{n-3} & 0 & 1 \\ 0 & 0 & 0 & 0 & \cdots & 24\binom{n}{n-4} & 0 & -6\binom{n}{n-2} & 0 \end{vmatrix}. \tag{2.2}$$

Proof These follow from taking $x \to 1$ and letting $(a, b) \to (1, 1)$ in Theorem 2.1, respectively. □

Theorem 2.2 *For $n \in \{0\} \cup \mathbb{N}$, bi-periodic Lucas polynomials $\ell_n(x)$ can be formulated determinantally as*

$$\ell_n(x) = \frac{(-1)^n}{n!}\begin{vmatrix} 2 & 1 & 0 & 0 & \cdots & 0 & 0 & 0 & 0 \\ ax & 0 & 1 & 0 & \cdots & 0 & 0 & 0 & 0 \\ -2(2+abx^2) & -2\binom{2}{0}(2+abx^2) & 0 & 1 & \cdots & 0 & 0 & 0 & 0 \\ 6ax & 0 & -2\binom{3}{1}(2+abx^2) & 0 & \cdots & 0 & 0 & 0 & 0 \\ 0 & 24\binom{4}{0} & 0 & -2\binom{4}{2}(2+abx^2) & \cdots & 0 & 0 & 0 & 0 \\ 0 & 0 & 24\binom{5}{1} & 0 & \cdots & 0 & 0 & 0 & 0 \\ 0 & 0 & 0 & 24\binom{6}{2} & \cdots & 0 & 0 & 0 & 0 \\ \vdots & \vdots & \vdots & \vdots & \ddots & \vdots & \vdots & \vdots & \vdots \\ 0 & 0 & 0 & 0 & \cdots & -2\binom{n-2}{n-4}(2+abx^2) & 0 & 1 & 0 \\ 0 & 0 & 0 & 0 & \cdots & 0 & -2\binom{n-1}{n-3}(2+abx^2) & 0 & 1 \\ 0 & 0 & 0 & 0 & \cdots & 24\binom{n}{n-4} & 0 & -2\binom{n}{n-2}(2+abx^2) & 0 \end{vmatrix}.$$

Proof Applying $u(t) = 2 + axt - \left(2 + abx^2\right)t^2 + axt^3$ and $v(t) = 1 - \left(2 + abx^2\right)$ $t^2 + t^4$ in (2.1) results in

$$
\begin{aligned}
\frac{\partial^n L(t;x)}{\partial t^n} &= \frac{(-1)^n}{[1-(2+abx^2)t^2+t^4]^{n+1}}\\
&\times\begin{vmatrix}
2+axt-(2+abx^2)t^2+axt^3 & 1-(2+abx^2)t^2+t^4 & 0 & 0 & 0 & \cdots & 0 & 0 & 0 & 0\\
ax-2(2+abx^2)t+3axt^2 & \binom{1}{0}[4t^3-2(2+abx^2)t] & 1-(2+abx^2)t^2+t^4 & 0 & 0 & \cdots & 0 & 0 & 0 & 0\\
-2(2+abx^2)+6axt & \binom{2}{0}[12t^2-2(2+abx^2)] & \binom{2}{1}[4t^3-2(2+abx^2)t] & 1-(2+abx^2)t^2+t^4 & 0 & \cdots & 0 & 0 & 0 & 0\\
6ax & 24\binom{3}{0}t & \binom{3}{1}[12t^2-2(2+abx^2)] & \binom{3}{2}[4t^3-2(2+abx^2)t] & 1-(2+abx^2)t^2+t^4 & \cdots & 0 & 0 & 0 & 0\\
0 & 24\binom{4}{0} & 24\binom{4}{1}t & \binom{4}{2}[12t^2-2(2+abx^2)] & \binom{4}{3}[4t^3-2(2+abx^2)t] & \cdots & 0 & 0 & 0 & 0\\
0 & 0 & 24\binom{5}{1} & 24\binom{5}{2}t & \binom{5}{3}[12t^2-2(2+abx^2)] & \cdots & 0 & 0 & 0 & 0\\
0 & 0 & 0 & 24\binom{6}{2} & 24\binom{6}{3}t & \cdots & 0 & 0 & 0 & 0\\
\vdots & \vdots & \vdots & \vdots & \vdots & \ddots & \vdots & \vdots & \vdots & \vdots\\
0 & 0 & 0 & 0 & 0 & \cdots & \binom{n-2}{n-4}[12t^2-2(2+abx^2)] & \binom{n-2}{n-3}[4t^3-2(2+abx^2)t] & 1-(2+abx^2)t^2+t^4 & 0\\
0 & 0 & 0 & 0 & 0 & \cdots & 24\binom{n-1}{n-4}t & \binom{n-1}{n-3}[12t^2-2(2+abx^2)] & \binom{n-1}{n-2}[4t^3-2(2+abx^2)t] & 1-(2+abx^2)t^2+t^4\\
0 & 0 & 0 & 0 & 0 & \cdots & 24\binom{n}{n-4} & 24\binom{n}{n-3}t & \binom{n}{n-2}[12t^2-2(2+abx^2)] & \binom{n}{n-1}[4t^3-2(2+abx^2)t]
\end{vmatrix}.
\end{aligned}
$$

Letting $t \to 0$ in the above equation yields

$$
\lim_{t\to 0}\frac{\partial^n L(t;x)}{\partial t^n}=(-1)^n
\begin{vmatrix}
2 & 1 & 0 & 0 & \cdots & 0 & 0 & 0 & 0\\
ax & 0 & 1 & 0 & \cdots & 0 & 0 & 0 & 0\\
-2(2+abx^2) & -2\binom{2}{0}(2+abx^2) & 0 & 1 & \cdots & 0 & 0 & 0 & 0\\
6ax & 0 & -2\binom{3}{1}(2+abx^2) & 0 & \cdots & 0 & 0 & 0 & 0\\
0 & 24\binom{4}{0} & 0 & -2\binom{4}{2}(2+abx^2) & \cdots & 0 & 0 & 0 & 0\\
0 & 0 & 24\binom{5}{1} & 0 & \cdots & 0 & 0 & 0 & 0\\
0 & 0 & 0 & 24\binom{6}{2} & \cdots & 0 & 0 & 0 & 0\\
\vdots & \vdots & \vdots & \vdots & \ddots & \vdots & \vdots & \vdots & \vdots\\
0 & 0 & 0 & 0 & \cdots & -2\binom{n-2}{n-4}(2+abx^2) & 0 & 1 & 0\\
0 & 0 & 0 & 0 & \cdots & 0 & -2\binom{n-1}{n-3}(2+abx^2) & 0 & 1\\
0 & 0 & 0 & 0 & \cdots & 24\binom{n}{n-4} & 0 & -2\binom{n}{n-2}(2+abx^2) & 0
\end{vmatrix}.
$$

The proof of Theorem 2.2 is thus complete. □

Corollary 2.2 *The bi-periodic Lucas numbers ℓ_n and the Lucas numbers L_n for $n \ge 0$ can be determinantally formulated by*

$$
\ell_n=\frac{(-1)^n}{n!}
\begin{vmatrix}
2 & 1 & 0 & 0 & \cdots & 0 & 0 & 0 & 0\\
a & 0 & 1 & 0 & \cdots & 0 & 0 & 0 & 0\\
-2(2+ab) & -2\binom{2}{0}(2+ab) & 0 & 1 & \cdots & 0 & 0 & 0 & 0\\
6a & 0 & -2\binom{3}{1}(2+ab) & 0 & \cdots & 0 & 0 & 0 & 0\\
0 & 24\binom{4}{0} & 0 & -2\binom{4}{2}(2+ab) & \cdots & 0 & 0 & 0 & 0\\
0 & 0 & 24\binom{5}{1} & 0 & \cdots & 0 & 0 & 0 & 0\\
0 & 0 & 0 & 24\binom{6}{2} & \cdots & 0 & 0 & 0 & 0\\
\vdots & \vdots & \vdots & \vdots & \ddots & \vdots & \vdots & \vdots & \vdots\\
0 & 0 & 0 & 0 & \cdots & -2\binom{n-2}{n-4}(2+ab) & 0 & 1 & 0\\
0 & 0 & 0 & 0 & \cdots & 0 & -2\binom{n-1}{n-3}(2+ab) & 0 & 1\\
0 & 0 & 0 & 0 & \cdots & 24\binom{n}{n-4} & 0 & -2\binom{n}{n-2}(2+ab) & 0
\end{vmatrix}.
$$

and

$$L_n = \frac{(-1)^n}{n!} \begin{vmatrix} 2 & 1 & 0 & 0 & \cdots & 0 & 0 & 0 & 0 \\ 1 & 0 & 1 & 0 & \cdots & 0 & 0 & 0 & 0 \\ -6 & -6\binom{2}{0} & 0 & 1 & \cdots & 0 & 0 & 0 & 0 \\ 6 & 0 & -6\binom{3}{1} & 0 & \cdots & 0 & 0 & 0 & 0 \\ 0 & 24\binom{4}{0} & 0 & -6\binom{4}{2} & \cdots & 0 & 0 & 0 & 0 \\ 0 & 0 & 24\binom{5}{1} & 0 & \cdots & 0 & 0 & 0 & 0 \\ 0 & 0 & 0 & 24\binom{6}{2} & \cdots & 0 & 0 & 0 & 0 \\ \vdots & \vdots & \vdots & \vdots & \ddots & \vdots & \vdots & \vdots & \vdots \\ 0 & 0 & 0 & 0 & \cdots & -6\binom{n-2}{n-4} & 0 & 1 & 0 \\ 0 & 0 & 0 & 0 & \cdots & 0 & -6\binom{n-1}{n-3} & 0 & 1 \\ 0 & 0 & 0 & 0 & \cdots & 24\binom{n}{n-4} & 0 & -6\binom{n}{n-2} & 0 \end{vmatrix}.$$

Proof These follow from taking $x \to 1$ and letting $(a, b) \to (1, 1)$ in Theorem 2.2, respectively. □

3 Recurrent Relations

In this section, we derive, by a different method, several recurrent relations.

Theorem 3.1 *For $n \geq 5$, we have the recurrent relations*

$$q_{n+4}(x) = \left(2 + abx^2\right)q_{n+2}(x) - q_n(x), \tag{3.1}$$

$$\ell_{n+4}(x) = \left(2 + abx^2\right)\ell_{n+2}(x) - \ell_n(x), \tag{3.2}$$

$$q_{n+4} = (2 + ab)q_{n+2} - q_n, \quad \ell_{n+4} = (2 + ab)\ell_{n+2} - \ell_n, \tag{3.3}$$

and

$$F_{n+4} = 3F_{n+2} - F_n, \quad L_{n+4} = 3L_{n+2} - L_n. \tag{3.4}$$

Proof Let $D_0 = 1$ and

$$D_n = \begin{vmatrix} e_{1,1} & e_{1,2} & 0 & \dots & 0 & 0 \\ e_{2,1} & e_{2,2} & e_{2,3} & \dots & 0 & 0 \\ e_{3,1} & e_{3,2} & e_{3,3} & \dots & 0 & 0 \\ \vdots & \vdots & \vdots & \vdots & \vdots & \vdots \\ e_{n-2,1} & e_{n-2,2} & e_{n-2,3} & \dots & e_{n-2,n-1} & 0 \\ e_{n-1,1} & e_{n-1,2} & e_{n-1,3} & \dots & e_{n-1,n-1} & e_{n-1,n} \\ e_{n,1} & e_{n,2} & e_{n,3} & \dots & e_{n,n-1} & e_{n,n} \end{vmatrix}$$

for $n \in \mathbb{N}$. In [3, p. 222, Theorem], it was proved that the sequence D_n for $n \geq 0$ satisfies $D_1 = e_{1,1}$ and

$$D_n = \sum_{r=1}^{n}(-1)^{n-r} e_{n,r}\left(\prod_{j=r}^{n-1} e_{j,j+1}\right) D_{r-1} \tag{3.5}$$

for $n \geq 2$, where the empty product is understood to be 1. See also [20, Lemma 5], [15, Lemma 2], [21, Lemma 2], and [26, Remark 3]. Applying the recurrent relation (3.5) to Theorem 2.1 reveals that

$$\begin{aligned} (-1)^{n-1}(n-1)!q_{n-1}(x) = 2\binom{n-1}{n-3}(2+abx^2)(-1)^{n-3}(n-3)!q_{n-3}(x) \\ - 24\binom{n-1}{n-5}(-1)^{n-5}(n-5)!q_{n-5}(x) \end{aligned}$$

for $n \geq 10$, which can be simplified as (3.1).

Similarly, applying the recurrent relation (3.5) to Theorem 2.2 and simplifying lead to the recurrent relation (3.2).

The recurrent relations in (3.3) and (3.4) follows from taking $x \to 1$ and letting $(a, b) \to (1, 1)$ in (3.1) and (3.2) respectively. The proof of Theorem 3.1 is complete. □

4 Remarks

In this section, we list several remarks on our main results and related things.

Remark 4.1 For more information on applications of the formula (2.1), please refer to the papers [7, 11, 13–17, 19, 22–26, 28, 30, 31, 33, 34, 36] and closely related references therein.

Remark 4.2 The determinantal formula (2.2) is different from those collected in [4, 12, 18, 27, 29, 32] and closely related references therein.

Remark 4.3 The recurrent relations (3.1), (3.2), (3.3), and (3.4) in Theorem 3.1 are different from corresponding ones in (1.1), (1.2), (1.3), (1.4), and (1.5).

Remark 4.4 Some of recurrent relations in Theorem 3.1 appeared in [1, Lemma 1], [5, p. 643], and [37, Eq. (7)].

Remark 4.5 One of the anonymous referees pointed out that a small misprint in the reference [6] has been detected.

Remark 4.6 This paper is a slightly revised version of the electronic preprint [35].

Acknowledgements The authors appreciate anonymous referees for their careful corrections to and valuable comments on the original version of this paper.

References

1. G. Bilgici, Two generalizations of Lucas sequence. Appl. Math. Comput. **245**, 526–538 (2014). https://doi.org/10.1016/j.amc.2014.07.111
2. N. Bourbaki, *Elements of Mathematics: Functions of a Real Variable: Elementary Theory* (Springer, Berlin, 2004). Translated from the 1976 French original by Philip Spain. Elements of Mathematics (Berlin). https://doi.org/10.1007/978-3-642-59315-4
3. N.D. Cahill, J.R. D'Errico, D.A. Narayan, J.Y. Narayan, Fibonacci determinants. College Math. J. **33**(3), 221–225 (2002). https://doi.org/10.2307/1559033
4. M.C. Dağlı, F. Qi, Several closed and determinantal forms for convolved Fibonacci numbers. J. Appl. Math. Stat. Inform. **17**(1) (2021). In press; OSF Preprints (2020). https://doi.org/10.31219/osf.io/e25yb
5. M. Edson, O. Yayenie,A new generalization of Fibonacci sequence and extended Binet's formula. Integers **9**, A48, 639–654 (2009). https://doi.org/10.1515/INTEG.2009.051
6. V. Hingins, C. Johnson, Inverse spectral problems for collections of leading principal submatrices of tridiagonal matrices. Linear Algebr. Appl. **489**, 104–122 (2016). https://doi.org/10.1016/j.laa.2015.10.004
7. C. Kızılateş, W.-S. Du, F. Qi, Several determinantal expressions of generalized Tribonacci polynomials and sequences. Tamkang J. Math. **53** (2022), In press; Available online at http://dx.doi.org/10.5556/j.tkjm.53.2022.3743
8. T. Koshy, Fibonacci and Lucas Numbers with Applications. Pure and Applied Mathematics (Wiley-Interscience, New York, 2001). https://doi.org/10.1002/9781118033067
9. R.S. Martin, J.H. Wilkinson, Handbook series linear algebra: similarity reduction of a general matrix to Hessenberg form. Numer. Math. **12**(5), 349–368 (1968). https://doi.org/10.1007/BF02161358
10. F. Qi, Derivatives of tangent function and tangent numbers. Appl. Math. Comput. **268**, 844–858 (2015). https://doi.org/10.1016/j.amc.2015.06.123
11. F. Qi, Determinantal expressions and recurrence relations for Fubini and Eulerian polynomials. J. Interdiscip. Math. **22**(3), 317–335 (2019). https://doi.org/10.1080/09720502.2019.1624063
12. F. Qi, Three closed forms for convolved Fibonacci numbers. Results Nonlinear Anal. **3**(4), 185–195 (2020)
13. F. Qi, A determinantal expression and a recursive relation of the Delannoy numbers. Acta Univ. Sapientiae Math. **13**(1) (2021). In press, arXiv:2003.12572
14. F. Qi, R.J. Chapman, Two closed forms for the Bernoulli polynomials, J. Number Theory **159**, 89–100 (2016). https://doi.org/10.1016/j.jnt.2015.07.021

15. F. Qi, B.-N. Guo, Some determinantal expressions and recurrence relations of the Bernoulli polynomials. Mathematics **4**(4), 11 pp. (2016). Article 65. https://doi.org/10.3390/math4040065
16. F. Qi, B.-N. Guo, A determinantal expression and a recurrence relation for the Euler polynomials. Adv. Appl. Math. Sci. **16**(9), 297–309 (2017)
17. F. Qi, B.-N. Guo, Explicit and recursive formulas, integral representations, and properties of the large Schröder numbers. Kragujevac J. Math. **41**(1), 121–141 (2017). https://doi.org/10.5937/KgJMath1701121F
18. F. Qi, B.-N. Guo, Expressing the generalized Fibonacci polynomials in terms of a tridiagonal determinant. Matematiche (Catania) **72**(1), 167–175 (2017). https://doi.org/10.4418/2017.72.1.13
19. F. Qi, B.-N. Guo, Two nice determinantal expressions and a recurrence relation for the Apostol–Bernoulli polynomials. J. Indones. Math. Soc. (MIHMI) **23**(1), 81–87 (2017). https://doi.org/10.22342/jims.23.1.274.81-87
20. F. Qi, B.-N. Guo, *A diagonal recurrence relation for the Stirling numbers of the first kind*. Appl. Anal. Discret. Math. **12**(1), 153–165 (2018). https://doi.org/10.2298/AADM170405004Q
21. F. Qi, J.-L. Zhao, Some properties of the Bernoulli numbers of the second kind and their generating function. Bull. Korean Math. Soc. **55**(6), 1909–1920 (2018). https://doi.org/10.4134/BKMS.b180039
22. F. Qi, B.-N. Guo, Several explicit and recursive formulas for generalized Motzkin numbers. AIMS Math. **5**(2), 1333–1345 (2020). https://doi.org/10.3934/math.2020091
23. F. Qi, C.-J. Huang, Computing sums in terms of beta, polygamma, and Gauss hypergeometric functions. Rev. R. Acad. Cienc. Exactas Fís. Nat. Ser. A Mat. RACSAM **114**, 9 pp. (2020). Article 191. https://doi.org/10.1007/s13398-020-00927-y
24. F. Qi, X.-T. Shi, B.-N. Guo, Two explicit formulas of the Schröder numbers. Integers **16**, 15 pp. (2016). Paper No. A23
25. F. Qi, X.-T. Shi, B.-N. Guo, Some properties of the Schröder numbers. Indian J. Pure Appl. Math. **47**(4), 717–732 (2016). https://doi.org/10.1007/s13226-016-0211-6
26. F. Qi, J.-L. Zhao, B.-N. Guo, Closed forms for derangement numbers in terms of the Hessenberg determinants. Rev. R. Acad. Cienc. Exactas Fís. Nat. Ser. A Mat. RACSAM **112**(4), 933–944 (2018). https://doi.org/10.1007/s13398-017-0401-z
27. F. Qi, V. Čerňanová, X.-T. Shi, B.-N. Guo, Some properties of central Delannoy numbers. J. Comput. Appl. Math. **328**, 101–115 (2018). https://doi.org/10.1016/j.cam.2017.07.013
28. F. Qi, J.-L. Wang, B.-N. Guo, A representation for derangement numbers in terms of a tridiagonal determinant. Kragujevac J. Math. **42**(1), 7–14 (2018). https://doi.org/10.5937/KgJMath1801007F
29. F. Qi, V. Čerňanová, Y. S. Semenov, Some tridiagonal determinants related to central Delannoy numbers, the Chebyshev polynomials, and the Fibonacci polynomials. Politehn. Univ. Bucharest Sci. Bull. Ser. A Appl. Math. Phys. **81**(1), 123–136 (2019)
30. F. Qi, C. Kızılateş, W.-S. Du, A closed formula for the Horadam polynomials in terms of a tridiagonal determinant. Symmetry **11**(6), 8 pp. (2019). https://doi.org/10.3390/sym11060782
31. F. Qi, D.-W. Niu, B.-N. Guo, Some identities for a sequence of unnamed polynomials connected with the Bell polynomials. Rev. R. Acad. Cienc. Exactas Fís. Nat. Ser. A Math. RACSAM **113**(2), 557–567 (2019). https://doi.org/10.1007/s13398-018-0494-z
32. F. Qi, J.-L. Wang, B.-N. Guo, A determinantal expression for the Fibonacci polynomials in terms of a tridiagonal determinant. Bull. Iran. Math. Soc. **45**(6), 1821–1829 (2019). https://doi.org/10.1007/s41980-019-00232-4
33. F. Qi, O. Kouba, I. Kaddoura, Computation of several Hessenberg determinants. Math. Slovaca **70**(6), 1521–1537 (2020). Available online at https://doi.org/10.1515/ms-2017-0445
34. F. Qi, M.C. Dağlı, W.-S. Du, Determinantal forms and recursive relations of the Delannoy two-functional sequence. Adv. Theory Nonlinear Anal. Appl. **4**(3), 184–193 (2020). https://doi.org/10.31197/atnaa.772734
35. F. Qi, E. Polatlı, B.-N. Guo, Determinantal formulas and recurrent relations for bi-periodic Fibonacci and Lucas polynomials. OSF Preprints (2020). https://doi.org/10.31219/osf.io/qvxd8

36. C.-F. Wei, F. Qi, Several closed expressions for the Euler numbers. J. Inequal. Appl., 8 pp. (2015). Paper No. 219. https://doi.org/10.1186/s13660-015-0738-9
37. N. Yılmaz, A. Coşkun, N. Taşkara, On properties of bi-periodic Fibonacci and Lucas polynomials. AIP Conf. Proc. **1863**, 310002 (2017). https://doi.org/10.1063/1.4992478

PBIB Designs and Their Association Schemes Arising from Minimum Bi-Independent Dominating Sets of Circulant Graphs

B. Chaluvaraju and S. A. Diwakar

Abstract A set $D \subseteq V(G)$ is a bi-independent dominating (BID) set of a graph $G = (V, E)$, if $\Delta(\langle D \rangle) = \Delta(\langle V - D \rangle) = 0$, where $\Delta(G)$ is the degree of the vertex with the greatest number of edges incident to it. The bi-independent domination number $\gamma_{bi}(G)$ is the smallest cardinality of a BID-set of G. A BID-set D of G with $|D| = \gamma_{bi}(G)$ is called γ_{bi}-set. In this paper, we obtain the total number of γ_{bi}-set along with partially balanced incomplete block (PBIB) designs with its association schemes arising from the γ_{bi}-sets in different jump sizes of some circulant graphs.

Keywords Association schemes · PBIB designs · Bi-independent dominating sets · Circulant graph

2010 Mathematics Subject Classification 05C51 · 05E30 · 05C69

1 Introduction

By a graph $G = (V, E)$, we mean a $V = V(G) \neq \phi$, $|V| = \infty$ and $E = E(G) = \{X \subseteq V(G) : |X| = 2\}$ called vertex set and edge set of G, respectively. For graph-theoretic terminology and notation not defined here, we follow [8].

For a given positive integer p, let $s_1, s_2, \ldots, s_t$ be a sequence of integers where

$$0 < s_1 < s_2 < \cdots < s_t < \frac{p+1}{2}.$$

B. Chaluvaraju (✉) · S. A. Diwakar
Department of Mathematics, Bangalore University, Jnana Bharathi Campus, Bengaluru 560056, India
e-mail: bchaluvaraju@gmail.com

S. A. Diwakar
e-mail: dsatumkur@gmail.com

S. K. Paikray et al. (eds.), *New Trends in Applied Analysis and Computational Mathematics*, Advances in Intelligent Systems and Computing 1356,
https://doi.org/10.1007/978-981-16-1402-6_19

Then the circulant graph $C_p(s_1, s_2, \ldots, s_t)$ for $1 \leq h \leq t$ is the graph on p vertices $v_1, v_2, \ldots, v_p$ with vertex v_h adjacent to each vertex $v_{h \pm s_l (mod\ p)}$. The values of s_t are called jump sizes. For more details, we refer to [17].

Partially Balanced Incomplete Block (PBIB) designs are the class of binary, equireplicate, and proper designs and these can be arranged into the varieties based on their association schemes, which was introduced by Bose and Nair [1]. By using, these can be grouped as first-, second-, third-, and m-associate classes of PBIB designs with $1 \leq k \leq m$. Further, based on the group and the type of underlying association scheme, there are other subgroups that can be made.

The following conditions satisfy m classes of association scheme on ν vertices (elements or objects) are

(i) If the associates are symmetric, then any two vertices are mth associates, where $1 \leq k \leq m$.
(ii) Each vertex x contains n_k kth associates, the number n_k being independent of vertex x.
(iii) If two vertices x and y are kth associates, then the number of vertices which are ath associates of x and bth associates of y is p^k_{ab} and is independent of the kth associates x and y. Hence, $p^k_{ab} = p^k_{ba}$.

A PBIB design contains ν vertices, ρ sets (blocks), g size with $g < \nu$ such that

(i) ν vertices having r blocks exactly and g distinct vertices contained in every block.
(ii) If x and y are of kth associates which are exactly in λ_k blocks with $1 \leq k \leq m$.

PBIB designs are used in wireless sensor networks particularly in key predistribution, visual cryptology via secret sharing schemes, traitor tracing in schemes with restricted access, one can refer [11, 15].

The parameters of the first kind are the numbers as $\nu, \rho, g, r, \lambda_1, \lambda_2, \ldots, \lambda_m$ and the parameters of the second kind are the numbers as $n_1, n_2, \ldots, n_m, p^k_{ab}$ with $1 \leq a, b, k \leq m$, respectively. For more details, we refer to [6].

A strongly regular graph with parameters (p, l, ω, μ) is a finite graph on p vertices, without loops or multiple edges, regular of degree l (with $0 < l < p - 1$, so that there are both edges and no edges), and such that any two distinct vertices have ω common neighbors when they are adjacent, and μ common neighbors when they are nonadjacent. For the related concepts of the strongly regular graph, we refer to [6].

A set $D \subseteq V$ is a dominating set of a graph G if every vertex not in D is adjacent to at least one in D. The domination number $\gamma(G)$ of a graph G is the cardinality of the smallest dominating set of G. The theory of domination and its related concepts, we refer to [7, 9, 10].

In [12], Kulli et al., initiate the study of Bi-independent dominating set is as, a dominating set D of vertices is a bi-independent dominating set (BID-set) of a graph G, if the induced subgraph of both $\Delta(\langle D \rangle) = \Delta(\langle D - V \rangle) = 0$. Further, a bi-independent domination number $\gamma_{bi}(G)$ of a graph G is the smallest BID-set. A minimum BID-set of a graph G with $|D| = \gamma_{bi}(G)$ is called γ_{bi}-set.

The concept of the total number of dominating set of graph G was introduced by Slater [16], where he denoted by $HED(G)$ in honor of Steve Hedetniemi. In this article, we will use $\tau_{bi}(G)$ to denote the total number of γ_{bi}-set of a graph G. For more details on related concepts, we refer to [2–5, 13, 14, 18, 19].

Generally, the PBIB designs obtained from the families of strongly regular graphs. Surprisingly, we found that the total number of γ_{bi}-sets in a certain class of circulant graphs apart from strongly regular graphs which are the blocks of PBIB design with almost $\lfloor \frac{p}{2} \rfloor$-association schemes.

1.1 Circulant Graph $C_p(1)$

The jump size of circulant graph is one, known as cycle C_p with $p \geq 3$ vertices. That is, $C_p(1) \cong C_p$; $p \geq 3$. The circulant graph $C_4(1)$ is the only strongly regular graphs.

Theorem 1.1 *For any circulant graph $C_p(1)$ with $p \geq 4$ vertices, where $p = 2n$; $n \geq 2$,*

(i) $\gamma_{bi}(C_p(1)) = \frac{p}{2}$,
(ii) $\tau_{bi}(C_p(1)) = 2$.

Proof Let $G_1 = C_p(1)$ be a circulant graph with $p \geq 4$ vertices, where $p = 2n$; $n \geq 2$,

(i) Since G_1 having jump size one and 2-regular. Thus, half of the vertices are nonadjacent to each other, there fore the vertex independence set $\beta(G)$, which is bi-independent. Hence, $\beta(G) = |D| = |V - D|$. Then $\gamma_{bi}(G_1) = \frac{p}{2}$.
(ii) If a $\gamma_{bi}(G_1) = 2n$, then γ_{bi}-set of G_1 consists n times of K_1's, which is co-independent and are fixed by the choice of the first K_1, then there exists exactly one γ_{bi}-set of G_1 containing the vertex v_1 and there is one γ_{bi}-set of G_1, neglecting the vertex v_1 such as γ_{bi}-set of G_1 containing the vertex v_2 and is containing the vertex v_p. Thus, $\tau_{bi}(G_1) = 2$.

□

Remark 1 Let G be a Circulant graph $C_p(1)$ with odd order. Then $\gamma_{bi}(C_p(1))$ does not exist.

1.2 Circulant Graph $C_p(\lfloor \frac{p}{2} \rfloor)$

The circulant graph with jump size $\lfloor \frac{p}{2} \rfloor$; $p \geq 4$ vertices, that is $C_p(\lfloor \frac{p}{2} \rfloor)$. The circulant graph $C_p(\lfloor \frac{p}{2} \rfloor)$ with $p = 2n$; $n \geq 2$ vertices contain n times of $K_2^{'}$s and they are disconnected.

Theorem 1.2 *For any circulant graph $C_p(\lfloor \frac{p}{2} \rfloor)$ with $p \geq 4$ verices,*

(i) $\gamma_{bi}\left(C_p\left(\lfloor \frac{p}{2} \rfloor\right)\right) = \lfloor \frac{p}{2} \rfloor$,
(ii) $\tau_{bi}\left(C_p\left(\lfloor \frac{p}{2} \rfloor\right)\right) = p$.

Proof (i) Let $G_2 = C_p(\lfloor \frac{p}{2} \rfloor)$ be any circulant graph with $p \geq 4$ verices. If D is the γ_{bi}-set and there exists $V - D$ vertices, which is greater than or equal to D, then $\gamma_{bi}(G_2) = \lfloor \frac{p}{2} \rfloor$

(ii) For total number of γ_{bi}-sets, there exists two cases as follows,
Case 1. If $p = 2n; n \geq 2$ verices and $\gamma_{bi}(G_2) = n$, then γ_{bi}-set of G_2 is bounded between $(n + 1)$ times of K_2's and they are disjoint unions. Locate exactly K_1 from each three times of K_2's and its induced subgraph becomes totally disconnected. Hence, $\tau_{bi}(G_2) = p$ follows.
Case 2. If $p = 2n + 1; n \geq 2$ vertices and $\gamma_{bi}(G_2) = n + 1$, then γ_{bi}-set of G_2 comprises $(2n + 1)$ times of K_1's which is connected. For totally disconnected induced subgraphs, choose three times of K_1's and repeat the process. Hence, $\tau_{bi}(G_2) = p$ follows.

□

1.3 Circulant Graph $C_p(1, 3, \ldots, \lfloor \frac{p}{2} \rfloor)$

The circulant graph with jump size $1, 3, \ldots, \lfloor \frac{p}{2} \rfloor$ with $p \geq 6$ vertices is known as a complete bipartite graph K_{p_1,p_2} with $p_1 = p_2$, that is $C_p(1, 3, \ldots, \lfloor \frac{p}{2} \rfloor) \cong K_{p_1,p_2}$. These graphs are all strongly regular graphs.

Theorem 1.3 *For any circulant graph $C_p(1, 3, \ldots, \lfloor \frac{p}{2} \rfloor)$ with $p = 4n - 2$; $n \geq 2$ vertices,*

(i) $\gamma_{bi}\left(C_p\left(1, 3, \ldots, \lfloor \frac{p}{2} \rfloor\right)\right) = \frac{p}{2}$,
(ii) $\tau_{bi}\left(C_p\left(1, 3, \ldots, \lfloor \frac{p}{2} \rfloor\right)\right) = 2$.

Proof Let $G_3 = C_p(1, 3, \ldots, \lfloor \frac{p}{2} \rfloor)$ be any circulant graph with $p = 4n - 2; n \geq 2$ vertices,

(i) Since G_3 having jump size of odd order and regular. Each vertex of the graph G_3 adjacent to the odd number of vertices, implies vertex independence number $\beta(G)$ and it is the half of the total number of vertices of G_3 which is bi-independent. Hence, $\beta(G) = |D| = |V - D|$. Then $\gamma_{bi}(G_3) = \frac{p}{2}$.
(ii) If a $\gamma_{bi}(G_3) = 2n - 1$, then γ_{bi}-set of G_3 consists $(2n - 1)$ times of K_1's, which are independent and are fixed by the choice of the first K_1, then there exists exactly one γ_{bi}-set of G_3 containing the vertex v_1 and there is one γ_{bi}-set of G_3, neglecting the vertex v_1 such as γ_{bi}-set of G_3 containing the vertex v_2 and is containing the vertex v_p. Thus, $\tau_{bi}(G_3) = 2$.

□

1.4 Circulant Graph $C_p(1, 3, \ldots, \lfloor\frac{p}{2}\rfloor - 1)$

A Circulant graph with jump size $1, 3, \ldots, \lfloor\frac{p}{2}\rfloor - 1$ with $p \geq 6$ vertices is a complete bipartite graph K_{p_1,p_2} with $p_1 = p_2$, that is $G_4 = C_p(1, 3, \ldots, \lfloor\frac{p}{2}\rfloor - 1) \cong K_{p_1,p_2}$. These graphs are all strongly regular graphs.

Theorem 1.4 *For any circulant graph $C_p(1, 3, \ldots, \lfloor\frac{p}{2}\rfloor - 1)$ with $p = 4n; n \geq 2$ vertices,*

(i) $\gamma_{bi}\left(C_p\left(1, 3, \ldots, \lfloor\frac{p}{2}\rfloor - 1\right)\right) = \frac{p}{2}$,
(ii) $\tau_{bi}\left(C_p\left(1, 3, \ldots, \lfloor\frac{p}{2}\rfloor - 1\right)\right) = 2$.

Proof The proof of (i) and (ii) is similar to Theorem 1.3. □

1.5 Circulant Graph $C_p(1, 2, \ldots, \lfloor\frac{p}{2}\rfloor)$

The jump size of circulant graph is $1, 2, \ldots, \lfloor\frac{p}{2}\rfloor$ known as complete graph K_p with $p \geq 3$ vertices, that is

$$C_p\left(1, 2, \ldots, \left\lfloor\frac{p}{2}\right\rfloor\right) \cong K_p.$$

The complete graph K_p is strongly regular graph for all $p \geq 3$ vertices. The status of the trivial singleton graph K_1 is unclear. Since the parameter μ is not well defined on K_2, it is difficult to analyze and conclude whether K_2 is a strongly regular graph or not.

Theorem 1.5 *For any circulant graph $C_p(1, 2, 3, \ldots, \lfloor\frac{p}{2}\rfloor)$, the γ_{bi}-set does not exists.*

Proof Let $G_6 = C_p(1, 2, 3, \ldots, \lfloor\frac{p}{2}\rfloor)$ be any circulant graph. Since it is a complete graph K_p and establish adjacency between every pair of two vertices. Hence, γ_{bi}-set of G_6 does not exists. □

1.6 Matrix Representation of Circulant Graphs via Association Schemes

By considering the certain classes of circulant graphs, we discuss the following observations:

Observation 1 *A circulant graph $G_1 = C_p(1)$ with $p = 2n; n \geq 1$ vertices labeled as $v_1, v_2, \ldots, v_p$ with the set of all γ_{bi}-sets by Theorem 1.1. Two distinct vertices are said to be first associates if its jump size is 1 and is adjacent; otherwise, it is kth associates $(2 \leq k \leq \lfloor\frac{p}{2}\rfloor)$ if its jump size is k and is nonadjacent. The $1 \leq a, b, k \leq \lfloor\frac{p}{2}\rfloor$ associates of each vertex, as shown in Table 1.*

Table 1 Association schemes of $C_p(s_1, s_2, \ldots, s_t)$; p is even

Elements	Association scheme					
	First	Second	...	k	...	$\frac{p}{2}$
v_1	v_p, v_2	v_{p-1}, v_3	...	$v_{(p-(k-1))(mod\ p)}$, $v_{(1+k)(mod\ p)}$	...	$v_{1+\frac{p}{2}}$
v_2	v_1, v_3	v_p, v_4	...	$v_{(p-(k-2))(mod\ p)}$, $v_{(2+k)(mod\ p)}$	...	$v_{2+\frac{p}{2}}$
v_3	v_2, v_4	v_1, v_5	...	$v_{(p-(k-3))(mod\ p)}$, $v_{(3+k)(mod\ p)}$	...	$v_{3+\frac{p}{2}}$
$\vdots$	$\vdots$	$\vdots$	$\vdots$	$\vdots$	$\vdots$	$\vdots$
v_a	$v_{(a-1)(mod\ p)}$, $v_{(a+1)(mod\ p)}$	$v_{(a-2)(mod\ p)}$, $v_{(a+2)(mod\ p)}$	...	$v_{(p-(k-a))(mod\ p)}$, $v_{(a+k)(mod\ p)}$	...	$v_{(a+\frac{p}{2})(mod\ p)}$
$\vdots$	$\vdots$	$\vdots$	$\vdots$	$\vdots$	$\vdots$	$\vdots$
v_p	v_{p-1}, v_1	v_{p-2}, v_2	...	v_{p-k}, v_k	...	$v_{\frac{p}{2}}$

Table 2 Association schemes of $C_p(s_1, s_2, \ldots, s_t)$; p is odd

Elements	Association scheme					
	First	Second	...	k	...	$\frac{p-1}{2}$
v_1	v_p, v_2	v_{p-1}, v_3	...	$v_{(p-(k-1))(mod\ p)}$, $v_{(1+k)(mod\ p)}$	...	$v_{1+\frac{p-1}{2}}$, $v_{1+\frac{p-1}{2}+1}$
v_2	v_1, v_3	v_p, v_4	...	$v_{(p-(k-2))(mod\ p)}$, $v_{(2+k)(mod\ p)}$	...	$v_{2+\frac{p-1}{2}}$, $v_{2+\frac{p-1}{2}+1}$
v_3	v_2, v_4	v_1, v_5	...	$v_{(p-(k-3))(mod\ p)}$, $v_{(3+k)(mod\ p)}$	...	$v_{3+\frac{p-1}{2}}$, $v_{3+\frac{p-1}{2}+1}$
$\vdots$	$\vdots$	$\vdots$	$\vdots$	$\vdots$	$\vdots$	$\vdots$
v_a	$v_{(a-1)(mod\ p)}$, $v_{(a+1)(mod\ p)}$	$v_{(a-2)(mod\ p)}$, $v_{(a+2)(mod\ p)}$	...	$v_{(p-(k-a))(mod\ p)}$, $v_{(a+k)(mod\ p)}$	...	$v_{(a+\frac{p-1}{2})(mod\ p)}$, $v_{(a+\frac{p-1}{2}+1)(mod\ p)}$
$\vdots$	$\vdots$	$\vdots$	$\vdots$	$\vdots$	$\vdots$	$\vdots$
v_p	v_{p-1}, v_1	v_{p-2}, v_2	...	v_{p-k}, v_k	...	$v_{\frac{p-1}{2}}$, $v_{\frac{p-1}{2}+1}$

Observation 2 *A circulant graph $G_2 = C_p(\lfloor \frac{p}{2} \rfloor)$, labeled as $v_1, v_2, \ldots, v_p$. The set of all γ_{ci}-sets is given by Theorem 1.2. Two distinct vertices are said to be kth associates, where $1 \le k \le \lfloor \frac{p-2}{2} \rfloor$ if its jump size is k and is non adjacent; otherwise, it is $\lfloor \frac{p}{2} \rfloor^{th}$ associates, if its jump size is $\lfloor \frac{p}{2} \rfloor$ and is adjacent. The $1 \le a, b, k \le \lfloor \frac{p}{2} \rfloor$ and $1 \le a, b, k \le \lfloor \frac{p-1}{2} \rfloor$ associates of each vertex, as shown in Tables 1 and 2.*

Observation 3 *A circulant graph $G_3 = C_p(1, 3, \ldots, \lfloor \frac{p}{2} \rfloor)$; $p = 4n - 2, n \ge 2$ vertices labeled as $v_1, v_2, \ldots, v_p$ with the set of all γ_{ci}-sets by Theorem 1.3. Two distinct vertices are said to be odd associates if its jump sizes are odd and are non adjacent;*

otherwise, it is even associates if its jump sizes are even and are adjacent. The $1 \leq a, b, k \leq \lfloor \frac{p}{2} \rfloor$ *associates of each vertex, as shown in Table 2.*

Observation 4 *A circulant graph* $G_5 = C_p(1, 3, \ldots, \lfloor \frac{p}{2} \rfloor - 1)$*;* $p = 4n, n \geq 2$ *vertices labeled as* $v_1, v_2, \ldots, v_p$ *with the set of all* γ_{ci}*-sets by Theorem 1.4. Two distinct vertices are said to be odd associates if its jump sizes are odd and are nonadjacent; otherwise, it is even associates if its jump sizes are even and are adjacent. The* $1 \leq a, b, k \leq \lfloor \frac{p}{2} \rfloor$ *associates of each vertex, as shown in Table 1.*

Observation 5 *A circulant graph* $G_6 = C_p(1, 2, \ldots, \lfloor \frac{p}{2} \rfloor)$ *be labeled as* $v_1, v_2, \ldots, v_p$*;* $p \geq 3$ *vertices. By Theorem 1.5, PBIB designs and association schemes does not exists.*

In view of the above observations, we have the following and they are

Type 1. The matrix representation through an association scheme of a circulant graph $C_p(s_1, s_2, \ldots, s_t)$; p (≥ 4) is even vertices.

By Table 1, the parameters of second kind are given by $n_a = 2$ for $1 \leq a \leq \frac{p}{2} - 1$ and $n_{\frac{p}{2}} = 1$.

With the association schemes for the Table 1, we have the matrix representation of circulant graph $C_p(s_1, s_2, \ldots, s_t)$; p (≥ 4) vertices are

$$P^k = \begin{pmatrix} p^k_{11} & p^k_{12} & \cdots & p^k_{1\frac{p}{2}} \\ p^k_{21} & p^k_{22} & \cdots & p^k_{2\frac{p}{2}} \\ \vdots & \vdots & \vdots & \vdots \\ p^k_{\frac{p}{2}1} & p^k_{\frac{p}{2}2} & \cdots & p^k_{\frac{p}{2}\frac{p}{2}} \end{pmatrix}$$

Therefore, the possible values of k in the matrix P^k are given below:

For $k = 1$,
$p^1_{ab} = 1$ for $1 \leq a \leq \frac{p}{2} - 1, b = a + 1$
$p^1_{ab} = 1$ for $a = 1 + b, 1 \leq b \leq \frac{p}{2} - 1$
For $2 \leq k \leq \frac{p}{2} - 1$,
$p^k_{ab} = 1$ for $1 \leq b \leq \frac{p}{2} - 1, a + b = k, b = k + a$ and $a + b = p - k$
$p^k_{ab} = 1$ for $1 \leq b \leq \frac{p}{2} - 1, a = k + b$ and $a + b = p - k$
For $k = \frac{p}{2}$,
$p^k_{ab} = 2$ for $1 \leq a \leq \frac{p}{2} - 1$ and $b = k - a$ and the other entries are all zero.

Type 2. The matrix representation through an association scheme of a circulant graph $C_p(s_1, s_2, \ldots, s_t)$; p (≥ 5) is odd vertices.

By Table 2, the parameters of second kind are given by $n_a = 2$ for $1 \leq a \leq \frac{p-1}{2}$.

With the association schemes for the Table 2, we have the matrix representation of circulant graph $C_p(s_1, s_2, \ldots, s_t)$; p (≥ 3) with odd vertices are

Table 3 Parameters of PBIB designs

Circulant graph	Parameters of PBIB design					
	ν	ρ	r	g	λ_m	
G_1	p	$\frac{p}{n}$	$\frac{p}{n}-1$	$\frac{p}{2}$	$\frac{1-(-1)^m}{2}$;$1 \le m \le \lfloor \frac{p}{2} \rfloor$	
G_2	p	p	$\lfloor \frac{p}{2} \rfloor$		$\frac{p}{2}-m$;$1 \le m \le \lfloor \frac{p}{2} \rfloor$	
G_3	p	2	$\frac{p}{2}$	1	0; if m is odd	1; if m is even
G_4	p	2	$\frac{p}{2}$	1		
G_5	p	Not exists				

$$P^k = \begin{pmatrix} p_{11}^k & p_{12}^k & \cdots & p_{1\,\frac{p-1}{2}}^k \\ p_{21}^k & p_{22}^k & \cdots & p_{2\,\frac{p-1}{2}}^k \\ \vdots & \vdots & \vdots & \vdots \\ p_{(\frac{p-1}{2})1}^k & p_{(\frac{p-1}{2})2}^k & \cdots & p_{(\frac{p-1}{2})\,(\frac{p-1}{2})}^k \end{pmatrix}$$

Therefore, the possible values of k in the matrix P^k are given below:
For $k = 1$,
$p_{ab}^1 = 1$ for $1 \le a \le \frac{p-1}{2} - 1, b = a + 1$
$p_{ab}^1 = 1$ for $1 \le b \le \frac{p-1}{2} - 1, a = 1 + b$
$p_{ab}^1 = 1$ for $a = \frac{p-1}{2}, b = \frac{p-1}{2}$
For $2 \le k \le \frac{p-3}{2}$,
$p_{ab}^k = 1$ for $1 \le a \le \frac{p-3}{2}, a + b = k, b = k + a$ and $a + b = p - k$
$p_{ab}^k = 1$ for $1 \le b \le \frac{p-3}{2}, a = k + b$ and $a + b = p - k$
For $k = \frac{p-1}{2}$,
$p_{ab}^k = 1$, for $1 \le a \le \frac{p-3}{2}, b = \frac{p-1}{2} - a$
$p_{ab}^k = 1$, for $1 \le a \le \frac{p-1}{2}, b = \frac{p+1}{2} - a$ and the other entries are all zero.

1.7 The Parameters of PBIB Designs

By considering the Theorems 1.1–1.5, Tables 1 and 2. Also the possible values of k in the matrix P^k by using above types, we have the parameters of PBIB designs as follows (Table 3).

2 Conclusion

Generally, the PBIB designs obtained from the families of strongly regular graphs. Interestingly, we determine the total number of γ_{bi}-set, the partially balanced incomplete block (PBIB) designs with its association schemes obtaining from the γ_{bi}-sets in some classes of circulant graph.

Finally, we pose the following open problem:

Open Problem: Obtain the PBIB-design associated with γ_{bi}-set of a/an integral circulant graphs, regular circulant graphs and Cayley graphs.

Acknowledgements The authors wish to thank Prof. N. D. Soner for his help and valuable suggestions in the preparation of this article.

References

1. R.C. Bose, K.R. Nair, Partially balanced incomplete block designs. Sankhyā: Indian J. Stat. **4**, 337–372 (1939)
2. B. Chaluvaraju, N. Manjunath, PBIB-designs with $m = 1, 2, 3, 4$ and $\lfloor \frac{p}{2} \rfloor$ associated classes arising from γ_{cc}-sets of some special kinds of graphs. South East Asian J. Math. Math. Sci. **12**(2), 69–80 (2016)
3. B. Chaluvaraju, N. Manjunath, PBIB-designs and association schemes arising from minimum bi-connected dominating sets of some special classes of graphs. Afr. Mat. (Springer) **29**(1–2), 47–63 (2018)
4. B. Chaluvaraju, S.A. Diwakar, N. Manjunath, PBIB designs emerging from minimum independent dominating sets of different jump sizes of circulant graphs. Manuscript (2020)
5. B. Chaluvaraju, V. Lokesha, S.A. Diwakar, Association scheme with PBIB designs for minimum co-independent dominating sets of circulant graphs. Manuscript (2020)
6. C.J. Colbourn, J.H. Dinitz, *Handbook of Combinatorial Designs* (CRC Press, Boca Raton, 1996)
7. E.J. Cockayne, S.T. Hedetniemi, Towards a theory of domination in graphs. Networks **7**, 247–261 (1977)
8. F. Harary, *Graph theory* (Addison-Wesley, Reading Mass, 1969)
9. T.W. Haynes, S.T. Hedetniemi, P.J. Slater, *Fundamentals of Domination in Graphs* (Marcel Dekker Inc., New York, 1998)
10. T.W. Haynes, S.T. Hedetniemi, P.J. Slater, *Domination in Graphs: Advanced Topics* (Marcel Dekker Inc., New York, 1998)
11. I. Kang, K. Sinha, Heung-Kyu Lee, New digital fingerprint code construction scheme using group-divisible design. IEICE Trans. Fundam. **E89-A**(12) 3732–3735 (2006)
12. V. R. Kulli, B. Chaluvaraju, C. Appajigowda, Bi-conditional domination related parameters of a graph-I. Bull. Int. Math. Virtual Inst. **7**, 451–464 (2017)
13. V.R. Kulli, B. Chaluvaraju, M. Kumara, PBIB-designs and association schemes from minimum split dominating sets of certain jump sizes of circulant graphs. South East Asian J. Math. & Math. Sci. **14**(1), 81–94 (2018)
14. N. Manjunath, G. Deepak, B. Chaluvaraju, PBIB-Designs and Association schemes arising from minimum connected dominating sets of a graph. South Asian J. Math. **4**(5), 225–232 (2014)
15. N.-K. Nguyen, Construction of optimal block designs by computer. Technometrics **36**(3), 300–307 (1994)

16. P.J. Slater, The Hedetniemi number of a graph. Congr. Numer. **139**, 65–75 (1999)
17. T. Tamizh Chelvam, I. Rani, Independent domination number of Cayley graphs on Z_n. J. Combin. Math. Combin. Comput. **69**, 251–255 (2009)
18. H.B. Walikar, H.S. Ramane, B.D. Acharya, H.S. Shekarappa, S. Arumugam, Partially balanced incomplete block design arising from minimum dominating sets of paths and cycles. AKCE J. Graphs Combin. **4**(2), 223–232 (2007)
19. E. Yi, Domination value in graphs. Contrib. Discret. Math. **7**(2), 30–43 (2012)

Skew-Harmonic and Skew-Sum Connectivity Energy of Some Digraphs

Y. Shanthakumari and V. Lokesha

Abstract The graph-theoretical matrices such as the adjacency matrix and the incidence matrix are used as sources of molecular descriptors. Graph energies are the backbone of chemical graph theory. It is well known that the notion of graph energy was firstly proposed by I. Gutman. In 2010, Adiga et al. introduced the skew energy of digraphs. In the present work, we introduce and calculate the skew-harmonic energy and skew-sum connectivity energy of some digraphs.

Keywords Digraph · Skew-harmonic energy · Skew-sum connectivity energy

AMS Subject Classification 05C50

1 Introduction

A graph is a mathematical entity, generally symbolized as G, which include two nonempty sets; one set, usually denoted by V, is a set of elements called vertices, and the other, usually denoted by E, is a set of unordered pairs of distinct elements of V called edges. Thus, $G = (V, E)$. The adjacency matrices (usually denoted by $A(G)$) are square $V \times V$ or $E \times E$ symmetric matrices that reveal the adjacencies between vertices or edges in graphs. The characteristic polynomial of the graph G is denoted by $Ch(G, \mu)$ and is defined as $Ch(G, \mu) =\mid \mu I - G \mid$, where μ is an eigenvalue of the graph G. Hence, by [2] the energy of G is defined as $\mathcal{E}(G) =$

Dedicated to Professor H. M. Srivastava on the Occasion of his 80th Birth Anniversary.

Y. Shanthakumari (✉) · V. Lokesha
Department of Studies in Mathematics, Vijayanagara Sri krishnadevaraya University, Ballari, Karnataka, India
e-mail: yskphd2019@gmail.com

V. Lokesha
e-mail: v.lokesha@gmail.com

S. K. Paikray et al. (eds.), *New Trends in Applied Analysis and Computational Mathematics*, Advances in Intelligent Systems and Computing 1356,
https://doi.org/10.1007/978-981-16-1402-6_20

$\sum_{i=1}^{n} \mid \mu_i \mid$. The molecular graph theory can make a topological representation of a molecule. Topological indices are well designed for tracing similarity or dissimilarity of molecules. In general, by [3], the topological index can be symbolized as

$$TI(G) = \sum_{i \sim j} F(d_i, d_j),$$

where $F = F(x, y)$ is a well-chosen function. For example, for the harmonic index and sum connectivity index, we have

$$F(x, y) = \frac{2}{x+y} \quad and \quad F(x, y) = \frac{1}{\sqrt{x+y}},$$

respectively. It is well known that a matching adjacency matrix $A = (a_{ij})$ of order $n \times n$ can be associated to each topological index as

$$a_{ij} = \begin{cases} F(d_i, d_j) & \text{if } i \sim j \\ 0 & \text{if otherwise} \end{cases}.$$

The concept of harmonic energy (HE) was deliberated by S. M. Hosamani et al. in 2017, [5] by considering the harmonic matrix $H(G) = (h_{ij})_{n \times n}$ which is defined as

$$h_{ij} = \begin{cases} \dfrac{2}{d_i + d_j} & \text{if } v_i \sim v_j \\ 0 & \text{if otherwise} \end{cases}.$$

In 2010, Zhou and Trinajstic [7, 8] have proposed the concept of the sum connectivity index $\chi(G)$ and the general sum connectivity index $\chi_\alpha(G)$, where

$$\chi(G) = \sum_{uv \in E(G)} (d_u + d_v)^{-\frac{1}{2}}$$

and

$$\chi_\alpha(G) = \sum_{uv \in E(G)} (d_u + d_v)^{\alpha} \tag{1}$$

where α is a real number.

The sum connectivity matrix of $SC(G) = (s_{ij})_{n \times n}$ is given as

$$s_{ij} = \begin{cases} \dfrac{1}{\sqrt{d_i + d_j}} & \text{if } v_i \sim v_j \\ 0 & \text{if otherwise} \end{cases}.$$

In 2010, Adiga et al. [1] introduced the skew energy of digraphs. Let G^σ be a directed graph of order n with the vertex set $V(G^\sigma)$ and the arc set $\Gamma(G^\sigma) \subset V(G^\sigma) \times V(G^\sigma)$. The skew-adjacent matrix of G^σ is the $n \times n$ matrix $S(G^\sigma) = [s_{ij}]$, where $s_{ij} = 1$ whenever $(v_i, v_j) \in \Gamma(G^\sigma)$, $s_{ij} = -1$ whenever $(v_j, v_i) \in \Gamma(G^\sigma)$, and $s_{ij} = 0$ otherwise. Thus, a corresponding skew-adjacency matrix $\mathcal{F} = (f_{ij})$ of order $n \times n$ can be associated to each topological indices as

$$f_{ij} = \begin{cases} F(d_i, d_j) & \text{if } v_i \sim v_j \\ -F(d_i, d_j) & \text{if } v_j \sim v_i \\ 0 & \text{if otherwise} \end{cases}. \tag{2}$$

As a generalization of $H(G)$ and $SC(G)$ into the directed graphs G^σ, we can define skew-harmonic matrix $SH(G^\sigma) = (h^\sigma_{ij})_{n\times n}$ and the skew-sum connectivity matrix $SSC(G^\sigma) = (s^\sigma_{ij})_{n\times n}$ of a digraph as

$$h^\sigma_{ij} = \begin{cases} \dfrac{2}{d_i + d_j} & \text{if } (v_i, v_j) \in \Gamma(G^\sigma) \\ -\dfrac{2}{d_i + d_j} & \text{if } (v_j, v_i) \in \Gamma(G^\sigma) \\ 0 & \text{if otherwise} \end{cases}$$

and

$$s^\sigma_{ij} = \begin{cases} \dfrac{1}{\sqrt{d_i + d_j}} & \text{if } (v_i, v_j) \in \Gamma(G^\sigma) \\ -\dfrac{1}{\sqrt{d_i + d_j}} & \text{if } (v_j, v_i) \in \Gamma(G^\sigma) \\ 0 & \text{if otherwise} \end{cases},$$

respectively.

Lemma 1.1 *Let G^σ be a directed graph containing n vertices and let $SH(G^\sigma) = (a_{ij})_{n\times n}$ be the skew-harmonic matrix. Then*

(i) $Tr(SH)^2 = -8\chi_{-2}(G)$.
(ii) $Tr(SH)^i = 0$, *for all odd* i.

Proof (i) Since the diagonal entries of $(SH)^2$ are

$$(SH)^2_{ii} = \sum_{j=1}^{n} a_{ij}a_{ji} = -\sum_{j=1}^{n} a^2_{ij},$$

we have

$$Tr(SH)^2 = -\sum_{i=1}^{n} \sum_{i \sim j,\, j \sim i}^{n} a_{ij}^2 = -\sum_{i=1}^{n} \sum_{i \sim j,\, j \sim i}^{n} \frac{4}{(d_i + d_j)^2} = -8 \sum_{i \sim j,\, j \sim i}^{n} \frac{1}{(d_i + d_j)^2}.$$

Hence by Eq. (1),

$$Tr(SH)^2 = -8\chi_{-2}(G)$$

as required.

(ii) Since for all odd i, $(SH)^i$ is a skew-symmetric matrix, we have $Tr(SH)^i = 0$. □

The following pre-results and definition will be needed in our results in the coming sections.

Proposition 1.2 ([6]) *Let* $C = \begin{pmatrix} 0 & Y \\ Y^T & 0 \end{pmatrix}$ *and* $D = \begin{pmatrix} 0 & Y \\ -Y^T & 0 \end{pmatrix}$ *be two real matrices. Then* $Spec(D) = iSpec(C)$.

Proposition 1.3 ([4]) *Let* $C \in M_{m \times m}$ *and* $D \in M_{n \times n}$, *and let* λ *and* μ *be the eigenvalues of* C *and* D, *respectively. Then* $\lambda\mu$ *is an eigenvalue of* $C \otimes D$.

In the following two main parts (part 2 and part 3), some peculiar digraphs (namely, complete, star, hyperoctahedral, crown, and complete bipartite) and their complements have been considered, we state and prove some results on the skew-harmonic energy (SHE) and skew-sum connectivity energy ($SSCE$). The orientations of all arcs of digraphs are taken from low labels to high labels.

2 New Energies over Special Digraphs

In this section, we will compute skew-harmonic energy (SHE) and skew-sum connectivity energy ($SSCE$) for some special cases of digraphs.

2.1 Case I: Complete Digraphs

Theorem 2.1 *Let* $V(K_n^\sigma) = \{v_1, \ldots, v_n\}$ *and* $\Gamma(K_n^\sigma) = \{(v_i, v_j) \mid 1 \le i, j \le n\}$ *be vertex and arc sets of a complete digraph* K_n^σ, *respectively. Then*

$$SHE(K_n^\sigma) = \frac{1}{n-1} \sum_{k=0}^{n-1} cot(2k+1)\frac{\pi}{2n} \text{ and } SSCE(K_n^\sigma) = \frac{1}{\sqrt{2(n-1)}} \sum_{k=0}^{n-1} cot(2k+1)\frac{\pi}{2n}.$$

Proof The skew-harmonic matrix of a complete digraph K_n^σ is given by

$$SH(K_n^\sigma) = \begin{bmatrix} 0 & \frac{1}{n-1} & \frac{1}{n-1} & \cdots & \frac{1}{n-1} & \frac{1}{n-1} \\ -\frac{1}{n-1} & 0 & \frac{1}{n-1} & \cdots & \frac{1}{n-1} & \frac{1}{n-1} \\ -\frac{1}{n-1} & -\frac{1}{n-1} & 0 & \cdots & \frac{1}{n-1} & \frac{1}{n-1} \\ \vdots & \vdots & \vdots & \ddots & \vdots & \vdots \\ -\frac{1}{n-1} & -\frac{1}{n-1} & -\frac{1}{n-1} & \cdots & 0 & \frac{1}{n-1} \\ -\frac{1}{n-1} & -\frac{1}{n-1} & -\frac{1}{n-1} & \cdots & -\frac{1}{n-1} & 0 \end{bmatrix},$$

which has the characteristic polynomial $\frac{(-1)^n[(\mu-\frac{1}{n-1})^n+(\mu+\frac{1}{n-1})^n]}{2}$ and the spectrum is $\mu = \frac{1}{n-1} i \cot(2k+1)\frac{\pi}{2n}$, where $k = 0, 1, 2, \ldots, (n-1)$. Hence, the skew-harmonic energy of K_n^σ is obtained as

$$\frac{1}{n-1} \sum_{k=0}^{n-1} \cot(2k+1)\frac{\pi}{2n}.$$

On the other hand, the skew-sum connectivity matrix of K_n^σ is defined by

$$SSC(K_n^\sigma) = \begin{bmatrix} 0 & \frac{1}{\sqrt{2(n-1)}} & \frac{1}{\sqrt{2(n-1)}} & \cdots & \frac{1}{\sqrt{2(n-1)}} & \frac{1}{\sqrt{2(n-1)}} \\ -\frac{1}{\sqrt{2(n-1)}} & 0 & \frac{1}{\sqrt{2(n-1)}} & \cdots & \frac{1}{\sqrt{2(n-1)}} & \frac{1}{\sqrt{2(n-1)}} \\ -\frac{1}{\sqrt{2(n-1)}} & -\frac{1}{\sqrt{2(n-1)}} & 0 & \cdots & \frac{1}{\sqrt{2(n-1)}} & \frac{1}{\sqrt{2(n-1)}} \\ \vdots & \vdots & \vdots & \ddots & \vdots & \vdots \\ -\frac{1}{\sqrt{2(n-1)}} & -\frac{1}{\sqrt{2(n-1)}} & -\frac{1}{\sqrt{2(n-1)}} & \cdots & 0 & \frac{1}{\sqrt{2(n-1)}} \\ -\frac{1}{\sqrt{2(n-1)}} & -\frac{1}{\sqrt{2(n-1)}} & -\frac{1}{\sqrt{2(n-1)}} & \cdots & -\frac{1}{\sqrt{2(n-1)}} & 0 \end{bmatrix}$$

such that the characteristic polynomial is $\frac{(-1)^n[(\mu-(\frac{1}{\sqrt{2(n-1)}}))^n+(\mu+(\frac{1}{\sqrt{2(n-1)}}))^n]}{2}$ and the spectrum is $\mu = \frac{1}{\sqrt{2(n-1)}} i \cot(2k+1)\frac{\pi}{2n}$, where $k = 0, 1, 2, \ldots, (n-1)$. Hence, the skew-sum connectivity energy of K_n^σ is

$$\frac{1}{\sqrt{2(n-1)}} \sum_{k=0}^{n-1} \cot(2k+1)\frac{\pi}{2n}.$$

□

2.2 Case II: Star Digraphs

Theorem 2.2 *Let $V(S_n^\sigma) = \{v_1, \ldots, v_n\}$ and $\Gamma(S_n^\sigma) = \{(v_1, v_j) \mid 2 \le j \le n\}$ be the vertex and arc sets of a star digraph S_n^σ, respectively. Then*

$$SHE(S_n^\sigma) = \frac{4\sqrt{n-1}}{n} \quad and \quad SSCE(S_n^\sigma) = \frac{2(n-1)}{n}.$$

Proof The skew-harmonic matrix of a star digraph S_n^σ is

$$SH(S_n^\sigma) = \begin{bmatrix} 0 & \frac{2}{n} & \cdots & \frac{2}{n} \\ -\frac{2}{n} & 0 & \cdots & 0 \\ \vdots & \vdots & \ddots & \vdots \\ -\frac{2}{n} & 0 & \cdots & 0 \end{bmatrix}$$

and so the characteristic polynomial is $\mu^{n-2}[\mu^2 + (n-1)\frac{4}{n^2}]$. Also the spectrum is

$$Spec(S_n^\sigma) = \begin{pmatrix} 0 & \frac{2i\sqrt{n-1}}{n} & -\frac{2i\sqrt{n-1}}{n} \\ n-2 & 1 & 1 \end{pmatrix}.$$

Thus $SHE(S_n^\sigma) = \frac{4\sqrt{n-1}}{n}$.

Secondly, the skew-sum connectivity matrix of the star digraph S_n^σ is presented by

$$SSC(S_n^\sigma) = \begin{bmatrix} 0 & \frac{1}{\sqrt{n}} & \cdots & \frac{1}{\sqrt{n}} \\ -\frac{1}{\sqrt{n}} & 0 & \cdots & 0 \\ \vdots & \vdots & \ddots & \vdots \\ -\frac{1}{\sqrt{n}} & 0 & \cdots & 0 \end{bmatrix}$$

and so the characteristic polynomial $\mu^{n-2}[\mu^2 + \frac{n-1}{n}]$ while the spectrum is

$$Spec(S_n^\sigma) = \begin{pmatrix} 0 & -\frac{i(n-1)}{n} & \frac{i(n-1)}{n} \\ n-2 & 1 & 1 \end{pmatrix}.$$

Therefore $SSCE(S_n^\sigma) = \frac{2(n-1)}{n}$. □

2.3 Case III: Hyperoctahedral Digraphs

Theorem 2.3 *Let* $V(H_{2n}^\sigma) = \{u_1, \ldots, u_n, v_1, \ldots, v_n\}$ *and* $\Gamma(H_{2n}^\sigma) = \{(u_i, v_j), (u_i, u_j), (v_i, v_j) \mid 1 \le i, j \le n\}$ *be the vertex and arc sets of hyperoctahedral digraph* H_{2n}^σ, *respectively. Then*

$$SHE(H_{2n}^\sigma) = \frac{2}{2n-1}\sum_{k=0}^{n-1} cot(2k+1)\frac{\pi}{2n} \quad and \quad SSCE(H_{2n}^\sigma) = \frac{2}{\sqrt{2(2n-1)}}\sum_{k=0}^{n-1} cot(2k+1)\frac{\pi}{2n}\sum_{k=0}^{n-1} cot(2k+1)\frac{\pi}{2n}.$$

Proof The skew-harmonic matrix of the hyperoctahedral digraph H_{2n}^σ is

$$SH(H_{2n}^{\sigma}) = \begin{bmatrix} 0 & \frac{1}{2n-1} & \frac{1}{2n-1} & \cdots & \frac{1}{2n-1} & 0 & \frac{1}{2n-1} & \frac{1}{2n-1} & \cdots & \frac{1}{2n-1} \\ -\frac{1}{2n-1} & 0 & \frac{1}{2n-1} & \cdots & \frac{1}{2n-1} & -\frac{1}{2n-1} & 0 & \frac{1}{2n-1} & \cdots & \frac{1}{2n-1} \\ -\frac{1}{2n-1} & -\frac{1}{2n-1} & 0 & \cdots & \frac{1}{2n-1} & -\frac{1}{2n-1} & -\frac{1}{2n-1} & 0 & \cdots & \frac{1}{2n-1} \\ \vdots & \vdots & \vdots & \ddots & \vdots & \vdots & \vdots & \vdots & \ddots & \vdots \\ -\frac{1}{2n-1} & -\frac{1}{2n-1} & -\frac{1}{2n-1} & \cdots & 0 & -\frac{1}{2n-1} & -\frac{1}{2n-1} & -\frac{1}{2n-1} & \cdots & 0 \\ 0 & \frac{1}{2n-1} & \frac{1}{2n-1} & \cdots & \frac{1}{2n-1} & 0 & \frac{1}{2n-1} & \frac{1}{2n-1} & \cdots & \frac{1}{2n-1} \\ -\frac{1}{2n-1} & 0 & \frac{1}{2n-1} & \cdots & \frac{1}{2n-1} & -\frac{1}{2n-1} & 0 & \frac{1}{2n-1} & \cdots & \frac{1}{2n-1} \\ -\frac{1}{2n-1} & -\frac{1}{2n-1} & 0 & \cdots & \frac{1}{2n-1} & -\frac{1}{2n-1} & -\frac{1}{2n-1} & 0 & \cdots & \frac{1}{2n-1} \\ \vdots & \vdots & \vdots & \ddots & \vdots & \vdots & \vdots & \vdots & \ddots & \vdots \\ -\frac{1}{2n-1} & -\frac{1}{2n-1} & -\frac{1}{2n-1} & \cdots & 0 & -\frac{1}{2n-1} & -\frac{1}{2n-1} & -\frac{1}{2n-1} & \cdots & 0 \end{bmatrix}$$

$$= \begin{bmatrix} 1 & 1 \\ 1 & 1 \end{bmatrix} \otimes \begin{bmatrix} 0 & \frac{1}{2n-1} & \frac{1}{2n-1} & \cdots & \frac{1}{2n-1} & \frac{1}{2n-1} \\ -\frac{1}{2n-1} & 0 & \frac{1}{2n-1} & \cdots & \frac{1}{2n-1} & \frac{1}{2n-1} \\ -\frac{1}{2n-1} & -\frac{1}{2n-1} & 0 & \cdots & \frac{1}{2n-1} & \frac{1}{2n-1} \\ \vdots & \vdots & \vdots & \ddots & \vdots & \vdots \\ -\frac{1}{2n-1} & -\frac{1}{2n-1} & -\frac{1}{2n-1} & \cdots & 0 & \frac{1}{2n-1} \\ -\frac{1}{2n-1} & -\frac{1}{2n-1} & -\frac{1}{2n-1} & \cdots & -\frac{1}{2n-1} & 0 \end{bmatrix}$$

in which the spectrum is $\mu = \frac{2}{2n-1} i cot(2k+1)\frac{\pi}{2n}$, where $k = 0, 1, 2, \ldots, (n-1)$ and 0 (n times). As a result of that, the skew harmonic of H_{2n}^{σ} is $\frac{2}{2n-1}\sum_{k=0}^{n-1} cot(2k+1)\frac{\pi}{2n}$.

The skew-sum connectivity matrix of the hyperoctahedral digraph H_{2n}^{σ} is

$$SSC(H_{2n}^{\sigma}) = \begin{bmatrix} 0 & \frac{1}{\sqrt{2(2n-1)}} & \cdots & \frac{1}{\sqrt{2(2n-1)}} & 0 & \frac{1}{\sqrt{2(2n-1)}} & \cdots & \frac{1}{\sqrt{2(2n-1)}} \\ -\frac{1}{\sqrt{2(2n-1)}} & 0 & \cdots & \frac{1}{\sqrt{2(2n-1)}} & -\frac{1}{\sqrt{2(2n-1)}} & 0 & \cdots & \frac{1}{\sqrt{2(2n-1)}} \\ \vdots & \vdots & \ddots & \vdots & \vdots & \vdots & \ddots & \vdots \\ -\frac{1}{\sqrt{2(2n-1)}} & -\frac{1}{\sqrt{2(2n-1)}} & \cdots & 0 & -\frac{1}{\sqrt{2(2n-1)}} & -\frac{1}{\sqrt{2(2n-1)}} & \cdots & 0 \\ 0 & \frac{1}{\sqrt{2(2n-1)}} & \cdots & \frac{1}{\sqrt{2(2n-1)}} & 0 & \frac{1}{\sqrt{2(2n-1)}} & \cdots & \frac{1}{\sqrt{2(2n-1)}} \\ -\frac{1}{\sqrt{2(2n-1)}} & 0 & \cdots & \frac{1}{\sqrt{2(2n-1)}} & -\frac{1}{\sqrt{2(2n-1)}} & 0 & \cdots & \frac{1}{\sqrt{2(2n-1)}} \\ \vdots & \vdots & \ddots & \vdots & \vdots & \vdots & \ddots & \vdots \\ -\frac{1}{\sqrt{2(2n-1)}} & -\frac{1}{\sqrt{2(2n-1)}} & \cdots & 0 & -\frac{1}{\sqrt{2(2n-1)}} & -\frac{1}{\sqrt{2(2n-1)}} & \cdots & 0 \end{bmatrix}$$

$$= \begin{bmatrix} 1 & 1 \\ 1 & 1 \end{bmatrix} \otimes \begin{bmatrix} 0 & \frac{1}{\sqrt{2(2n-1)}} & \cdots & \frac{1}{\sqrt{2(2n-1)}} \\ -\frac{1}{\sqrt{2(2n-1)}} & 0 & \cdots & \frac{1}{\sqrt{2(2n-1)}} \\ \vdots & \vdots & \ddots & \vdots \\ -\frac{1}{\sqrt{2(2n-1)}} & -\frac{1}{\sqrt{2(2n-1)}} & \cdots & 0 \end{bmatrix}$$

such that the spectrum is $\mu = \frac{2}{\sqrt{2(2n-1)}} i cot(2k+1)\frac{\pi}{2n}$, where $k = 0, 1, 2, \ldots, (n-1)$ and 0 (n times). Therefore, the skew-sum connectivity energy of H_{2n}^{σ} is $\frac{2}{\sqrt{2(2n-1)}} \sum_{k=0}^{n-1} cot(2k+1)\frac{\pi}{2n}$. □

2.4 Case IV: Crown Digraphs

Theorem 2.4 *Let $V(C_{2n}^{\sigma}) = \{u_1, \ldots, u_n, v_1, \ldots, v_n\}$ and $\Gamma(C_{2n}^{\sigma}) = \{(u_i, v_j) \mid 1 \leq i, j \leq n\}$ be the vertex and arc sets of a crown digraph C_{2n}^{σ}, respectively. Then*

$$SHE(C_{2n}^{\sigma}) = \frac{2}{n-1}\sum_{k=0}^{n-1} cot(2k+1)\frac{\pi}{2n} \text{ and } SSCE(C_{2n}^{\sigma}) = \sqrt{\frac{2}{n-1}\sum_{k=0}^{n-1} cot(2k+1)\frac{\pi}{2n}}.$$

Proof The skew-harmonic matrix of the crown digraph C_{2n}^{σ} is

$$SH(C_{2n}^{\sigma}) = \begin{bmatrix} 0 & SH(K_n^{\sigma}) \\ SH(K_n^{\sigma}) & 0 \end{bmatrix}$$

$$= \begin{bmatrix} 0 & 1 \\ 1 & 0 \end{bmatrix} \otimes SH(K_n^{\sigma}),$$

where $SH(K_n^{\sigma})$ is adjacency matrix of K_n^{σ}. The spectrum is $\mu = \pm\frac{1}{n-1} i \cot(2k+1)\frac{\pi}{2n}$, where $k = 0, 1, 2, \ldots, (n-1)$. So the skew harmonic of C_{2n}^{σ} is $\frac{2}{n-1}\sum_{k=0}^{n-1}$ $cot(2k+1)\dfrac{\pi}{2n}$.

For the skew-sum connectivity energy case, the skew-sum connectivity matrix of the crown digraph C_{2n}^{σ} is given by

$$SSC(C_{2n}^{\sigma}) = \begin{bmatrix} 0 & SSC(K_n^{\sigma}) \\ SSC(K_n^{\sigma}) & 0 \end{bmatrix}$$

$$= \begin{bmatrix} 0 & 1 \\ 1 & 0 \end{bmatrix} \otimes SSC(K_n^{\sigma}).$$

In here since the spectrum is $\mu = \pm\frac{1}{\sqrt{2(n-1)}} i \cot(2k+1)\frac{\pi}{2n}$ $(k = 0, 1, 2, \ldots, (n-1))$, we obtain the skew-sum connectivity energy of C_{2n}^{σ} is $\sqrt{\frac{2}{n-1}\sum_{k=0}^{n-1}}$ $cot(2k+1)\dfrac{\pi}{2n}$. □

2.5 Case V: Complete Bipartite Digraphs

Theorem 2.5 *Let $V(K_{m,n}^{\sigma}) = \{u_1, \ldots, u_m, v_1, \ldots, v_n\}$ and $\Gamma(K_{m,n}^{\sigma}) = \{(u_i, v_j) \mid 1 \leq i \leq m, 1 \leq j \leq n\}$ be the vertex and arc sets of a complete bipartite digraph $K_{m,n}^{\sigma}$, respectively. Then*

$$SHE(K_{m,n}^{\sigma}) = \frac{4\sqrt{mn}}{m+n} \text{ and } SSCE(K_{m,n}^{\sigma}) = 2\sqrt{\frac{mn}{m+n}}.$$

Proof Similarly, as in the previous cases, the skew-harmonic matrix of the complete bipartite digraph $K^{\sigma}_{m,n}$ is defined by

$$SH(K^{\sigma}_{m,n}) = \begin{bmatrix} 0 & 0 & \cdots & \frac{2}{m+n} & \frac{2}{m+n} \\ 0 & 0 & \cdots & \frac{2}{m+n} & \frac{2}{m+n} \\ \vdots & \vdots & \ddots & \vdots & \vdots \\ -\frac{2}{m+n} & -\frac{2}{m+n} & \cdots & 0 & 0 \\ -\frac{2}{m+n} & -\frac{2}{m+n} & \cdots & 0 & 0 \end{bmatrix} \tag{3}$$

and so the characteristic polynomial is $(\mu)^{m+n-2}(\mu - \frac{2i\sqrt{mn}}{m+n})(\mu + \frac{2i\sqrt{mn}}{m+n})$ while the spectrum is

$$Spec(K^{\sigma}_{m,n}) = \begin{pmatrix} 0 & -\frac{2i\sqrt{mn}}{m+n} & \frac{2i\sqrt{mn}}{m+n} \\ m+n-2 & 1 & 1 \end{pmatrix}. \tag{4}$$

Therefore, the skew harmonic of $K^{\sigma}_{m,n}$ is $\frac{4\sqrt{mn}}{m+n}$.

On the other hand, the skew-sum connectivity matrix of the complete bipartite digraph $K^{\sigma}_{m,n}$ is defined by

$$SSC(K^{\sigma}_{m,n}) = \begin{bmatrix} 0 & 0 & \cdots & \frac{1}{\sqrt{m+n}} & \frac{1}{\sqrt{m+n}} \\ 0 & 0 & \cdots & \frac{1}{\sqrt{m+n}} & \frac{1}{\sqrt{m+n}} \\ \vdots & \vdots & \ddots & \vdots & \vdots \\ -\frac{1}{\sqrt{m+n}} & -\frac{1}{\sqrt{m+n}} & \cdots & 0 & 0 \\ -\frac{1}{\sqrt{m+n}} & -\frac{1}{\sqrt{m+n}} & \cdots & 0 & 0 \end{bmatrix} \tag{5}$$

and so the characteristic polynomial is

$$\mu^{m+n-2}\left(\mu - i\sqrt{\frac{mn}{m+n}}\right)\left(\mu + i\sqrt{\frac{mn}{m+n}}\right)$$

and the spectrum is

$$Spec(K^{\sigma}_{m,n}) = \begin{pmatrix} 0 & i\sqrt{\frac{mn}{m+n}} & -i\sqrt{\frac{mn}{m+n}} \\ m+n-2 & 1 & 1 \end{pmatrix}$$

This means that the skew-sum connectivity energy of $K^{\sigma}_{m,n}$ is $2\sqrt{\frac{mn}{m+n}}$. □

3 New Energies over Complements of Special Digraphs

3.1 Case I: Complement of a Complete Digraph

Theorem 3.1 *For the complement of a complete digraph $\overline{K_n^\sigma}$ having the vertex set $V(\overline{K_n^\sigma}) = \{v_1, \ldots, v_n\}$ and the arc set $\Gamma(\overline{K_n^\sigma}) = \{(v_i, v_j) \mid 1 \leq i, j \leq n\}$, respectively, we have*

$$SHE(\overline{K_n^\sigma}) = 0 \text{ and } SSCE(\overline{K_n^\sigma}) = 0.$$

Proof It actually follows from the fact that the complement of complete digraph is a null graph and the definition of $SSCE$. □

3.2 Case II: Complement of a Star Digraph

Theorem 3.2 *For the complement of a star digraph $\overline{S_n^\sigma}$ having the vertex set $V(\overline{S_n^\sigma}) = \{v_1, \ldots, v_n\}$ and the arc set $\Gamma(\overline{S_n^\sigma}) = \{(v_1, v_j) \mid 2 \leq j \leq n\}$, respectively, we have*

$$SHE(\overline{S_n^\sigma}) = \frac{1}{n-1}\sum_{k=0}^{n-2} cot(2k+1)\frac{\pi}{2n} \text{ and } SSCE(\overline{S_n^\sigma}) = \frac{1}{\sqrt{2(n-1)}}\sum_{k=0}^{n-2} cot(2k+1)\frac{\pi}{2n},$$

where $n \geq 2$.

Proof Since the skew-harmonic matrix of $\overline{S_n^\sigma}$ is defined by

$$SH(\overline{S_n^\sigma}) = \begin{bmatrix} 0 & 0 & 0 & \cdots & 0 & 0 \\ 0 & 0 & \frac{1}{n-1} & \cdots & \frac{1}{n-1} & \frac{1}{n-1} \\ \vdots & \vdots & \vdots & \ddots & \vdots & \vdots \\ 0 & -\frac{1}{n-1} & -\frac{1}{n-1} & \cdots & 0 & \frac{1}{n-1} \\ 0 & -\frac{1}{n-1} & -\frac{1}{n-1} & \cdots & -\frac{1}{n-1} & 0 \end{bmatrix},$$

the characteristic polynomial is obtained as $\frac{(-1)^{n-1}\mu[(\mu-\frac{1}{n-1})^{n-1}+(\mu+\frac{1}{n-1})^{n-1}]}{2}$ while the spectrum is

$$Spec(\overline{S_n^\sigma}) = \begin{pmatrix} 0 & \frac{1}{n-1}icot(\frac{\pi}{2n}) & \frac{1}{n-1}icot(\frac{3\pi}{2n}) & \cdots & \frac{1}{n-1}icot(\frac{2n-3}{2n})\pi \\ 1 & 1 & 1 & \cdots & 1 \end{pmatrix}.$$

Hence, the skew-harmonic energy of $\overline{S_n^\sigma}$ is $\frac{1}{n-1}\sum_{k=0}^{n-2} cot(2k+1)\frac{\pi}{2n}$, where $n \geq 2$.

Secondly, since the skew-sum connectivity matrix of $\overline{S_n^\sigma}$ is given by

$$SSC(\overline{S_n^{\sigma}}) = \begin{bmatrix} 0 & 0 & 0 & \cdots & 0 & 0 & 0 \\ 0 & 0 & \frac{1}{\sqrt{2(n-1)}} & \cdots & \frac{1}{\sqrt{2(n-1)}} & \frac{1}{\sqrt{2(n-1)}} & \frac{1}{\sqrt{2(n-1)}} \\ 0 & -\frac{1}{\sqrt{2(n-1)}} & 0 & \cdots & \frac{1}{\sqrt{2(n-1)}} & \frac{1}{\sqrt{2(n-1)}} & \frac{1}{\sqrt{2(n-1)}} \\ \vdots & \vdots & \vdots & \ddots & \vdots & \vdots & \vdots \\ 0 & -\frac{1}{\sqrt{2(n-1)}} & -\frac{1}{\sqrt{2(n-1)}} & \cdots & -\frac{1}{\sqrt{2(n-1)}} & 0 & \frac{1}{\sqrt{2(n-1)}} \\ 0 & -\frac{1}{\sqrt{2(n-1)}} & -\frac{1}{\sqrt{2(n-1)}} & \cdots & -\frac{1}{\sqrt{2(n-1)}} & -\frac{1}{\sqrt{2(n-1)}} & 0 \end{bmatrix},$$

the characteristic polynomial is obtained by $\frac{(-1)^{n-1}\mu[(\mu-\frac{1}{\sqrt{2(n-1)}})^{n-1}+(\mu+\frac{1}{\sqrt{2(n-1)}})^{n-1}]}{2}$ while the spectrum is

$$Spec(\overline{S_n^{\sigma}}) = \begin{pmatrix} 0 & i\frac{1}{\sqrt{2(n-1)}}cot(\frac{\pi}{2n}) & i\frac{1}{\sqrt{2(n-1)}}cot(\frac{3\pi}{2n}) & \cdots & i\frac{1}{\sqrt{2(n-1)}}cot(\frac{2n-3}{2n})\pi \\ 1 & 1 & 1 & \cdots & 1 \end{pmatrix}.$$

Thus, the skew-sum connectivity energy of $\overline{S_n^{\sigma}}$ is $\frac{1}{\sqrt{2(n-1)}}\sum_{k=0}^{n-2} cot(2k+1)\frac{\pi}{2n}$, $n \geq 2$. □

3.3 *Case III: Complement of a Hyperoctahedral Digraph*

Theorem 3.3 *For the complement of a hyperoctahedral digraph $\overline{H_{2n}^{\sigma}}$ having the vertex set $V(\overline{H_{2n}^{\sigma}}) = \{u_1, \ldots, u_n, v_1, \ldots, v_n\}$ and the arc set $\Gamma(\overline{H_{2n}^{\sigma}}) = \{(u_i, v_j) \mid 1 \leq i, j \leq n\}$, respectively, we have*

$$SHE(\overline{H_{2n}^{\sigma}}) = 2n \text{ and } SSCE(\overline{H_{2n}^{\sigma}}) = \sqrt{2}n\,.$$

Proof The skew-harmonic matrix of $\overline{H_{2n}^{\sigma}}$ is

$$SH(\overline{H_{2n}^{\sigma}}) = \begin{bmatrix} 0 & 0 & 0 & \cdots & 0 & 1 & 0 & \cdots & 0 & 0 \\ 0 & 0 & 0 & \cdots & 0 & 0 & 1 & \cdots & 0 & 0 \\ 0 & 0 & 0 & \cdots & 0 & 0 & 0 & \cdots & 0 & 0 \\ \vdots & \vdots & \vdots & \ddots & \vdots & \vdots & \vdots & \ddots & \vdots & \vdots \\ 0 & 0 & 0 & \cdots & 0 & 0 & 0 & \cdots & 0 & 1 \\ -1 & 0 & 0 & \cdots & 0 & 0 & 0 & \cdots & 0 & 0 \\ 0 & -1 & 0 & \cdots & 0 & 0 & 0 & \cdots & 0 & 0 \\ \vdots & \vdots & \vdots & \ddots & \vdots & \vdots & \vdots & \ddots & \vdots & \vdots \\ 0 & 0 & 0 & \cdots & 0 & 0 & 0 & \cdots & 0 & 0 \\ 0 & 0 & 0 & \cdots & -1 & 0 & 0 & \cdots & 0 & 0 \end{bmatrix}$$

$$= \begin{bmatrix} 0 & 1 \\ -1 & 0 \end{bmatrix} \otimes \begin{bmatrix} 1 & 0 & \cdots & 0 \\ 0 & 1 & \cdots & 0 \\ \vdots & \vdots & \ddots & \vdots \\ 0 & 0 & \cdots & 1 \end{bmatrix}.$$

So the characteristic polynomial is $(\mu - i)^n(\mu + i)^n$ and the spectrum is $spec(\overline{H_{2n}^{\sigma}}) = \begin{pmatrix} i & -i \\ n & n \end{pmatrix}$. Hence, the skew-harmonic energy of $\overline{H_{2n}^{\sigma}}$ is only $2n$.

Nevertheless, the skew-sum connectivity matrix of $\overline{H_{2n}^{\sigma}}$ is

$$SSC(\overline{H_{2n}^{\sigma}}) = \begin{bmatrix} 0 & 0 & 0 & \cdots & 0 & \frac{1}{\sqrt{2}} & 0 & \cdots & 0 & 0 \\ 0 & 0 & 0 & \cdots & 0 & 0 & \frac{1}{\sqrt{2}} & \cdots & 0 & 0 \\ 0 & 0 & 0 & \cdots & 0 & 0 & 0 & \cdots & 0 & 0 \\ \vdots & \vdots & \vdots & \ddots & \vdots & \vdots & \vdots & \ddots & \vdots & \vdots \\ 0 & 0 & 0 & \cdots & 0 & 0 & 0 & \cdots & 0 & \frac{1}{\sqrt{2}} \\ -\frac{1}{\sqrt{2}} & 0 & 0 & \cdots & 0 & 0 & 0 & \cdots & 0 & 0 \\ 0 & -\frac{1}{\sqrt{2}} & 0 & \cdots & 0 & 0 & 0 & \cdots & 0 & 0 \\ \vdots & \vdots & \vdots & \ddots & \vdots & \vdots & \vdots & \ddots & \vdots & \vdots \\ 0 & 0 & 0 & \cdots & 0 & 0 & 0 & \cdots & 0 & 0 \\ 0 & 0 & 0 & \cdots & -\frac{1}{\sqrt{2}} & 0 & 0 & \cdots & 0 & 0 \end{bmatrix}$$

$$= \begin{bmatrix} 0 & 1 \\ -1 & 0 \end{bmatrix} \otimes \begin{bmatrix} \frac{1}{\sqrt{2}} & 0 & \cdots & 0 \\ 0 & \frac{1}{\sqrt{2}} & \cdots & 0 \\ \vdots & \vdots & \ddots & \vdots \\ 0 & 0 & \cdots & \frac{1}{\sqrt{2}} \end{bmatrix}.$$

Hence, the characteristic polynomial $(\mu - \frac{1}{\sqrt{2}}i)^n(\mu + \frac{1}{\sqrt{2}}i)^n$ and the spectrum is $spec(\overline{H_{2n}^{\sigma}}) = \begin{pmatrix} \frac{1}{\sqrt{2}}i & -\frac{1}{\sqrt{2}}i \\ n & n \end{pmatrix}$. As a result, the skew-sum connectivity energy of $\overline{H_{2n}^{\sigma}}$ is just $\sqrt{2}n$. □

3.4 Case IV: Complement of a Complete Bipartite Digraph

Theorem 3.4 *For the complement of a complete bipartite digraph $\overline{K_{m,n}^{\sigma}}$ having the vertex set $V(\overline{K_{m,n}^{\sigma}}) = \{u_1, \ldots, u_m, v_1, \ldots, v_n\}$ and the arc set $\Gamma(\overline{K_{m,n}^{\sigma}}) = \{(u_i, v_j) \mid 1 \le i \le m, 1 \le j \le n\}$, respectively, we have*

$$SHE(\overline{K^{\sigma}_{m,n}}) = \frac{1}{m-1}\sum_{k=0}^{m-1} cot(2k+1)\frac{\pi}{2m} + \frac{1}{n-1}\sum_{k=0}^{n-1} cot(2k+1)\frac{\pi}{2n} \quad and$$

$$SSCE(\overline{K^{\sigma}_{m,n}}) = \frac{1}{\sqrt{2(m-1)}}\sum_{k=0}^{m-1} cot(2k+1)\frac{\pi}{2m} + \frac{1}{\sqrt{2(n-1)}}\sum_{k=0}^{n-1} cot(2k+1)\frac{\pi}{2n}.$$

Proof The skew-harmonic matrix of $\overline{K^{\sigma}_{m,n}}$ is

$$SH(\overline{K^{\sigma}_{m,n}}) = \begin{bmatrix} 0 & \frac{1}{m-1} & \frac{1}{m-1} & \cdots & \frac{1}{m-1} & 0 & 0 & 0 & \cdots & 0 \\ -\frac{1}{m-1} & 0 & \frac{1}{m-1} & \cdots & \frac{1}{m-1} & 0 & 0 & 0 & \cdots & 0 \\ -\frac{1}{m-1} & -\frac{1}{m-1} & 0 & \cdots & \frac{1}{m-1} & 0 & 0 & 0 & \cdots & 0 \\ \vdots & \vdots & \vdots & \ddots & \vdots & \vdots & \vdots & \vdots & \ddots & \vdots \\ -\frac{1}{m-1} & -\frac{1}{m-1} & -\frac{1}{m-1} & \cdots & 0 & 0 & 0 & 0 & \cdots & 0 \\ 0 & 0 & 0 & \cdots & 0 & 0 & \frac{1}{n-1} & \frac{1}{n-1} & \cdots & \frac{1}{n-1} \\ 0 & 0 & 0 & \cdots & 0 & -\frac{1}{n-1} & 0 & \frac{1}{n-1} & \cdots & \frac{1}{n-1} \\ 0 & 0 & 0 & \cdots & 0 & -\frac{1}{n-1} & -\frac{1}{n-1} & 0 & \cdots & \frac{1}{n-1} \\ \vdots & \vdots & \vdots & \ddots & \vdots & \vdots & \vdots & \vdots & \ddots & \vdots \\ 0 & 0 & 0 & \cdots & 0 & -\frac{1}{n-1} & -\frac{1}{n-1} & -\frac{1}{n-1} & \cdots & 0 \end{bmatrix},$$

which gives the spectrum as $\mu_1 = \frac{1}{m-1} i cot(2k+1)\frac{\pi}{2m}$ $(k = 0, 1, 2, \ldots, m-1)$ and $\mu_2 = \frac{1}{n-1} i cot(2k+1)\frac{\pi}{2n}$ $(k = 0, 1, 2, \ldots, n-1$. So the skew-harmonic energy of $\overline{K^{\sigma}_{m,n}}$ is

$$SHE(\overline{K^{\sigma}_{m,n}}) = \frac{1}{m-1}\sum_{k=0}^{m-1} cot(2k+1)\frac{\pi}{2m} + \frac{1}{n-1}\sum_{k=0}^{n-1} cot(2k+1)\frac{\pi}{2n}.$$

With a similar approximation, the skew-sum connectivity matrix of $\overline{K^{\sigma}_{m,n}}$ is given by

$$SSC(\overline{K^{\sigma}_{m,n}}) = \begin{bmatrix} 0 & \frac{1}{\sqrt{2(m-1)}} & \cdots & \frac{1}{\sqrt{2(m-1)}} & 0 & 0 & \cdots & 0 \\ -\frac{1}{\sqrt{2(m-1)}} & 0 & \cdots & \frac{1}{\sqrt{2(m-1)}} & 0 & 0 & \cdots & 0 \\ -\frac{1}{\sqrt{2(m-1)}} & -\frac{1}{\sqrt{2(m-1)}} & \cdots & \frac{1}{\sqrt{2(m-1)}} & 0 & 0 & \cdots & 0 \\ \vdots & \vdots & \ddots & \vdots & \vdots & \vdots & \ddots & \vdots \\ -\frac{1}{\sqrt{2(m-1)}} & -\frac{1}{\sqrt{2(m-1)}} & \cdots & 0 & 0 & 0 & \cdots & 0 \\ 0 & 0 & \cdots & 0 & 0 & \frac{1}{\sqrt{2(n-1)}} & \cdots & \frac{1}{\sqrt{2(n-1)}} \\ 0 & 0 & \cdots & 0 & -\frac{1}{\sqrt{2(n-1)}} & 0 & \cdots & \frac{1}{\sqrt{2(n-1)}} \\ 0 & 0 & \cdots & 0 & -\frac{1}{\sqrt{2(n-1)}} & -\frac{1}{\sqrt{2(n-1)}} & \cdots & \frac{1}{\sqrt{2(n-1)}} \\ \vdots & \vdots & \ddots & \vdots & \vdots & \vdots & \ddots & \vdots \\ 0 & 0 & \cdots & 0 & -\frac{1}{\sqrt{2(n-1)}} & -\frac{1}{\sqrt{2(n-1)}} & \cdots & 0 \end{bmatrix}$$

and so the spectrum is $\mu_1 = \frac{1}{\sqrt{2(m-1)}} i cot(2k+1)\frac{\pi}{2m}$ $(k = 0, 1, 2, \ldots, m-1)$ and $\mu_2 = \frac{1}{\sqrt{2(n-1)}} i cot(2k+1)\frac{\pi}{2n}$ $(k = 0, 1, 2, \ldots, n-1)$. Thus, the skew-sum connectivity energy of $\overline{K^{\sigma}_{m,n}}$ is

$$\frac{1}{\sqrt{2(m-1)}}\sum_{k=0}^{m-1} cot(2k+1)\frac{\pi}{2m} + \frac{1}{\sqrt{2(n-1)}}\sum_{k=0}^{n-1} cot(2k+1)\frac{\pi}{2n}.$$

□

4 Conclusion

In this article, by considering some special digraphs (namely, complete, star, hyperoctahedral, crown, and complete bipartite) and their complements, the skew-harmonic and skew-sum connectivity energies have been calculated.

References

1. C. Adiga, R. Balakrishnan, W. So, The skew energy of a digraph. Linear Algebr. Appl. **432**, 1825–1835 (2010)
2. I. Gutman, The energy of a graph. Ber. Math. Stat. Sekt. Forschungsz. Graz **103**, 1–22 (1978)
3. I. Gutman, Degree-based topological indices. Croat Chem. Acta **8G**, 351–361 (2013)
4. R.A. Horn, C.R. Johnson, *Topics in Matrix Analysis* (Cambridge University Press, Cambridge, 1991)
5. S.M. Hosamani, B.B. Kulkarni, R.G. Boli, V.M. Gadag, QSPR analysis of certain graph theoretical matrices and their corresponding energy. Appl. Math. Nonlinear Sci **2**, 131–150 (2017)
6. B. Shader, W. So, Skew spectra of oriented graphs. Electron. J. Combin. **16**, N32 (2009)
7. B. Zhou, N. Trinajstic, On a novel connectivity index. J. Math. Chem. **46**, 1252–1270 (2009)
8. B. Zhou, N. Trinajstic, On general sum-connectivity index. J. Math. Chem **47**, 210–218 (2010)

Multi-valued Analysis of *CR* Iterative Process in Banach Spaces

Nisha Sharma and Lakshmi Narayan Mishra

Abstract The main objective of this paper is to define multi-valued *CR* iteration scheme involving quasi-nonexpansive mapping by relaxing the compactness of domain. We also established multi-valued *CR* iteration scheme in whole new sense which removes a restrictive condition ($i.e.\ \wp(\sigma) = \{\sigma\}$) for any $\sigma \in \mathcal{P}_{\wp}$ where $\mathcal{P}_{\wp}$ is the set of all fixed points of $\wp$.

Keywords Multi-valued map · *CR* iteration scheme · Fixed points · Uniformly convex Banach spaces

AMS Subject Classification 47H10 · 54H25

1 Introduction

Let $\mathcal{B}_s$ be nonempty convex subset of a Banach space $\mathcal{B} := (\mathcal{B}, ||.||)$. The set $\mathcal{B}_s$ is known to be proximinal, wherein, there is in existence, an element $\jmath \in \mathcal{B}_s$ for every $\ell \in \mathcal{B}$, as

$$||\ell - \jmath|| = d(\ell, \mathcal{B}_s) = \inf\{d(\ell, \jmath) : \jmath \in \mathcal{B}_s\}.$$

The family of all bounded proximinal subsets of $\mathcal{B}_s$ in $\mathcal{B}$ and those of nonempty bounded and closed subsets of $\mathcal{B}$ are denoted by $P(\mathcal{B}_s)$ and $CB(\mathcal{B}_s)$, respectively. The Hausdorff metric on $CB(\mathcal{B}_s)$ is defined by

N. Sharma
Department of Mathematics, Pt. J. L. N Govt College, Faridabad, India
e-mail: nnishaa.bhardwaj@gmail.com

L. N. Mishra (✉)
Department of Mathematics, School of Advanced Sciences, Vellore Institute of Technology (VIT) University, Vellore 632014, Tamil Nadu, India
e-mail: lakshminarayanmishra04@gmail.com

S. K. Paikray et al. (eds.), *New Trends in Applied Analysis and Computational Mathematics*, Advances in Intelligent Systems and Computing 1356,
https://doi.org/10.1007/978-981-16-1402-6_21

$$\mathcal{H}(A, B) = \max\left\{\sup_{\ell \in A} d(\ell, B), \sup_{\jmath \in B} d(A, \jmath)\right\}$$

for $A, B \in CB(\mathcal{B}_s)$. An element $\sigma \in \mathcal{B}_s$ is called a fixed point of $\wp : \mathcal{B}_s \to \mathcal{B}_s$ (respectively, $\wp : \mathcal{B}_s \to CB(\mathcal{B}_s)$) if $\sigma = \wp(\sigma)$ (respectively, $\sigma \in \wp(\sigma)$). It is important to note that approximation of common fixed points of two quasi-nonexpansive multi-valued maps in Banach spaces and an implicit iteration process for solving a fixed point problem of a finite family of multi-valued mappings is very precisely studied by Suantai et al. [1]. Notice that Cholamjiak et al. proposed many applications in the framework of multi-valued mappings like modified inertial shrinking projection method for solving inclusion problems (see [2, 3]).

The theory of multi-valued mappings is challenging than the corresponding theory of single-valued mappings. To approximate the fixed points of the multi-valued mappings, different iterative procedures have been established. It is important to note that Sastry and Babu [4] established some Mann [5] and Ishikawa [6] iteration-based convergence results for multi-valued nonexpansive mappings in the framework of Hilbert spaces. Panyanak [7] extended the results of Sastry and Babu [4] to the framework of uniformly convex Banach spaces. Actually, Panyanak proved some results using Ishikawa-type iterative procedure without the endpoint condition. Song and Wang [8] have shown that Panyanak's result was not well defined without the endpoint condition. They reconstructed the procedure using the endpoint condition which was more well-defined manner. After this, results studied by Shahzad and Zegeye [9] are notable generalizations of Mann and Ishikawa iteration process especially in the case of multi-valued mappings. There is now extensive literature on iterative fixed points for the different class of mappings. We refer readers to Berinde [10] for an almost all the recent account of the literature on the iterative fixed points. There shall be a reference made to carve out a detailed analysis and review of literature with respect to *CR* iterates by taking recourse to the Chug et al. [11]. Transformation from single-valued map to multi-valued map, thereby extending the convergence results of single-valued mapping with the aid of *CR* iteration scheme shall be the focal point of this paper. In process, several relevant results, especially those contained in [12–17].

It is not been long that a single-valued iterate scheme was introduced by Chug et al. [11] which provided for an iteration convergence rate, which was faster than many preexisting iteration schemes and the aforementioned iteration scheme is as follows:

$$\begin{cases} \ell_n \in \mathcal{B}_s, \\ \ell_{n+1} = (1 - \eta_n^0)\jmath_n + \eta_n^0 \wp \jmath_n, \\ \jmath_n = (1 - \eta_n^1)\wp \ell_n + \eta_n^1 \wp \varepsilon_n, \\ \varepsilon_n = (1 - \eta_n^2)\ell_n + \eta_n^2 \wp \ell_n, \ n \in \mathbb{N}(\textit{set of natural numbers}), \end{cases}$$

where real sequences $\{\eta_n^0\}$, $\{\eta_n^1\}$ and $\{\eta_n^2\}$ satisfy η_n^0, η_n^1 and η_n^2 in [0, 1] and satisfying $\sum_{n=0}^{\infty} \eta_n^0 < \infty$.

The presumption of compactness is quite significant, because it is simple to find a sequence in defined domain that converges to the mapping's fixed point. After this, without the need for a compactness hypothesis, we are going to establish a convergence theorem. In order to achieve this purpose, we define the iterations of *CR* iteration scheme in a new context, which is a slight modification from the one originally defined by Bancha Panyanak (see [7]).

Let $\mathcal{B}_s$ be a nonempty convex subset of a Banach space $\mathcal{B}$ and the real sequences $\{\eta_n^0\}$, $\{\eta_n^1\}$ and $\{\eta_n^2\}$ satisfy η_n^0, η_n^1 and η_n^2 in [0, 1], then the multi-valued *CR* iteration scheme is defined as

$$\begin{cases} \ell_1 \in \mathcal{B}_s, \\ \varepsilon_n = (1-\eta_n^1)\ell_n + \eta_n^2 z_n', \\ \jmath_n = (1-\eta_n^1)z_n' + \eta_n^1 z_n''', \\ \ell_{n+1} = (1-\eta_n^0)\jmath_n + \eta_n^0 z_n'', \end{cases} \qquad (\mathcal{C}),$$

where $z_n' \in \wp\ell_n,\ z_n'' \in \wp\jmath_n,\ z_n''' \in \wp\varepsilon_n \forall\ n \in \mathbb{N}$.

2 Preliminaries

The proof of theorems is studied by us using lemma and definitions as well as various results and iteration processes to make this presentation more closed and self-contained.

Definition 2.1 A Banach space $\mathcal{B}$ is said to be uniformly convex if for each $\varepsilon \in (0, 2]$, there is a $\mu > 0$ such that for every $\ell, \jmath \in \mathcal{B}$, if for $||\ell|| \leq 1, ||\jmath|| \leq 1$ and $||\ell - \jmath|| > \varepsilon$ implies

$$\frac{1}{2}||\ell + \jmath|| \leq (1-\mu).$$

Definition 2.2 A multi-valued map $\wp : \mathcal{B}_s \to \mathcal{B}_s$ is known as nonexpansive if

$$\mathcal{H}(\wp(\ell), \wp(\jmath)) \leq \|\ell - \jmath\| \quad \text{for all } \ell, \jmath \in \mathcal{B}_s.$$

Definition 2.3 ([18]) A multi-valued map $\wp : \mathcal{B}_s \to \mathcal{B}_s$ is known as quasi-nonexpansive if $\mathcal{P}_\wp \neq \emptyset$ and

$$\mathcal{H}(\wp(\ell), \wp(\sigma)) \leq \|\ell - \sigma\| \quad \text{for all } \ell \in \mathcal{B}_s.$$

Definition 2.4 ([14]) A Banach space $\mathcal{B}$ is said to have Opial's property if for each sequence $\{\ell_n\}$ in $\mathcal{B}$ which weakly converges to $\ell \in \mathcal{B}$ and for every $\jmath \in \mathcal{B}$, it follows the following:

$$\lim sup_{n\to\infty}||\ell_n - \ell|| < \lim sup_{n\to\infty}||\ell_n - \jmath||.$$

The following lemmas will be useful in our subsequent discussion and are easy to establish.

Lemma 2.5 ([4]) *Considering $\{\eta_n^0\}$, $\{\eta_n^1\}$ being real sequences, wherein*

(1) $0 \leq \eta_n^0, \eta_n^1 < 1$
(2) $\eta_n^1 \to 0$ *as* $n \to \infty$
(3) $\sum \eta_n^0 \eta_n^1 = \infty$.

Let there be some real sequence $\{\eta_n^2\}$ which is non-negative and exists in such a manner that $\sum \eta_n^0 \eta_n^1 (1 - \eta_n^1)\gamma_n$ is bounded, then the sequence η_n^2 has a null subsequence.

Lemma 2.6 ([17] *Let a real number $\mathcal{R} > 1$ be a fixed number and $\mathcal{B}$ a Banach space. Then $\mathcal{B}$ is uniformly convex if and only if there exists a continuous, strictly increasing and convex function $g : [0,) \to [0, \infty)$ with $g(0) = 0$ such that*

$$||\mu\ell + (1-\mu)\jmath||^2 \leq \mu||\ell||^2 + (1-\mu)||\jmath||^2 - \mu(1-\mu)g(||\ell - \jmath||) \tag{2.1}$$

for all $\ell, \jmath \in B_{\mathcal{R}}(0) = \{\ell \in \mathcal{B} : ||\ell|| \leq \mathcal{R}\}$ and $\in [0, 1]$.

Definition 2.7 ([13]) The mapping $\wp : \mathcal{B}_s \to CB(\mathcal{B}_s)$ is called hemicompact if, for any sequence $\{\ell_n\}$ in $\mathcal{B}_s$ such that $d(\ell_n, \wp(\ell_n)) \to 0$ as $n \to \infty$, there exists a subsequence $\{\ell_{n_k}\}$ of $\{\ell_n\}$ such that $\{\ell_{n_k}\} \to p \in \mathcal{B}_s$. We note that if $\mathcal{B}_s$ is compact, then every multi-valued mapping $\wp : \mathcal{B}_s \to CB(\mathcal{B}_s)$ is hemicompact.

Definition 2.8 ([19]) A mapping $\wp : \mathcal{B}_s \to CB(\mathcal{B}_s)$ is said to satisfy Condition (I) if there is a nondecreasing function $f : [0, \infty) \to [0, \infty)$ with $f(0) = 0, f(r) > 0$ for $r \in (0, \infty)$ such that

$$d(\ell, \wp(\ell)) \geq f(d(\ell, \mathcal{P}_\wp)$$

for all $\ell \in \mathcal{B}_s$.

A mapping $\wp$ is said to be demiclosed if, for any sequence $\{\ell_n\}$ which weakly converges to $\jmath$, and if the sequence $\{\wp\ell_n\}$ strongly converges to ε, we have $\wp(\jmath) = \varepsilon$, (see [20]).

Definition 2.9 (*see* [21, 22]) Let $\mathcal{B}_s$ be a nonempty closed convex subset of a real Banach space and $\wp$ a mapping from $\mathcal{B}_s \to \mathcal{B}_s$. The mapping $\wp$ is called zero-demiclosed if $\{\ell_n\}$ in $\mathcal{B}_s$ satisfying $||\ell_n - \wp\ell_n|| \to 0$ and $\ell_n \rightharpoonup \varepsilon \in \mathcal{B}_s$ implies $\wp\varepsilon = \varepsilon$.

The following lemma is useful in our subsequent discussion and are easy to establish.

Lemma 2.10 ([15]) *Let $\mathcal{B}_s$ be a nonempty closed convex subset of a uniformly convex Banach space $\mathcal{B}$ and $\wp$ a quasi-nonexpansive mapping on $\mathcal{B}_s$. Then $I - \wp$ is demiclosed at* 0.

3 Main Results

Lemma 3.1 *Let $\mathcal{B}$ be a uniformly convex Banach space, a nonempty $\mathcal{B}_s$ be a closed and convex subset of $\mathcal{B}$. Assume that $\wp : \mathcal{B}_s \to CB(\mathcal{B}_s)$ be a multi-valued quasi-nonexpansive mapping with $\mathcal{P}_\wp \neq \emptyset$ and for which $\wp(\sigma) = \{\sigma\}$ for all $\sigma \in \mathcal{P}_\wp$. Let $\{\ell_n\}$ be the sequence of CR iterates defined by $(\mathcal{A})$ with $\eta_n^0, \eta_n^1, \eta_n^2 \in [a, b] \subset [0, 1]$. Then $\lim_{n\to\infty} ||\ell_n - \sigma||$ exists for every $\sigma \in \mathcal{P}_\wp$.*

Proof Since $\wp$ is a multi-valued quasi-nonexpansive mapping, for an arbitrary $\sigma \in \mathcal{P}_\wp$ and $n \in \mathbb{N}$, we have

$$
\begin{aligned}
||\varepsilon_n - \sigma|| &= ||(1 - \eta_n^2)\ell_n + \eta_n^2 z_n' - \sigma|| \\
&\leq (1 - \eta_n^2)||\ell_n - \sigma|| + \eta_n^2||z_n' - \sigma|| \\
&\leq (1 - \eta_n^2)||\ell_n - \sigma|| + \eta_n^2 d(z_n', \wp(\sigma)) \\
&\leq (1 - \eta_n^2)||\ell_n - \sigma|| + \eta_n^2 \mathcal{H}(\wp(\ell_n), \wp(\sigma) \\
&\leq (1 - \eta_n^2)||\ell_n - \sigma|| + \eta_n^2||\ell_n - \sigma|| \\
||\varepsilon_n - \sigma|| &\leq ||\ell_n - \sigma||
\end{aligned}
$$

and

$$
\begin{aligned}
||J_n - \sigma|| &= ||(1 - \eta_n^1)z_n' + \eta_n^1 z_n''' - \sigma|| \\
&\leq (1 - \eta_n^1)||z_n' - \sigma|| + \eta_n^1||z_n''' - \sigma|| \\
&\leq (1 - \eta_n^1)d(z_n', \wp(\sigma)) + \eta_n^1 d(z_n''', \wp(\sigma)) \\
&\leq (1 - \eta_n^1)\mathcal{H}(\wp\ell_n, \wp\sigma) + \eta_n^1 \mathcal{H}(\wp\varepsilon_n, \wp\sigma) \\
&\leq (1 - \eta_n^1)||\ell_n - \sigma|| + \eta_n^1||\varepsilon_n - \sigma||.
\end{aligned}
$$

On substituting the value of $||\varepsilon_n - \sigma||$, we have

$$
\begin{aligned}
||J_n - \sigma|| &\leq (1 - \eta_n^1)||\ell_n - \sigma|| + \eta_n^1||\ell_n - \sigma|| \\
&\leq ||\ell_n - \sigma||.
\end{aligned}
$$

Now, similarly, we have

$$
\begin{aligned}
||\ell_{n+1} - \sigma|| &= ||(1 - \eta_n^0)J_n + \eta_n^0 z_n'' - \sigma|| \\
&\leq (1 - \eta_n^0)||J_n - \sigma|| + \eta_n^0||z_n'' - \sigma|| \\
&\leq (1 - \eta_n^0)||J_n - \sigma|| + \eta_n^0 d(z_n'', \wp(\sigma)) \\
&\leq (1 - \eta_n^0)||J_n - \sigma|| + \eta_n^0 \mathcal{H}(\wp(J_n), \wp\sigma) \\
&\leq (1 - \eta_n^0)||J_n - \sigma|| + \eta_n^0||J_n - \sigma|| \\
||\ell_{n+1} - \sigma|| &\leq ||J_n - \sigma||.
\end{aligned}
$$

Using the value of $||\jmath_n - \sigma||$, we have

$$||\ell_{n+1} - \sigma|| \leq ||\ell_n - \sigma||.$$

It results in $\{||\ell_n - \sigma||\}$ which is a non-increasing sequence, which implies $\lim_{n\to\infty} ||\ell_n - \sigma||$ exists for all $\sigma \in \mathcal{P}_{\wp}$. Also $\{\ell_n\}$ is bounded.
First, we approximate weak convergence of the sequence $\{\ell_n\}$ generated by $\mathcal{B}_s$.

Theorem 3.2 *Let $\mathcal{B}$ be a uniformly convex Banach space with Opial's property $\mathcal{B}_s = \emptyset$ which is a closed and convex subset of $\mathcal{B}$. Assume that $\wp : \mathcal{B}_s \to CB(\mathcal{B}_s)$ be a multi-valued quasi-nonexpansive mapping with $\mathcal{P}_{\wp} \neq \emptyset$ and $\wp(\sigma) = \{\sigma\}$. If $\{\ell_n\}$ is the sequence of $(\mathcal{C})$ iterates defined by $\eta_n^0, \eta_n^1, \eta_n^2 \in [a, b] \subset [0, 1]$, then the sequence $\{\ell_n\}$ converges weakly to an element of $\wp$.*

Proof Consider arbitrarily chosen $\sigma \in \mathcal{P}_{\wp}$, since $\mathcal{B}$ is uniformly convex, by Lemma 2.6, there exists a strictly increasing continuous function $g : [0, \infty) \to [0, \infty)$ with $g(0) = 0$ such that

$$\begin{aligned}
||\varepsilon_n - \sigma||^2 &= ||(1 - \eta_n^2)\ell_n + \eta_n^2 z_n' - \sigma||^2 \\
&\leq (1 - \eta_n^2)||\ell_n - \sigma||^2 + \eta_n^2||z_n' - \sigma||^2 - \eta_n^2(1 - \eta_n^2)g(||\ell_n - z_n'||) \\
&\leq (1 - \eta_n^2)||\ell_n - \sigma||^2 + \eta_n^2 d^2(z_n', \wp(\sigma)) - \eta_n^2(1 - \eta_n^2)g(||\ell_n - z_n'||) \\
&\leq (1 - \eta_n^2)||\ell_n - \sigma||^2 + \eta_n^2\mathcal{H}^2(\wp(\ell_n), \wp(\sigma)) - \eta_n^2(1 - \eta_n^2)g(||\ell_n - z_n'||) \\
&\leq (1 - \eta_n^2)||\ell_n - \sigma||^2 + \eta_n^2||\ell_n - \sigma||^2 \\
||\varepsilon_n - \sigma||^2 &\leq ||\ell_n - \sigma||^2
\end{aligned}$$

and

$$\begin{aligned}
||\jmath_n - \sigma||^2 &= ||(1 - \eta_n^1)z_n' + \eta_n^1 z_n''' - \sigma||^2 \\
&\leq (1 - \eta_n^1)||z_n' - \sigma||^2 + \eta_n^1||z_n''' - \sigma||^2 - \eta_n^1(1 - \eta_n^1)g(||z_n''' - z_n'||) \\
&\leq (1 - \eta_n^1)d^2(z_n', \wp(\sigma)) + \eta_n^1 d^2(z_n''', \wp(\sigma)) - \eta_n^1(1 - \eta_n^1)g(||\ell_n - z_n'||) \\
&\leq (1 - \eta_n^1)\mathcal{H}^2(\wp\ell_n, \wp\sigma) + \eta_n^1\mathcal{H}^2(\wp\varepsilon_n, \wp\sigma) - \eta_n^1(1 - \eta_n^1)g(||\ell_n - z_n'||) \\
&\leq (1 - \eta_n^1)||\ell_n - \sigma||^2 + \eta_n^1||\varepsilon_n - \sigma||^2 - \eta_n^1(1 - \eta_n^1)g(||\ell_n - z_n'||).
\end{aligned}$$

On substituting the value of $||\varepsilon_n - \sigma||$, we have

$$\begin{aligned}
||\jmath_n - \sigma||^2 &\leq (1 - \eta_n^1)||\ell_n - \sigma||^2 + \eta_n^1||\ell_n - \sigma||^2 - \eta_n^2\eta_n^1(1 - \eta_n^1)g(||\ell_n - z_n'||) \\
&\leq ||\ell_n - \sigma||^2.
\end{aligned}$$

Now, similarly, we have

$$\begin{aligned}
||\ell_{n+1}-\sigma||^2 &= ||(1-\eta_n^0)\jmath_n+\eta_n^0 z_n''-\sigma||^2\\
&\le (1-\eta_n^0)||\jmath_n-\sigma||^2+\eta_n^0||z_n''-\sigma||^2-\eta_n^0(1-\eta_n^0)g(||\jmath_n-z_n''||)\\
&\le (1-\eta_n^0)||^2\jmath_n-\sigma||+\eta_n^0 d^2(z_n'',\wp(\sigma))-\eta_n^0(1-\eta_n^0)g(||\ell_n-z_n'||)\\
&\le (1-\eta_n^0)||\jmath_n-\sigma||^2+\eta_n^0\mathcal{H}^2(\wp(\jmath_n),\wp\sigma)-\eta_n^0(1-\eta_n^0)g(||\jmath_n-z_n''||)\\
&\le (1-\eta_n^0)||\jmath_n-\sigma||^2+\eta_n^0||\jmath_n-\sigma||^2-\eta_n^0(1-\eta_n^0)g(||\jmath_n-z_n''||)\\
||\ell_{n+1}-\sigma||^2 &\le ||\jmath_n-\sigma||^2-\eta_n^0(1-\eta_n^0)g(||\jmath_n-z_n''||).
\end{aligned}$$

Using the value of $||\jmath_n-\sigma||$, we have

$$||\ell_{n+1}-\sigma||^2 \le ||\ell_n-\sigma||^2-\eta_n^2\eta_n^1\eta_n^0(1-\eta_n^0)g(||\jmath_n-z_n''||).$$

On substituting the value of $||\varepsilon_n-\sigma||$, we have

$$||\jmath_n-\sigma||^2 \le ||\ell_n-\sigma||^2-\eta_n^1\eta_n^2(1-\eta_n^2)g(||\ell_n-z_n'||.$$

Now, similarly, we have

$$\begin{aligned}
||\ell_{n+1}-\sigma||^2 &= ||(1-\eta_n^0)\jmath_n+\eta_n^0 z_n''-\sigma||^2\\
&\le (1-\eta_n^0)||\jmath_n-\sigma||^2+\eta_n^0||z_n''-\sigma||^2-\eta_n^0(1-\eta_n^0)g(||\jmath_n-z_n''||)\\
&\le (1-\eta_n^0)||\jmath_n-\sigma||^2+\eta_n^0 d^2(z_n'',\wp\sigma)-\eta_n^0(1-\eta_n^0)g(||\jmath_n-z_n''||)\\
&\le (1-\eta_n^0)||\jmath_n-\sigma||^2+\eta_n^0\mathcal{H}^2(\wp\jmath_n,\wp\sigma)-\eta_n^0(1-\eta_n^0)g(||\jmath_n-z_n''||)\\
&\le (1-\eta_n^0)||\jmath_n-\sigma||^2+\eta_n^0||\jmath_n-\sigma||^2-\eta_n^0(1-\eta_n^0)g(||\jmath_n-z_n''||)\\
&\le ||\jmath_n-\sigma||-\eta_n^0(1-\eta_n^0)g(||\jmath_n-z_n''||)\\
&\le ||\ell_n-\sigma||-\eta_n^0\eta_n^1\eta_n^2(1-\eta_n^2)g(||\ell_n-z_n'||).
\end{aligned} \tag{3.1}$$

This implies

$$\sum_{n=1}^{\infty} a^3(1-b)g(\|\ell_n-z_n'\|) \le \sum_{n=1}^{\infty}\eta_n^0\eta_n^1\eta_n^2(1-\eta_n^2)g(\|\ell_n-z_n'\|), \tag{3.2}$$

which implies that $\{||\ell_n-z_n'||\}$ is a null sequence. But the function g is strictly increasing and continuous so

$$\lim_{n\to\infty}||\ell_n-z_n'||=0,$$

and hence we have

$$\lim_{n\to\infty} d(\ell_n,\wp\ell_n)=\lim_{n\to\infty}||\ell_n-z_n'||=0. \tag{3.3}$$

Now to prove that $\{\ell_n\}$ converges weakly to the element of $\mathcal{P}_{\wp}$. To prove the same, we need to prove that $\{\ell_n\}$ has a unique weak subsequential limit in $\mathcal{P}_{\wp}$. For the same result, let us make an assumption of two convergent subsequences $\{\ell_{n_u}\}$ and $\{\ell_{n_v}\}$ of $\{\ell_n\}$, in such a manner that $\{\ell_{n_u}\} \rightharpoonup u$ and $\{\ell_{n_v}\} \rightharpoonup v$, $\lim_{n\to\infty} d(\ell_{n_u}, u) = 0$ pre-defined lemmas, $u \in \mathcal{P}_{\wp}$. Similarly, $v \in \mathcal{P}_{\wp}$ it is obvious that $v \in \mathcal{P}_{\wp}$. Now, we need to prove equality of u and v. On contrary, assume that $u \neq v$, then by Lemma 3.1 together with *Opial*'s property, we have

$$\begin{aligned}
\lim_{n\to\infty} \|\ell_n - u\| &= \lim_{n\to\infty} \|\ell_{n_u} - u\| \\
&< \lim_{n\to\infty} \|\ell_{n_u} - v\| \\
&= \lim_{n\to\infty} \|\ell_n - v\| \\
&= \lim_{n\to\infty} \|\ell_{n_v} - v\| \\
&< \lim_{n\to\infty} \|\ell_{n_v} - u\| \\
&= \lim_{n\to\infty} \|\ell_n - u\|
\end{aligned}$$

which is a contradiction. So $u = v$ and it proves that $\{\ell_n\}$ converges weakly to a fixed point of $\wp$. □

We are now all set to go for our strong convergence theorem.

Theorem 3.3 *Let $\mathcal{B}$ be a uniformly convex Banach space with $\mathcal{B}_s = \emptyset$ which is a closed and convex subset of $\mathcal{B}$. Assume that $\wp : \mathcal{B}_s \to CB(\mathcal{B}_s)$ be a multi-valued quasi-nonexpansive map with $\mathcal{P}_{\wp} \neq \emptyset$ and $\wp\sigma = \{\sigma\}$ for every $\sigma \in \mathcal{P}_{\wp}$. Assuming $\{\ell_n\}$ to be the sequence of CR iterates defined by (C) satisfying $\eta_n^0, \eta_n^1, \eta_n^2 \in [a, b] \subset [0, 1]$. If $\wp$ satisfies condition (I), then the sequence $\{\ell_n\}$ converges strongly to an element of $\wp$.*

Proof $\sigma \in \mathcal{P}_{\wp}$. Then as in the proof of Lemma 3.1, $\{\ell_n\}$ is bounded and so $\{\jmath_n\}$ is bounded. Therefore, there is an existence of a real number $\mathcal{R} > 0$ in such a way that $\ell_n - 0$ and $\jmath_n - \sigma \in \mathcal{B}_{\wp}(0)$ for every $n \geq 0$. On applying Lemma 2.6, we have

$$||\ell_{n+1} - \sigma||^2 \leq ||\ell_n - \sigma||^2 - \eta_n^2\eta_n^1\eta_n^0(1 - \eta_n^0)g(||\jmath_n - z_n''||).$$

$$\sum_{n=1}^{\infty} a^3(1 - b)g(\|\ell_n - z_n'\|) \leq \sum_{n=1}^{\infty} \eta_n^0\eta_n^1\eta_n^2(1 - \eta_n^2)g(\|\ell_n - z_n'\|). \tag{3.4}$$

This implies that $\lim_{n\to\infty} g(||\jmath_n - z_n''||) = 0$. But g is strictly increasing and continuous, and it follows that $\lim_{n\to\infty} ||\jmath_n - z_n''$. Also, $d(\ell_n, \wp(\ell_n)) \leq ||\jmath_n - z_n''||$ approaches to 0 as $n \to \infty$. Therefore, it is given that $\wp$ satisfies condition (I), we have

$$\lim_{n\to\infty} d(\ell_n, \wp\ell_n) \leq ||\ell_n - z_n'|| \to 0$$

as $n \to \infty$. Since, $\wp$ satisfies condition (I), we have $\lim_{n\to\infty} d(\ell_n, \mathcal{P}_\wp) = 0$. Thus, there is an existence of a subsequence $\{\ell_{n_\kappa}\}$ of $\{\ell_n\}$ such that

$$||\ell_{n_{\kappa+1}} - \sigma_\kappa|| < \frac{1}{2^\kappa}$$

for some $\{\sigma_\kappa\} \subset \mathcal{P}_\wp)$ and all κ. Note that by Lemma 3.1, we obtain

$$||\ell_{n_{\kappa+1}} - \sigma_\kappa|| \leq ||\ell_{n_\kappa} - \sigma_\kappa|| < \frac{1}{2^\kappa}.$$

To show that $\{\sigma\}_\kappa$ is a Cauchy sequence in $\mathcal{B}_s$. Note that

$$\begin{aligned} ||\sigma_{\kappa+1} - \sigma_\kappa|| &\leq ||\sigma_{\kappa+1} - \ell_{n_{\kappa+1}}|| + ||\ell_{n_{\kappa+1}} - \sigma_\kappa|| \\ &< \frac{1}{2^{\kappa+1}} + \frac{1}{2^\kappa} \\ &< \frac{1}{2^{\kappa-1}}. \end{aligned}$$

This shows that σ_κ is a Cauchy sequence in $\mathcal{B}_s$ and thus converges to $\sigma' \in \mathcal{B}_s$. Since

$$d(\sigma_\kappa, \wp(\sigma')) \leq \mathcal{H}(\wp(\sigma'), \wp(\sigma_\kappa)) \qquad \leq ||\sigma' - \sigma_\kappa||$$

and σ_κ approaches to σ' as $\kappa \to \infty$, which implies that $d(\sigma', \wp(\sigma)) = 0$ and hence $\sigma' \in \wp$ and $\{\ell_{n_\kappa}\}$ converges strongly to σ'. It is also given that there is an existence of $\lim_{n\to\infty} ||\sigma_\kappa - \sigma'||$. It follows that $\{\ell_n\}$ converges strongly to σ', which completes the proof. □

Theorem 3.4 *Let $\mathcal{B}$ be a uniformly convex Banach space, $\mathcal{B}_s$ a nonempty closed convex subset of $\mathcal{B}$, and $\wp : \mathcal{B}_s \to CB(\mathcal{B}_s)$ a quasi-nonexpansive multi-valued map with $\mathcal{P}_\wp \neq \emptyset$ and for which $\wp(\sigma) = \{\sigma\}$ for each $\sigma \in \mathcal{P}_\wp$. Let $\{\ell_n\}$ be the CR iterates defined by C. Assume that $\wp$ is hemicompact and continuous, and $0 \leq \eta_n^0, \eta_n^1, \eta_n^2 < 1$, $\eta_n^2 \to 0$ and $\sum \eta_n^0\eta_n^1\eta_n^2 = \infty$. Then $\{\ell_n\}$ converges strongly to a fixed point of $\wp$.*

Proof Let σ is an arbitrarily chosen element of $\mathcal{P}_\wp$. Then, as in the proof of Theorem 3.3

$$\sum_{n=1}^{\infty} \eta_n^0\eta_n^1\eta_n^2(1-\eta_n^2)g(\|\ell_n - z_n'\|) \leq ||\ell_n - \sigma||^2 - ||\ell_{n+1} - \sigma||^2,$$

and hence

$$\sum_{n=1}^{\infty} \eta_n^0 \eta_n^1 \eta_n^2 (1 - \eta_n^2) g(||\ell_n - z_n'||) < \infty.$$

On the other hand, $\wp$ satisfies hemicompactness, and we may choose a subsequence $\{\ell_{n_k}\}$ of $\{\ell_n\}$, such that $\{\ell_{n_k}\} \to \sigma'$ for some $\sigma' \in \mathcal{B}_s$. Since $\wp$ is continuous, we have

$$\lim_{n\to\infty} d(\ell_{n_k}, \wp(\ell_{n_k})) \to d(\sigma', \wp(\sigma')).$$

As a result, we have $d(\sigma', \wp(\sigma')) = 0$ and so $q \in \mathcal{P}_\wp$. By Lemma 3.1, we have $\lim_{n\to\infty} ||\ell_n - \sigma||$ which exists for each $\sigma \in \mathcal{P}$ and it follows that $\{\ell_n\}$ converges strongly to σ'. This completes the proof. □

Corollary 3.5 *Let $\mathcal{B}$ be a uniformly convex Banach space, $\mathcal{B}_s$ a nonempty closed convex subset of $\mathcal{B}$, and $\wp : \mathcal{B}_s \to CB(\mathcal{B}_s)$ a quasi-nonexpansive multi-valued map with $\mathcal{P}_\wp \neq \emptyset$ and for which $\wp(\sigma) = \{\sigma\}$ for each $\sigma \in \mathcal{P}_\wp$. Let $\{\ell_n\}$ be the CR iterates defined by $\mathcal{C}$. Assume that $\wp$ is hemicompact, and $0 \le \eta_n^0, \eta_n^1, \eta_n^2 < 1, \eta_n^2 \to 0$ and $\sum \eta_n^0 \eta_n^1 \eta_n^2 = \infty$. Then $\{\ell_n\}$ converges strongly to a fixed point of $\wp$.*

The hemicompactness of $\wp$ to conditions (I) of Theorem 3.3 is interesting to compare. The following iteration scheme is defined by eliminating the restriction on $\wp$ $(i.e.\ \wp(\sigma) = \{\sigma\})$ for all $\sigma \in \mathcal{P}_\wp$ and nonempty proximinal bounded subsets of $\mathcal{B}_s$ and denoted by $\mathcal{P}^s$ also let $\wp : \mathcal{B}_s \to \mathcal{P}^s(\mathcal{B}_s)$ and $\mathcal{P}_\wp^s(\ell) = \{y \in \wp(\ell) : ||\ell - \jmath|| = d(\ell, \wp(\ell))\}$. Then the sequence of *CR* iterates is defined by $x_0 \in \mathcal{B}_s$,

$$\begin{cases} \varepsilon_n = (1 - \eta_n^2)\ell_n + \eta_n^2 z_n', \\ \jmath_n = (1 - \eta_n^1)z_n' + \eta_n^1 z_n''', \\ \ell_{n+1} = (1 - \eta_n^0)\jmath_n + \eta_n^0 z_n'', \end{cases} \qquad (\mathcal{C}^m)$$

where $z_n' \in \mathcal{P}_\wp^s(\ell_n)$, $z_n'' \in \mathcal{P}_\wp^s(\jmath_n)$ and $z_n''' \in \mathcal{P}_\wp^s(\varepsilon_n)$.

It is to note that if $\wp$ is single valued, then this iteration scheme reduced to the well-known *CR* iteration scheme.

Theorem 3.6 *Let $\mathcal{B}$ be a uniformly convex Banach space, $\mathcal{B}_s \neq \emptyset$ be a closed and convex subset of $\mathcal{B}$. Also, $\mathcal{P}_\wp^s$ is nonexpansive with $\mathcal{P}_\wp \neq \emptyset$ and $\sigma \in \mathcal{P}_\wp$. Let $\{\ell_n\}$ be the sequence of CR iterates defined by $(\mathcal{C}^m)$. Assume that $\wp$ satisfies condition (I) and $\eta_n^0, \eta_n^1, \eta_n^2 \in [a, b] \subset (0, 1)$. Then the sequence $\{\ell_n\}$ converges strongly to an element of $\wp$.*

Proof Since $\mathcal{P}_\wp \neq \emptyset$, let $\sigma \in \mathcal{P}_\wp$. Then $\sigma \in \mathcal{P}_\wp(\sigma) = \sigma$ and thus we obtain

$$
\begin{aligned}
||\varepsilon_n - \sigma|| &= ||(1-\eta_n^2)\ell_n + \eta_n^2 z_n' - \sigma|| \\
&\leq (1-\eta_n^2)||\ell_n - \sigma|| + \eta_n^2||z_n' - \sigma|| \\
&\leq (1-\eta_n^2)||\ell_n - \sigma|| + \eta_n^2 d(z_n', \mathcal{P}_{\wp}^s(\sigma)) \\
&\leq (1-\eta_n^2)||\ell_n - \sigma|| + \eta_n^2 \mathcal{H}(\mathcal{P}_{\wp}^s(\ell_n), \mathcal{P}_{\wp}^s(\sigma)) \\
&\leq (1-\eta_n^2)||\ell_n - \sigma|| + \eta_n^2||\ell_n - \sigma|| \\
||\varepsilon_n - \sigma|| &\leq ||\ell_n - \sigma|| \qquad (3.5)
\end{aligned}
$$

and

$$
\begin{aligned}
||\jmath_n - \sigma|| &= ||(1-\eta_n^1)z_n' + \eta_n^1 z_n''' - \sigma|| \\
&\leq (1-\eta_n^1)||z_n' - \sigma|| + \eta_n^1||z_n''' - \sigma|| \\
&\leq (1-\eta_n^1)d(z_n', \mathcal{P}_{\wp}^s(\sigma)) + \eta_n^1 d(z_n''', \mathcal{P}_{\wp}(\sigma)) \\
&\leq (1-\eta_n^1)\mathcal{H}(\mathcal{P}_{\wp}^s(\ell_n), \mathcal{P}_{\wp}^s(\sigma)) + \eta_n^1 \mathcal{H}(\mathcal{P}_T(\varepsilon_n), \mathcal{P}_{\wp}^s(\sigma)) \\
&\leq (1-\eta_n^1)||\ell_n - \sigma|| + \eta_n^1||\varepsilon_n - \sigma||.
\end{aligned}
$$

On substituting the value of $||\varepsilon_n - \sigma||$, we have

$$
\begin{aligned}
||\jmath_n - \sigma|| &\leq (1-\eta_n^1)||\ell_n - \sigma|| + \eta_n^1||\ell_n - \sigma|| \\
&\leq ||\ell_n - \sigma||. \qquad (3.6)
\end{aligned}
$$

Now, similarly, we have

$$
\begin{aligned}
||\ell_{n+1} - \sigma|| &= ||(1-\eta_n^0)\jmath_n + \eta_n^0 z_n'' - \sigma|| \\
&\leq (1-\eta_n^0)||\jmath_n - \sigma|| + \eta_n^0||z_n'' - \sigma|| \\
&\leq (1-\eta_n^0)||\jmath_n - \sigma|| + \eta_n^0 d(z_n'', \mathcal{P}_{\wp}^s(\sigma)) \\
&\leq (1-\eta_n^0)||\jmath_n - \sigma|| + \eta_n^0 \mathcal{H}(\mathcal{P}_{\wp}^s(\jmath_n), \mathcal{P}_{\wp}^s(\sigma)) \\
&\leq (1-\eta_n^0)||\jmath_n - \sigma|| + \eta_n^0||\jmath_n - \sigma|| \\
||\ell_{n+1} - \sigma|| &\leq ||\jmath_n - \sigma||.
\end{aligned}
$$

Consequently, the sequence $\{||\ell_n - \sigma||\}$ is decreasing and bounded below and thus $\lim_{n\to\infty} ||\ell_n - \sigma||$ exists for any arbitrarily chosen $\sigma \in \mathcal{P}_{\wp}$. Applying Lemma 2.6, we have

$$\begin{aligned}
||\varepsilon_n - \sigma||^2 &= ||(1-\eta_n^2)\ell_n + \eta_n^2 z_n' - \sigma||^2\\
&\le (1-\eta_n^2)||\ell_n - \sigma||^2 + \eta_n^2||z_n' - \sigma||^2 - \eta_n^2(1-\eta_n^2)g(||\ell_n - z_n'||)\\
&\le (1-\eta_n^2)||\ell_n - \sigma||^2 + \eta_n^2 d^2(z_n', \mathcal{P}_{\wp}^s(\sigma)) - \eta_n^2(1-\eta_n^2)g(||\ell_n - z_n'||)\\
&\le (1-\eta_n^2)||\ell_n - \sigma||^2 + \eta_n^2\mathcal{H}^2((\mathcal{P}_{\wp}^s(\ell_n), \mathcal{P}_{\wp}^s(\sigma)) - \eta_n^2(1-\eta_n^2)g(||\ell_n - z_n'||)\\
&\le (1-\eta_n^2)||\ell_n - \sigma||^2 + \eta_n^2||\ell_n - \sigma||^2 - \eta_n^2(1-\eta_n^2)g(||\ell_n - z_n'||)\\
||\varepsilon_n - \sigma||^2 &\le ||\ell_n - \sigma||^2 - \eta_n^2(1-\eta_n^2)g(||\ell_n - z_n'||)
\end{aligned}$$

and

$$\begin{aligned}
||\jmath_n - \sigma||^2 &= ||(1-\eta_n^1)z_n' + \eta_n^1 z_n''' - \sigma||^2\\
&\le (1-\eta_n^1)||z_n' - \sigma||^2 + \eta_n^1||z_n''' - \sigma||^2 - \eta_n^1(1-\eta_n^1)g(||z' - z_n'''||)\\
&\le (1-\eta_n^1)d^2(z_n', \mathcal{P}_{\wp}^s(\sigma)) + \eta_n^1 d^2(z_n''', \mathcal{P}_{\wp}^s(\sigma)) - \eta_n^1(1-\eta_n^1)g(||z' - z_n'''||)\\
&\le (1-\eta_n^1)\mathcal{H}^2(\mathcal{P}_{\wp}^s(\ell_n), \mathcal{P}_{\wp}^s(\sigma)) + \eta_n^1\mathcal{H}^2(\mathcal{P}_{\wp}^s(\varepsilon_n), \mathcal{P}_{\wp}^s(\sigma)) - \eta_n^1(1-\eta_n^1)g(||z' - z_n'''||)\\
&\le (1-\eta_n^1)||\ell_n - \sigma||^2 + \eta_n^1||\varepsilon_n - \sigma||^2 - \eta_n^1(1-\eta_n^1)g(||z' - z_n'''||)\\
&\le (1-\eta_n^1)||\ell_n - \sigma||^2 + \eta_n^1||\varepsilon_n - \sigma||^2.
\end{aligned}$$

On substituting the value of $||\varepsilon_n - \sigma||$, we have

$$||\jmath_n - \sigma||^2 \le ||\ell_n - \sigma||^2 - \eta_n^1\eta_n^2(1-\eta_n^2)g(||\ell_n - z_n'||.$$

Now, similarly, we have

$$\begin{aligned}
||\ell_{n+1} - \sigma||^2 &= ||(1-\eta_n^0)\jmath_n + \eta_n^0 z_n'' - \sigma||^2\\
&\le (1-\eta_n^0)||\jmath_n - \sigma||^2 + \eta_n^0||z_n'' - \sigma||^2 - \eta_n^0(1-\eta_n^0)g(||\jmath_n - z_n''||)\\
&\le (1-\eta_n^0)||\jmath_n - \sigma||^2 + \eta_n^0\mathcal{H}^2(\mathcal{P}_{\wp}^s(\jmath_n), \mathcal{P}_{\wp}^s(\sigma)) - \eta_n^0(1-\eta_n^0)g(||\jmath_n - z_n''||)\\
&\le (1-\eta_n^0)||\jmath_n - \sigma||^2 + \eta_n^0||\jmath_n - \sigma||^2 - \eta_n^0(1-\eta_n^0)g(||\jmath_n - z_n''||)\\
&\le ||\jmath_n - \sigma|| - \eta_n^0(1-\eta_n^0)g(||\jmath_n - z_n''||)\\
||\ell_{n+1} - \sigma|| &\le ||\ell_n - \sigma|| - \eta_n^0\eta_n^1\eta_n^2(1-\eta_n^2)g(||\ell_n - z_n'||)
\end{aligned}$$

so

$$\begin{aligned}
a^3(1-b)g(\|\ell_n - z_n'\|) &\le \eta_n^0\eta_n^1\eta_n^2(1-\eta_n^2)g(\|\ell_n - z_n'\|)\\
&\le ||\ell_n - \sigma||^2 - ||\ell_{n+1} - \sigma||^2
\end{aligned} \tag{3.7}$$

and so

$$\sum_{n=1}^{\infty} a^3(1-b)g(\|\ell_n - z_n'\|) \le \infty.$$

This implies that $\lim_{n\to\infty} ||\ell_n - z_n'||$. It is given that g is strictly increasing and is continuous at 0, we have$\lim_{n\to\infty} ||\ell_n - z'_n||$. Also $d(\ell_n, \wp(\ell_n)) \leq ||\ell_n - z'_n||$ as $n \to \infty$. Since $\wp$ satisfies Condition (*I*), we have $\lim_{n\to\infty} d(\ell_n, \mathcal{P}_\wp) = 0$. Thus, there is a subsequence $\{\ell_{n_k}\}$ of $\{\ell_n\}$ such that $||\ell_{n_k} - \sigma_k|| < \frac{1}{2^k}$ for some $\{\sigma_k\} \subset \mathcal{P}_\wp$ for all k. As in the proof of Theorem 3.2, $\{\sigma_k\}$ is a Cauchy sequence in $\mathcal{B}_s$ and thus converges to $\sigma' \in \mathcal{B}_s$. Since,

$$\begin{aligned} d(\sigma_k, \wp(\sigma') &\leq d(\sigma_k, \mathcal{P}^s_\wp(\sigma')) \\ &\leq \mathcal{H}(\mathcal{P}^s_\wp(\sigma_k), \mathcal{P}^s_\wp(\sigma')) \\ &\leq ||\mathcal{P}_k - \sigma'||, \end{aligned}$$

which implies that $\mathcal{P}_k \to \sigma'$ as $k \to \infty$, and hence $d(\sigma', \wp(\sigma')) = 0$ which results in $\sigma' \in \mathcal{P}_\wp$ and $\{\ell_{n_k}\}$ converges strongly to σ'. Also, because of existence of $\lim_{n\to\infty} ||\ell_n - \sigma'||$, it follows that $\{\ell_n\}$ converges strongly to σ', which completes the proof. □

Theorem 3.7 *Let $\mathcal{B}$ be a uniformly convex Banach space, $\mathcal{B}_s \neq \emptyset$ be a closed and convex subset of $\mathcal{B}$. Also, $\mathcal{P}^s_\wp$ is nonexpansive with $\mathcal{P}_\wp \neq \emptyset$ and $\sigma \in \mathcal{P}_\wp$. Let $\{\ell_n\}$ be the sequence of CR iterates defined by (C^m). Assume that $\wp$ is hemicompact and* $\eta^0_n, \eta^1_n, \eta^2_n \in [a, b] \subset (0, 1)$. *Then the sequence $\{\ell_n\}$ converges strongly to an element of $\wp$.*

Proof Let σ be an arbitrarily chosen element of $\mathcal{P}_\wp$. Then, as in the proof of Theorem 3.6, there is an existence of $\lim_{n\to\infty} ||\ell_n - \sigma||$ for any arbitrarily chosen $\sigma \in \mathcal{P}_s$. According to Theorem 3.6, it is obvious that $d(\ell_n, \wp\ell_n) \to 0$ as $\kappa \to \infty$. Also, the map $\wp$ is hemicompact, and it can be assumed that ℓ_{n_κ} for any arbitrarily chosen $\sigma' \in \mathcal{B}_s$. Also, it is to note that

$$\begin{aligned} d(\sigma', \wp(\sigma')) &\leq ||\sigma' + \ell_{n_\kappa}|| + ||\ell_{n_\kappa} - z'_{n_\kappa}|| + d(z'_{n_\kappa}, \wp(\sigma')) \\ &\leq ||\sigma' + \ell_{n_\kappa}|| + ||\ell_{n_\kappa} - z'_{n_\kappa}|| + \mathcal{H}(\mathcal{P}^s_\wp(\ell_{n_\kappa}), (\mathcal{P}^s_\wp(\sigma')) \\ &\leq ||\sigma' + \ell_{n_\kappa}|| + ||\ell_{n_\kappa} - z'_{n_\kappa}|| \to 0. \end{aligned}$$

This implies that $d(\sigma, \wp(\sigma))$ and thus $\sigma' \in \mathcal{P}_\wp$. Since $\lim_{n\to\infty}$ exists for each $\sigma \in \mathcal{P}_\wp$, it follows that $\ell_n \to \sigma'$ as $n \to 1$. This completes the proof. □

4 Conclusion

We have defined multi-valued version of *CR* iteration scheme involving quasi-nonexpansive mapping by relaxing the compactness of domain along with weak and strong convergence. We also established multi-valued *CR* iteration scheme in whole new sense which removes a restrictive condition (*i.e.* $\wp(\sigma) = \{\sigma\}$) for any $\sigma \in \mathcal{P}_\wp$. Our findings extend many existing results.

Acknowledgements The authors are thankful to the learned anonymous referee(s) for the valuable suggestions. Particularly, for introduction and providing some relevant references to make this manuscript more effective.

References

1. S. Suantai, W. Cholamjiak, P. Cholamjiak, An implicit iteration process for solving a fixed point problem of a finite family of multi-valued mappings in Banach spaces. Appl. Math. Lett. **25**(11), 1656–1660 (2012)
2. W. Cholamjiak, N. Pholasa, S. Suantai, A modified inertial shrinking projection method for solving inclusion problems and quasi-nonexpansive multi-valued mappings. Comput. Appl. Math. **37**(5), 5750–5774 (2018)
3. S. Suantai, W. Cholamjiak, P. Cholamjiak, Strong convergence theorems of a finite family of quasi-nonexpansive and Lipschitz multi-valued mappings. Afr. Mat. **26**, 345–355 (2015)
4. K.P.R. Sastry, G.V.R. Babu, Convergence of Ishikawa iterates for a multi-valued mappings with a fixed point. Czechoslovak Math. J. **55**, 817–826 (2005)
5. W.R. Mann, Mean value methods in iteration. Proc. Am. Math. Soc. **4**, 506–510 (1953)
6. S. Ishikawa, Fixed points by a new iteration method. Proc. Am. Math. Soc. **44**, 147–150 (1974)
7. B. Panyanak, Mann and Ishikawa iterative processes for multi-valued mappings in Banach spaces. Comput. Math. Appl. **54**, 872–877 (2007)
8. Y. Song, H. Wang, Erratum to Mann and Ishikawa iterative processes for multi-valued mappings in Banach spaces. Comput. Math. Appl. Comput. Math. Appl. **55**, 2999–3002 (2008)
9. N. Shahzad, H. Zegeye, On Mann and Ishikawa iteration schemes for multi-valued maps in Banach spaces. Nonlinear Anal. **71**, 838–844 (2009)
10. V. Berinde, *Iterative Approximation of Fixed Points*, vol. 1912, Lecture Notes in Mathematics (Springer, Berlin, 2007)
11. R. Chugh, V. Kumar, S. Kumar, Strong convergence of a new three step iterative scheme in Banach spaces. Am. J. Comput. Math. **2**, 345–357 (2012)
12. T. Hussain, A. Latif, Fixed points of multi-valued quasi-nonexpansive maps. Math. Jpn. **33**, 385–391 (1988)
13. S. Reich, Weak convergence theorems for quasi-nonexpansive mappings in Banach spaces. J. Math. Anal. Appl. **67**, 274–276 (1979)
14. H.F. Senter, W.G. Dotson, Approximating fixed points of quasi-nonexpansive mappings. Proc. Am. Math. Soc. **44**, 375–380 (1974)
15. K.K. Tan, H.K. Xu, Approximating fixed points of quasi-nonexpansive mappings by the Ishikawa iteration process. J. Math. Anal. Appl. **178**, 301–308 (1993)
16. H.K. Xu, Inequalities in Banach spaces with applications. Nonlinear Anal. **16**, 1127–1138 (1991)
17. H.K. Xu, On weakly quasi-nonexpansive and quasi-nonexpansive multi-valued mappings. Math. Jpn. **36**, 441–445 (1991)
18. C. Shiau, K.K. Tan, C.S. Wong, Quasi-nonexpansive multi-valued maps and selections. Fundam. Math. **87**, 109–119 (1975)
19. Z. Opial, Weak and strong convergence of the sequence of successive approximations for quasi-nonexpansive mappings. Bull. Am. Math. Soc. **73**, 591–598 (1967)
20. A. Moudafi, A note on the split common fixed-point problem for quasi-nonexpansive operators. Nonlinear Anal. **74**, 4083–4087 (2011)
21. Z. He, W.S. Du, Nonlinear algorithms approach to split common solution problems. Fixed Point Theory Appl. **2012**, 130 (2012)
22. W.S. Du Z. He, Feasible iterative algorithms for split common solution problems. J. Nonlinear Convex Anal. (in press)

Author Index

S. K. Paikray et al. (eds.), *New Trends in Applied Analysis and Computational Mathematics*, Advances in Intelligent Systems and Computing 1356,
https://doi.org/10.1007/978-981-16-1402-6

The manufacturer's authorised representative in the EU is Springer Nature Customer Service Centre GmbH, Europaplatz 3, 69115 Heidelberg, Germany. If you have any concerns regarding our products, please contact ProductSafety@springernature.com

Printed and bound by CPI Group (UK) Ltd, Croydon, CR0 4YY
30/09/2026
02225292-0003